VEHICLE MAINTENANCE AND REPAIR SERIES

Level 3

TRANSMISSION, CHASSIS and RELATED SYSTEMS

John Whipp ● Edited by Roy Brooks

D1098374

THOMSON

Australia • Canada • Mexico • Singapore • Spain • United Kingdom • United S

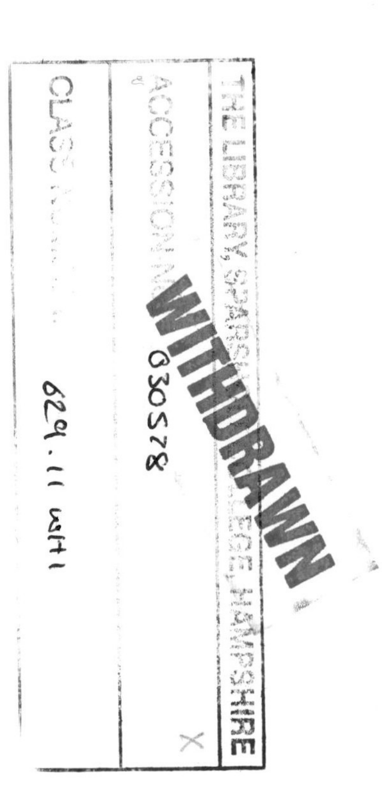

THOMSON

Vehicle Maintenance and Repair, Level 3 – 3rd Edition
Transmission, chassis and related systems

Copyright © Roy Brooks, John Whipp 1982, 1990, 1994, 2002
© Original illustrations and layout Thomson Learning 2002

The Thomson Learning logo is a registered trademark used herein under licence.

For more information, contact Thomson Learning, High Holborn House, 50–51 Bedford Row, London, WC1R 4LR or visit us on the World Wide Web at:
http://www.thomsonlearning.co.uk

All rights reserved by Thomson Learning 2003. The text of this publication, or any part thereof, may not be reproduced or transmitted in any form or by any means, electronic or mechanical, including photocopying, recording, storage in an information retrieval system, or otherwise, without prior permission of the publisher.

While the publisher has taken all reasonable care in the preparation of this book the publisher makes no representation, express or implied, with regard to the accuracy of the information contained in this book and cannot accept any legal responsibility or liability for any errors or omissions from the book or the consequences thereof.

Products and services that are referred to in this book may be either trademarks and/or registered trademarks of their respective owners. The publisher and author/s make no claim to these trademarks.

British Library Cataloguing-in-Publication Data
A catalogue record for this book is available from the British Library

---

ISBN 1-86152-806-X

First published 1994 Macmillan Press Ltd
(Vehicle Mechanical and Electronic Systems – Transmission, chassis and related systems)
Reprinted 2000 by Thomson Learning
This edition by Thomson Learning 2002
Reprinted 2003 by Thomson Learning

Typeset by Photoprint, Torquay, Devon UK
Printed in Croatia by Zrinski dd

Contents

	Preface	iv
	Acknowledgements	v
	Basic Essentials: Health, Safety and Relationships in the Workplace	vi
1.	Clutches	1
2.	Manual gearboxes	18
3.	Automatic transmission systems	48
4.	Drive line	71
5.	Final drive and differentials	81
6.	Suspension	93
7.	Steering	119
8.	Tyres and wheels	144
9.	Braking	163
10.	Bodywork	206
11.	Electrical and electronic systems	211
12.	Lighting systems	220
13.	Auxiliary electrical systems	231
14.	Enhance vehicle system features	245

Preface

Welcome to this workbook, which along with its companion volume, *Engines, Electronics and Related Systems* is one of the final books in the Thomson Learning, Vehicle Maintenance and Repair Series. It is great to know that you are probably on the last lap towards gaining Level 3 NVQ/SVQ or similar important qualifications.

Very likely, as with thousands of other students, you have already been helped on your way by using the earlier books in this long established series. Consequently you will know that by properly completing the text and diagrams you will achieve a high standard of essential knowledge. What is more the book will then become a valuable source of reference and provide vital evidence of achievement for your portfolio. One quick point though, don't forget to complete the answers in pencil – just in case of mistakes!

It is worth stressing that although you will now be working at quite an advanced level and often using equipment that is complex and dedicated, it is most important that you understand the basic principles of such items. For this reason the book concentrates where possible on fundamentals rather than the latest fancy variations on what may well be fairly simple designs. This helps you to understand and cope with the many and rapid technological changes that occur in the automobile world. No doubt very wisely you will consult manufacturer's data for any more detailed information that might be required.

The author and editor wish you every success with your studies and progression within the motor industry. If you have any comments you wish to pass on to us, please do so via the publisher.

Roy Brooks
Series Editor

Acknowledgements

The editor, author and publisher would like to thank all who helped so generously with information, assistance, illustrations and inspiration. In particular the book's principal illustrator, Harvey Dearden (previously principal lecturer in Motor Vehicle Subjects, Moston College of Further Education); and the North Manchester College; Andrea Whipp for her dedication in preparing the manuscript; and the persons, firms and organisations listed below. Should there by any omissions, they are completely unintentional.

A-C Delco Division of General Motors Ltd
Alfa Romeo (Great Britain) Ltd
AP Racing
Automotive Products plc
Robert Bosch Ltd
British Standards Institution
Castrol (UK) Ltd
Champion Automotive UK Ltd
Citroën UK Ltd
City & Guilds of London Institute
Clayton Dewandre Ltd
Delphi Lockheed Automotive
Dunlop Tyres Ltd
Fiat Auto (UK) Ltd
Ford Motor Co. Ltd
Girling Ltd
GKN Group
Honda UK Ltd
Land Rover
Lucas Industries plc
Luminetion Ltd
MAN Truck & Bus UK Ltd
Mercedes Benz UK Ltd
Meritor HVS (ROR)
Metalistic Ltd
Michelin Tyre plc
Mitsubishi Motors (The Colt Car Co. Ltd)
Peugeot Talbot Motor Co. Ltd
Renault UK Ltd
Ripaults Ltd
Rover Cars
Sachs Automotive Components Ltd
Scania (Great Britain) Ltd
Seddon Atkinson Vehicles Ltd
Subaru (UK) Ltd
Suzuki GB plc
Telma Retarder Ltd
Alfred Teves GmbH
Unipart Group of Companies
Vauxhall Motors Ltd
Volkswagen Group United Kingdom Ltd
Volvo Car UK Ltd
Westinghouse CVB Ltd
ZF Great Britain Ltd

Coverage of Standards

QUICK CHECK UNIT GRID

VEHICLE MAINTENANCE and REPAIR SERIES LEVEL 3

The subject material in chapters covers
Basic Essential Knowledge
for the unit areas indicated.

UNIT NUMBERS and TITLES	1 Contribute to Good Housekeeping	2 Ensure Your Own Actions Reduce Risks to Health and Safety	3 Maintain Positive Working Relationships	11 Carry Out Routine Vehicle Maintenance	13 Diagnose Complex System Faults	14 Rectify Complex System Faults	15 Enhance Vehicle System Features	17 Overhaul Units	18 Identify and Agree Customer Vehicle Needs	19 Inspect Vehicles
TRANSMISSION, CHASSIS and RELATED SYSTEMS										
BASIC ESSENTIALS Health, Safety and Relationships in the Workplace	•	•	•						•	
1. Clutches				•				•		•
2. Manual Gearboxes				•	•	•		•		•
3. Automatic Transmission Systems				•	•	•		•		•
4. Drive Line				•				•		•
5. Final Drive and Differentials				•	•	•		•		•
6. Suspension				•	•	•		•		•
7. Steering				•	•	•		•		•
8. Tyres and Wheels				•				•		•
9. Braking				•	•	•		•		•
10. Bodywork				•				•		•
11. Electrical and Electronic Systems				•	•	•		•		•
12. Lighting Systems				•				•		•
13. Auxiliary Electrical Systems				•	•	•		•		•
14. Enhance Vehicle System Features							•			•

For complete syllabus coverage see also the other books in this series – Maintenance and Repair of Road Vehicles Level2 and Engines, Electronics and Related Systems Level3.

Basic Essentials

Health, Safety and Relationships in the Workplace

Whichever subjects, at whatever level you are studying to obtain NVQ/SVQ qualifications in Motor Vehicle Work, you must be successful in the first three units:

1 Contribute to Good Housekeeping

2 Ensure Your Own Actions Reduce Risks to Health and Safety

3 Maintain Positive Working Relationships

Even if you are already familiar with this area of work, the pages in this section revise and reinforce these vital units. Their contents consist of items/actions which you should observe and carry out every day of your working career.

Good Housekeeping – keeping the workshop clean, tidy, and safe viii
Workshop Resources – looking after equipment, power, and time x
Health and Safety at Work – what to be aware of xi
Accidents and First Aid – what to do if something happens xi
Hazards – what to watch for xii
Personal Protection – clothes and equipment xii
Safe Handling – of loads, of equipment, and of harmful substances xiv
Health and Safety Signs – what they mean xvi
Emergency Procedures – fire alarms and fire extinguishers xvi
Positive Working Relationships – how to develop good relationships with colleagues xviii
Working as a Team – knowing your own job and what other people do xviii
Organisational Structure – knowing who does what in the company xix
Communication – spoken, written, telephone, and non-verbal xx

GOOD HOUSEKEEPING

Maintaining a clean work area

We are all impressed when we see a clean and shiny car, even though we know it would work just as well dirty. In the same way customers will be impressed if you keep your workshop clean and tidy. No one wants to see a dirty workshop, a cluttered parts department, an untidy forecourt, or a patch of oil on the floor.

Remember – there are benefits in keeping the workshop clean:

- Passing customers who need work done may be attracted in.
- Regular customers will be happy to keep coming back.
- Staff will work better.
- Accidents are less likely to happen.
- Work will be completed faster.
- You are less likely to lose tools and parts.

Housekeeping routines

Each workshop will have a system to keep the place clean and tidy. As an employee you may have to do some simple cleaning or tidying up as part of your duties.

A large company may employ cleaners; in a small one the work will be shared between the staff. All technicians are likely to be responsible for cleaning up after their work. Some jobs are particularly dirty – for example, jobs on exhausts or suspensions always leave dust and dirt on the floor.

Make sure you know your firm's housekeeping routines. Here is a typical daily housekeeping routine for a small garage workshop.

OPENING

- Move parked vehicles away from the work areas.
- Check that the lifts and floor area are clean and free from obstructions.
- Check that all tools and equipment are clean and tidy.
- Check the reception counter is tidy, and that the sales computer equipment is working.
- Check that the reception area and seating are straight and tidy.
- Check the special workshop tools and equipment such as the air line, the tyre-remover and the wheel balancer are working.

DURING THE DAY

- Keep the lift and the floor clean.
- After use, put special tools back where they are normally kept.
- Make sure that items such as wheels and removed tyres do not obstruct work areas or pathways.
- Make sure that the reception area is kept tidy, with ashtrays empty and magazines straight.

CLOSING

- Put away neatly all special tools.
- Lock up your personal toolbox.
- Check that any customers' vehicles are secure.
- Clean and tidy the work area.
- Switch off power to equipment.
- Tidy reception and empty the till.

Delivery of goods

When goods are delivered, do not place or leave them where they would block walkways or exits.

Cleaning

Cleaning equipment should be kept in a separate store, as many chemicals are highly concentrated.

Always read the instructions on the labels before using them (see COSHH regulations). If specialised cleaning is required, your employer will provide protective clothing.

When you have finished cleaning, put the cleaning equipment and unused chemicals back in the store.

HEALTH & SAFETY

If you need to use a hazardous cleaning material, read the label on the container. This will tell you how to use it safely, and what to do if you *do* have an accident – for example, if the cleaning material touches your eyes or skin.

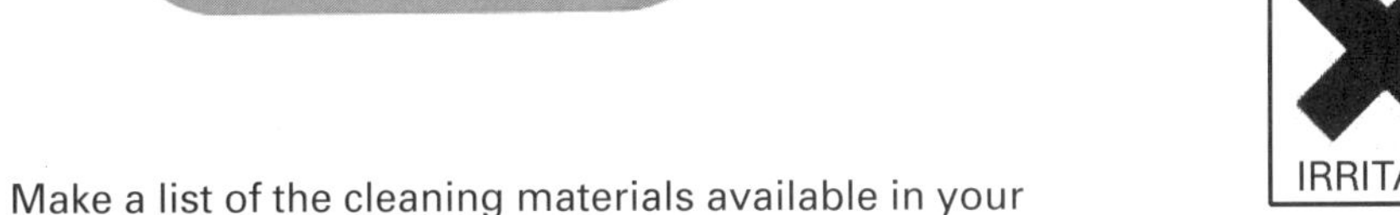

Make a list of the cleaning materials available in your garage. Think about:

- *Personal cleanliness* – e.g. barrier cream, hand-cleaning material.
- *Equipment* – e.g. lifts, trolley jacks, wheel-balancing machines, special and personal tools.
- *The workshop* – e.g. the floor area, car bays, walls, windows, lighting.

Emergency cleaning

Sometimes you will need to clean up after a breakage or spillage. Some oil might be spilt, for example, or some glass broken on the floor.

Breakages and spillages must be cleaned up immediately. If this is not done, someone may be injured. Also, the firm could be in breach of the Health and Safety at Work Act (see page xi). If an accident happens, the firm may be fined.

Disposing of dangerous waste material

All workshops produce dangerous waste materials – dirty engine oil and filters, scrap exhausts and batteries, broken or scrap plastic or metal components, and waste paper.

Items must be disposed of in different ways. Usually this is decided by the local council, who pass by-laws. Refuse disposal requirements differ from place to place.

Some types of dangerous material must be kept separate. They will be collected by specialist agencies, or taken to the local refuse collection point.

HEALTH & SAFETY
While cleaning, place cones and notices to warn others. Section off areas that could be dangerous, such as slippery floors.
DANGER
Cleaning in progress

All must take note of the Environmental Protection Act.

WORKSHOP RESOURCES

Every workshop has many resources. You need to be aware of what they are, and how you can make best use of them. Resources include:

- *Special workshop tools and equipment fixtures*, such as vehicle lifts, the air compressor, the wheel-balancer, steering alignment tools, the headlamp aligner, and so on.
- *Stock* in the stores – different types of tyres, exhausts, shock absorbers, batteries, oil, and so on.
- *Fixtures and fittings* in the reception area, the staff dining area, and elsewhere.
- *Utilities*, such as electricity, gas, water and the telephone.
- *Space* available – to work, to store parts, to park and display vehicles.
- *Time* – often the firm's most valuable resource.

Use resources to the best advantage

TOOLS

Use tools and equipment safely and properly. Avoid damage, and don't risk your own health and safety, or anyone else's.

UTILITIES

Electricity, gas and telephone calls are costly. Waste will reduce the firm's profit. To save energy, your firm will probably have some automatic timing controls fitted.

CONSUMABLE ITEMS

Do not waste consumable items, even if they are small. After fitting an exhaust system, for instance, return to stock any unused components such as nuts, brackets or rubber mounting rings.

Security

Make sure that parts are kept as safe as possible. If visitors wander around the workshop and are not observed, they could steal things. Theft by staff may also occur.

Taking goods, equipment or money without permission is always theft. Theft by staff is gross misconduct, and can lead to dismissal.

Do not leave keys in unattended cars. It is not uncommon to have cars driven away from the forecourt.

If you see someone acting suspiciously, ask them what they are doing. If the answer is not satisfactory, tell a senior member of staff immediately. Do not put yourself at risk.

Using resources economically

AVOID WASTING POWER

- Turn off lights when they are not needed.
- Keep workshop doors shut, to keep heat in.
- Report faulty components – for example, a leaking air line that causes the compressor to keep switching on.
- Turn off water when you are not using it, especially when washing cars.

USE SPACE SENSIBLY

Workshop space is expensive. Costs include rates, taxes, heating and lighting. Do not waste this space!

- Place vehicles so that you can work properly, but so that they take up minimum space.
- Keep the working space around ramps and in gangways clear of obstructions.
- Clear away quickly when a job is finished.

TIME

You rightly expect to be paid for your time at work. So you must play your part and help the firm to make a profit.

- As a trade trainee to the trade you are unlikely to work as quickly as a skilled technician. But aim to build up your speed.
- Work steadily, but not slowly. Time wasted is almost impossible to make up. You may have a bonus at stake!
- Organise yourself. Gather tools, information and equipment before you start a job.
- Complete jobs properly. Putting things right later is costly.

HEALTH AND SAFETY AT WORK

You must always take care of your own health, hygiene and safety. Here we look at health and safety matters that affect you at work and explain some of the regulations. These exist to make sure that you work in good, safe conditions. All major regulations have developed from the legislation in the **Health and Safety at Work Act 1974**.

Regulations

In the United Kingdom and Europe, health and safety matters are covered by the **Management of Health and Safety at Work Regulations 1992**.

As you train you will need to learn and understand the safety regulations that apply to your job. You are as responsible as your employer for following workshop safety regulations. You will find copies of the regulations displayed in the workshop. For your own safety, **read them**.

Duties

Under the Health and Safety at Work Regulations, **you** have certain duties. Bear these in mind as you work.

- Take reasonable care of your own health and safety.
- Take reasonable care for the safety of other people who may be affected by your actions.
- Work with your employer to keep safety rules.
- Report any accidents, hazards, or damage to equipment.

Warning

If you do not follow the Health and Safety at Work Regulations, you could be taken to court. For example, suppose you were welding an exhaust pipe. If you did not wear the goggles provided, you would be breaking the law, because you would not be taking reasonable care of your own health and safety.

ACCIDENTS AND FIRST AID

An accident involves something that is unexpected and unplanned. One or more people may be injured.

If the accident is minor, it may just be inconvenient. If it is serious, though, it could affect you for the rest of your life.

Accidents may be caused if you:

- Do not know the dangers involved in what you are doing.
- Daydream.
- Do not take enough precautions.
- Fool around.

Accidents may also occur because of faulty equipment or bad work conditions:

- Unsafe tools.
- Unguarded machinery.
- Poor ventilation.
- Poor lighting.

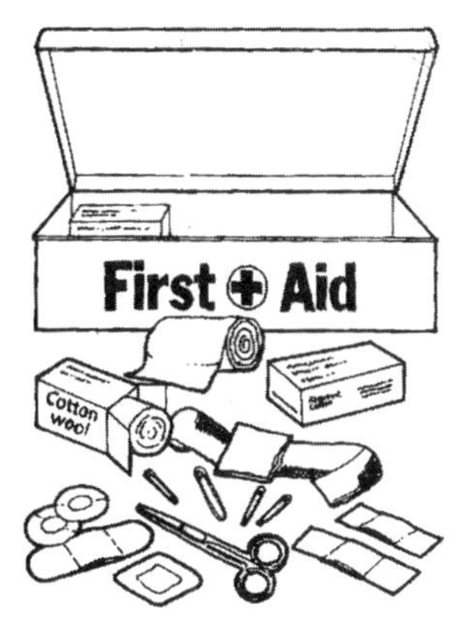

In your place of work, someone will have been specially trained as a first-aider. Normally this person will give first-aid. If the accident is minor, you may be able to help.

Whenever an accident happens, it is important to record the details in a special accident book.

Every first-aid kit should contain a card giving advice on what to do if anyone is injured.

HAZARDS

A hazard is anything that might cause an accident or injury.

Look around your workshop. You will probably see at least one hazard, perhaps more. Some can quickly be removed. Examples are:

- Wheels left lying on the floor, after being taken off a vehicle.
- Oil spilt on the floor.
- A trolley jack handle left lying where someone could trip over it.

Some hazards are always present. Warning notices or guards, or both, keep the risk to a minimum. Examples are:

- Brake tester rollers.
- A wheel balancer.
- A stand drill.

HEALTH & SAFETY

Hazard
"a hazard is something with potential to cause harm"

Risk
"a risk is the likelihood of the hazard's potential being realised"

PERSONAL PROTECTION

Various equipment is available which can be worn or held by people at work, protecting them from risks to health and safety. This equipment is covered by the **Personal Protection Equipment at Work Regulations 1992**. All Personal Protective Equipment (PPE) in use at work should carry the CE mark and where appropriate should comply with a European Norm (EN) standard. The Regulations don't include ordinary working clothes that do not specifically protect the health and safety of the wearer.

Personal presentation

Take care in what you wear.

To protect yourself and your clothes, it is sensible to wear:

- One-piece overalls (a boiler suit).
- Stout footwear (preferably with steel toe-caps).
- A suitable cap or bump cap.

Do **not** wear:

- Loose or torn overalls (especially if the sleeves are loose or torn).
- Rings or watches.
- Trainers or similar.
- Long hair (unless protected by suitable headgear).

HEALTH & SAFETY

Your personal presentation at work should:

- help to ensure the health and safety of yourself and others
- meet legal requirements
- be in accordance with workplace policies

Specialised personal protection equipment

In your work you will need specialised equipment to protect you. Examples include dark-tinted glass goggles when welding, masks when painting, waterproof clothing when steam cleaning, or high-visibility clothing when going out to vehicle breakdowns.

As you do vehicle repair and maintenance work you will sometimes need special equipment to protect the top of your head, your eyes, your ears, your hands and feet, and your breathing.

SAFETY CAPS

Bump caps protect your head from banging on the underside of a vehicle when you work under a ramp.

Soft caps keep your head and hair clean. They also prevent long hair from catching in revolving parts, such as drills on the bench, or engine drive belts under the bonnet.

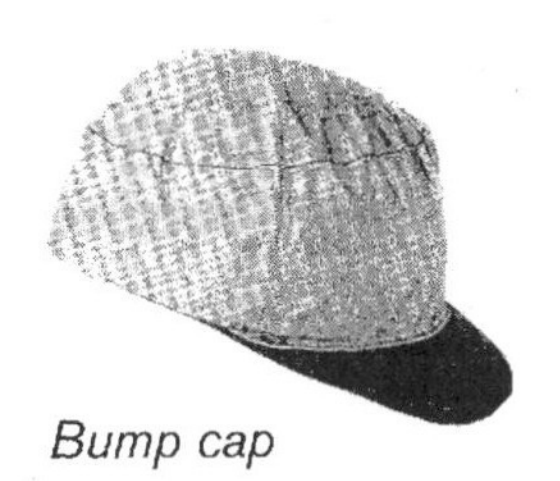

Bump cap

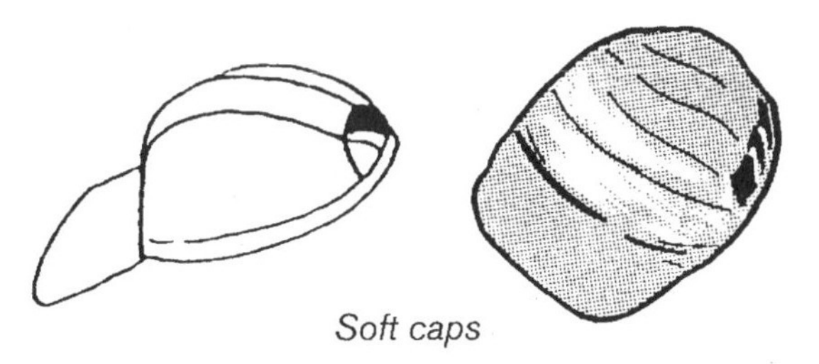

Soft caps

EYE PROTECTORS

An accident to the eyes can be very painful and may result in blindness.

Spectacles protect you from rust or dirt falling off the car. Some have side shields and can be adjusted to fit your face.

Goggles protect you from dust and chemicals. They are used, for example, when sanding body filler off bumped car wings.

Welding goggles protect your eyes from the bright glare of the welding flame.

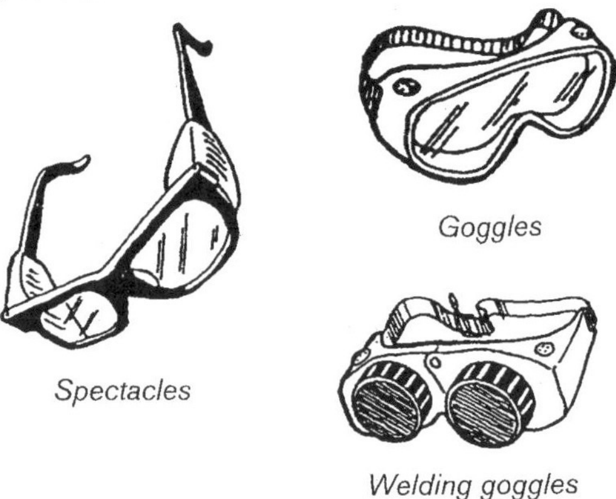

Spectacles

Goggles

Welding goggles

EAR PROTECTORS

Ear muffs protect your ears from damage when there is a loud, continuous noise.

Ear plugs are as effective as, and in some cases more effective than, ear muffs!

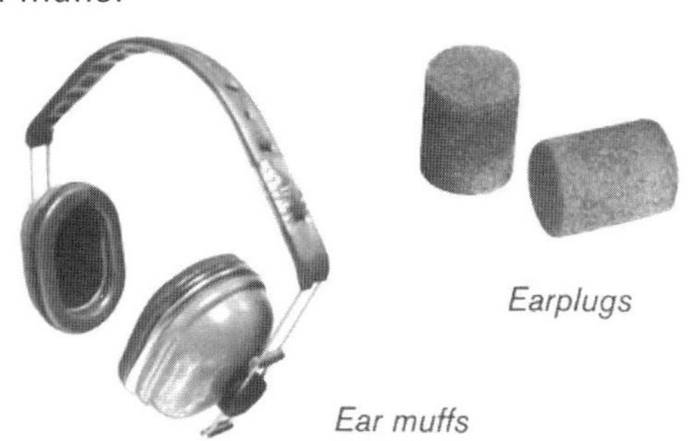

Earplugs

Ear muffs

MASKS

Face masks protect your lungs from dust (some of the dust may be toxic.) They use special moulded pads made from cotton gauze or special filter paper.

Gas respirators are used in vehicle paint shops. The paint fumes may be toxic.

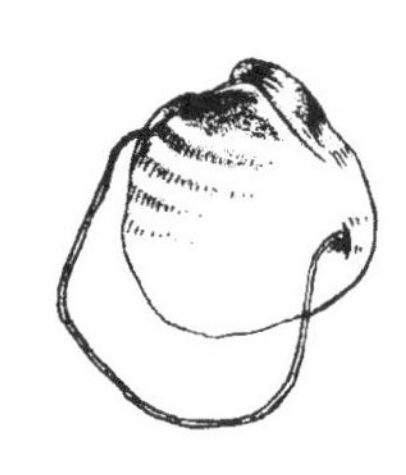

Face mask

HAND PROTECTION

Industrial gloves should be used when moving rough or heavy parts. They protect your hands and wrists from cuts, scratches and burns.

Heat-resistant gloves should be worn when working on items such as a hot exhaust or radiator.

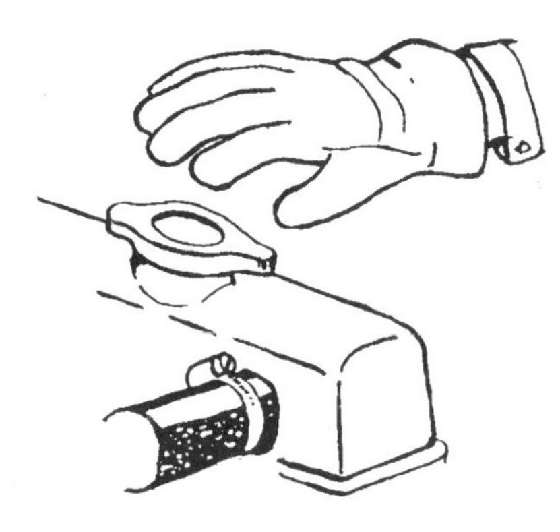

Heat-resistant gloves

SKIN PROTECTION

Before you start work, put barrier cream on your hands. If you have sensitive skin you might easily develop an infection such as dermatitis. Thin plastic gloves (like surgical gloves) can be worn to prevent contact with fuel and oil.

After work, clean your hands with an antiseptic hand cleaner. Rub this on your hands before you get them wet.

Keep your overalls clean, and do not put dirty rags in your pocket. Oil might pass through your clothes and onto your body, and you might develop a skin infection.

FOOT PROTECTION

Safety boots protect your feet and toes from falling objects. In a workshop there is also a risk that a car might run over your feet!

TOTAL PROTECTION

Total waterproof protection will sometimes be needed.

- When working at a car wash or valeting firm, wear waterproof clothing.
- When working on breakdowns, wear high-visibility clothing.
- When paint-spraying a car, wear Tyvek overalls. These have elasticated hoods, cuffs and ankles. Wear a gas respirator, too.

SAFE HANDLING

Moving loads

A load is any heavy object that must be moved, whether by hand or by lifting equipment.

CORRECT HANDLING TECHNIQUES

When you have to lift something big or heavy, you need to lift it in the right way. Look at the drawings and read the numbered instructions.

LIFTING A HEAVY PART FROM THE FLOOR

1 Stand as close to the load as possible. Spread your feet.
2 Bend your knees and keep your back in a straight line. Do not bend your knees fully, as this would leave you with little lifting power.
3 Grip the load firmly.
4 Raise your head.
5 Lift by straightening your legs. Keep the action smooth.
6 Hold the load close to the centre of your body.

UNLOADING ONTO A BENCH

1. Bend your knees to lower the load. Keep your back straight, and weight close to your body.
2. Be careful with your fingers as you set the load down.
3. *Slide* the load into position on the bench. Push with your body.
4. Make sure that the load is secure, and that it won't tip, fall or roll over.

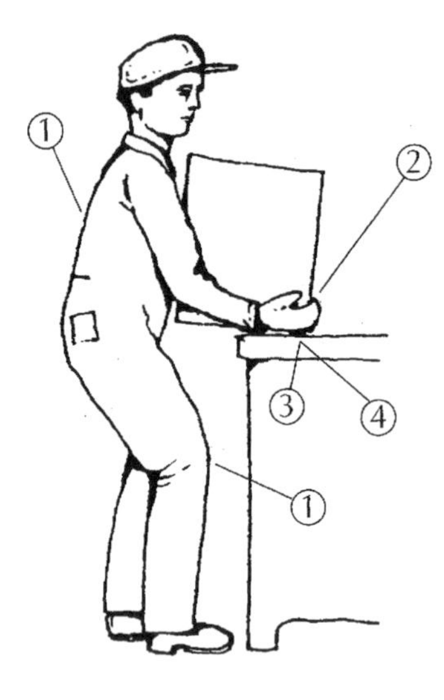

HEALTH & SAFETY

Avoid personal injury. Do not lift anything too heavy for you – about 20 kg is a recommended amount.

If using lifting gear, never exceed the Safe Working Load (SWL).

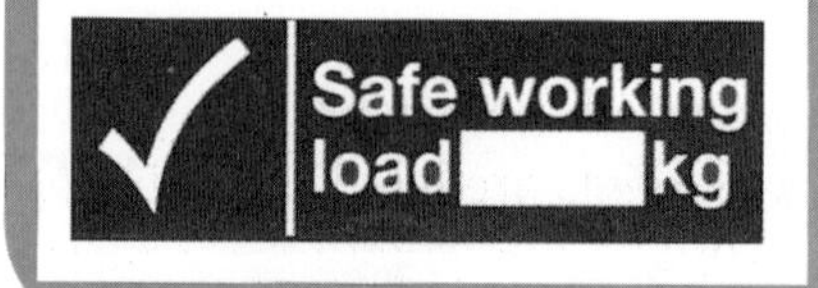

Handling harmful substances

Workshops often store dangerous chemicals. Some could catch fire; some could even explode when handled. Others are corrosive or caustic, and could damage your skin.

There are regulations about the **Control of Substances Hazardous to Health** (**COSHH**). These state that every hazardous substance must be described on a health and safety data sheet. The sheet gives details of safe handling, and says whether protective equipment should be worn.

Safe use of garage machinery and equipment

RAMPS AND JACKS

In the workshop you will often use vehicle lifts and trolley jacks. Make sure you know how to use them safely.

The diagram opposite shows basic precautions you should take when working under a lift. Here are three extra precautions:

- Do not exceed the lift's safe working load (SWL).
- Before raising a car, check that the radio aerial, bonnet and boot lid are down. They could hit lights, beams or the roof.
- Before lowering, make sure that all tools and old parts have been removed.

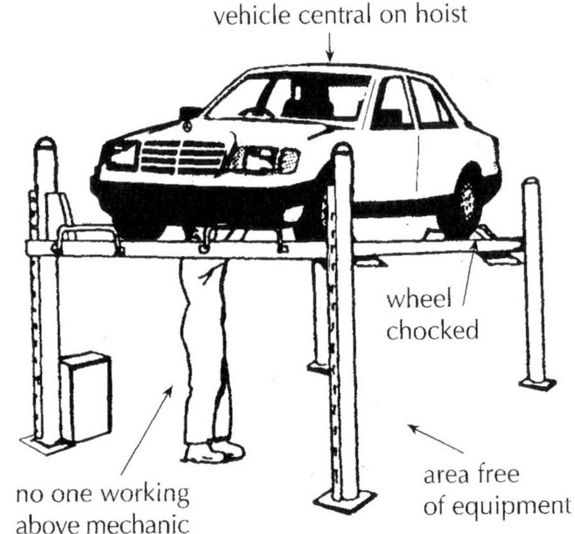

COMBUSTIBLE MATERIALS

Some liquids or chemicals found in a garage catch fire very easily. Petrol is the most obvious example. Vapour from such chemicals could be ignited by a spark – even the tiny spark when a light switch is operated.

Such fluids must be stored in fireproof containers. These are designed to prevent leakage and evaporation.

HEALTH & SAFETY

For your own safety, make sure the equipment you use has been checked and is safe.

ELECTRICAL SAFETY

The main dangers caused by electricity are:

- Fire – due to cables being overloaded, overheating, or loose connections.
- Electric shock – due to touching a live circuit.

Unguarded cables or connections, like those in the diagram opposite, could cause a fire or a shock.

In industrial premises, all electrical equipment must be checked regularly by a qualified person.

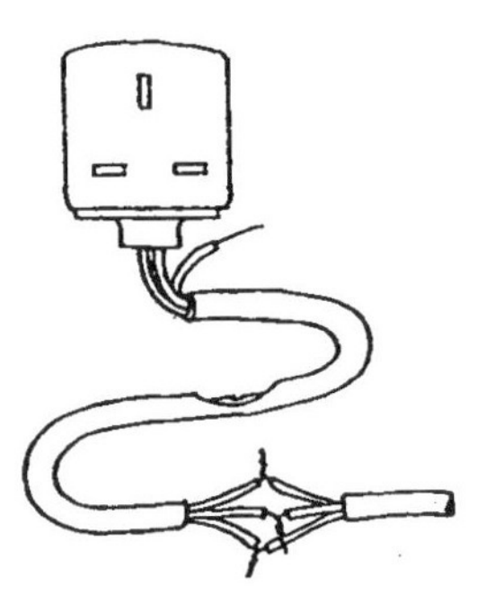

HAND TOOLS

Hand tools are spanners, sockets, screwdrivers, pliers, hammers, chisels and files. To work safely with them use your common sense, know which tool to use, and follow safe procedures.

In a workshop the most common small injuries are cut fingers or skinned knuckles and fingers. Usually these are due to the misuse of a hand tool.

ROTATING MACHINERY

All high-speed rotating machinery, such as wheel balancers, drills and grindstones, should have guards fitted.

COMPRESSED-AIR EQUIPMENT

Compressed air is dangerous if misused. Before you use flexible pipe extensions, make sure that the quick-release couplings are fully engaged.

When working with compressed air:

- Never direct it onto any part of your body.
- Never use it to blow away brake dust (or any other type of dust).
- Never use it to clear dirt or filings off benches.
- Never use it to clean ball and roller bearings (by spinning them).

HEALTH AND SAFETY SIGNS

All public places, including workshops, must display safety signs to warn people of dangers. If you look around buildings you will see such signs. Some will be so familiar that you hardly notice them.

By law (BS 5378 and BS 5449) safety signs must clearly show what they mean. There are different shapes, colours, and symbols or words.

There are four types of safety signs, and fire signs.

Prohibition

A red circular band and a cross bar.

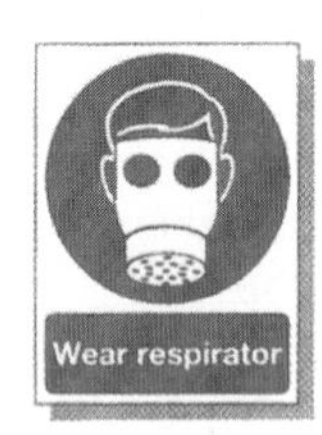

Mandatory

A blue circle with a symbol inside.

Warning

A yellow triangle with a thick black border.

Safe condition

A green square or rectangle with a symbol inside.

EMERGENCY PROCEDURES

Emergencies may be caused by many things: a workshop accident, a fire, a spillage of a flammable or hazardous substance – even a bomb scare.

If you are sure there *is* an emergency, sound the alarm, evacuate the building and ring **999** for the emergency services. (See next page.)

Fire alarm

In a small workshop, the fire alarm may be simply shouting '*Fire*'. In a large workshop an automatic alarm may be linked to the fire station, for immediate action.

If a workshop employs more than five people, it must have an emergency evacuation procedure. It must also have a building plan, and signs that show where to find:

- Fire extinguishers.
- Fire exits.
- Assembly points.
- First-aid points.

Fire extinguishers

There are several types of fire extinguisher, suitable for different kinds of fire. Your garage should have extinguishers to fight fuel fires and electrical fires, as well as ordinary water extinguishers.

Fire equipment

A red square or rectangle with a symbol or text.

Extinguisher contents

The different kinds of extinguisher have different contents.

- **Water extinguisher** Water will kill the heat and put out the fire. This should be used for wood and paper fires. Do not use it for electrical fires: you could get an electric shock. Do not use it for petrol or oil fires; burning fuel will float on the water.
- **Foam extinguisher** The foam is water-based. It smothers the fire which goes out because there is no oxygen. This can be used with flammable liquids. Do not use this in a garage.
- **CO_2 extinguisher** This produces carbon dioxide gas, which removes the oxygen around the fire. However, because CO_2 does not remove the heat, wood and paper could re-ignite later.
- **Dry powder extinguisher** The powder is a fire-retardant dust. This covers the fire like a blanket.
- **BCF extinguisher** This smothers the fire with a blanket of heavy vapour. It is very clean, and leaves no deposit.

New colours

Since January 1997 the British Standard for fire extinguishers has been BS EN 3. Under this standard, all fire extinguishers must be coloured *red*. However, 5% of the surface area may be colour-coded using the colours many people are already familiar with. The table shows types of fires and the extinguishers recommended for fighting them.

Classification of Fire Risk	WATER	FOAM	CO_2 GAS	POWDER	BCF
A Paper, Wood Textile and Fabric	✓			✓	✓
B Flammable Liquids		✓	✓	✓	✓
C Flammable Gases			✓	✓	✓
Electrical Hazards			✓	✓	✓
Vehicle Protection				✓	✓

HEALTH & SAFETY

A fire blanket can be wrapped around someone who is burning. It smothers the fire.

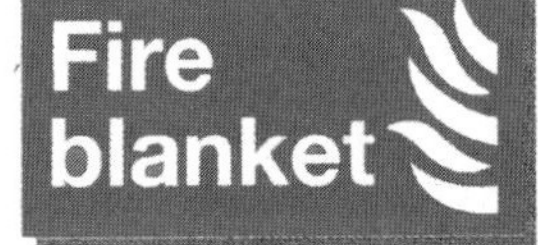

Emergencies

In an emergency, for **fire, police** or **accident**, press or dial **999**.

1. Tell the operator which service you want.
2. Wait for the operator to connect you.
3. Tell the emergency service:
   a. Where the trouble is
   b. What the trouble is
   c. Where you are
   d. The number of the phone you are using.
4. Let the person at the other end of the telephone ask the questions.
5. Don't hang up until the service has all the information it needs.

WARNING

Never make a false emergency call. It is against the law, and you could risk the lives of others who *really* need help.

You can be traced immediately to the telephone where the call came from.

POSITIVE WORKING RELATIONSHIPS

Working with others

There are several different kinds of relationships. You have a family relationship with a relative – a parent, a child, an aunt or a cousin. With people you like at school or college, you have a friendly relationship. You may have a romantic relationship with a girlfriend or boyfriend.

Working relationships develop with the people who work alongside you. As you interact in the workshop with the manager, other mechanics and fitters, or your supervisor, you build working relationships with them.

Good working relationships are very important to the success of every business. You may not be friends with all your colleagues; occasionally you may even dislike some of them. But to help the business run smoothly, you must get on with all of them professionally.

WORKING AS A TEAM

To make the company successful, all of its employees must work together. They must co-operate, like members of a football team: this is teamwork.

If the firm does well, and employees get on with each other and trust each other, there will be a good feeling in the workplace, and people will be enthusiastic about their jobs. This feeling is called good morale.

When everyone works hard, and no one wastes time or resources, the firm will be efficient. By being efficient, employees will get a lot done – they will be productive.

If a team with good morale works efficiently and productively, customers will be satisfied and pleased to come again. They are also likely to recommend the firm to others, so it will gain a good reputation. And this in turn will bring more business, and the firm will become even more successful, and will grow. It will gain a good company image.

The company's aim

It is important to be aware of the aim of the company for which you work. This may be set out in what is known as a mission statement. This explains the main objectives of the company: the reasons why it is trading, and what it hopes to achieve.

All firms are in business to make a profit. This is shared between the owners or any shareholders. A profitable business is likely to be a successful one, and a successful business can offer its workforce job security.

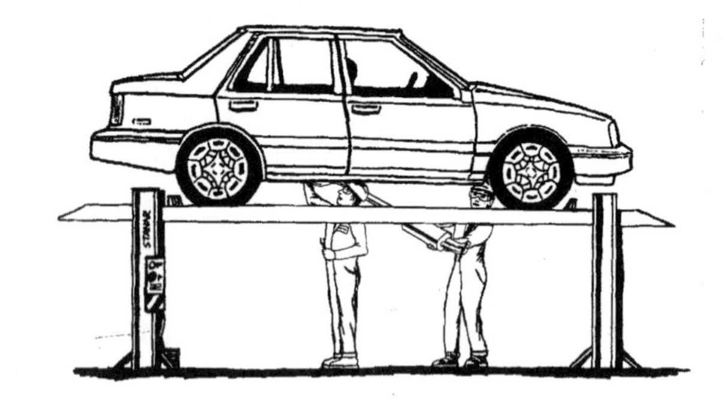

ORGANISATIONAL STRUCTURE

The organisational structure describes who does what in the company. It describes each person's duties and level of authority. It also shows everyone's working relationships, from the managing director to the most junior employee.

Job description

A job description names a job title and states the duties and responsibilities of a particular job.

Each employee should be aware of their own role and the roles of others. Disputes may arise if anyone is uncertain about who does what.

An organisation chart such as the one below shows levels of authority and responsibilities. Vertical links show the line of authority from senior staff downwards. Horizontal links indicate people with equal status and authority in the company.

general manager
service manager
sales manager
parts manager
reception
works foreman
sales admin.
senior storeman
works staff
sales staff
parts staff

Building good relationships

Here are some ways in which you can build good working relationships with your colleagues.

- Recognise that there are differences in personality and temperament.
- Treat colleagues politely and with respect.
- Co-operate, and assist willingly with requests.
- Talk with colleagues about problems, changes or proposals.
- If something you are doing goes wrong, or if you break something, tell your supervisor straight away.
- When a working relationship breaks down, be honest and fair, and try to put things right.

Here are a few examples of the kind of things that can upset good working relationships

- In everyday conversation you may discuss your social life, sports, films and so on. You may find that a workmate has opinions on some topics which are different from yours. Never allow these differences to spoil your working relationship.
- Sometimes you may find colleagues who lack interest or enthusiasm, who are lazy or incompetent, who keep being absent, and so on. If people do not 'pull their weight' in the workplace, this can cause anger and frustration among other members of the team.
- Beware of anyone who ignores company rules and regulations, or safe working practices. This behaviour can create problems or even dangers for everyone else in the company.
- Personal appearance and hygiene are important. Employees who do not bother with these may upset colleagues, and this may affect the performance of a workforce.
- Managers and supervisors should not show favouritism and should pay everyone fairly.

The company image depends in part on the way its workers are judged by customers. Customers will see your work, and also how you behave and interact with your colleagues.

COMMUNICATION

When problems and misunderstandings do arise, it is often because of poor communication. Good communication is an essential part of every working day, for employees at all levels.

In a typical garage, there will be daily communication between colleagues within a department, between departments, between management, supervisors and shop-floor staff, and between customers and reception staff. The company will also communicate with suppliers, subcontractors, vehicle manufacturers, advertising agencies, banks, accountants, lawyers, the council, and so on.

Trust and goodwill with customers will be very much helped if you practise good, clear communication. Misunderstandings can easily make customers go elsewhere. In Britain, customers who are not satisfied tend not to complain – they just don't come back. So you may never find out that something is wrong – until, perhaps, there is no work!

INCORRECT COMMUNICATION

Problems can also arise if information given is wrong. A good example of this is when a spoken message is passed on and wrongly remembered – always write down messages for people, while the information is clear in your mind. Similarly, if a job sheet lacks some detail this causes confusion.

If a message is passed orally via several people, it may become totally changed!

Non-verbal communication

It is not only your words that communicate – so too do your facial expression and the way you stand and move. Customers will notice your body language: for example whether you smile and look directly at them or slouch and avoid their eyes.

Problems in communication

Some things can cause problems in communication.

LACK OF COMMUNICATION

Problems can arise if people are not told things they need to know. Changes to systems, safety issues, shop-floor problems and the like must be passed on to the right people without delay.

Delays in getting information to staff may be caused by pressure of work or different hours of work (such as part-time, full-time, or shift work).

Methods of communication

This section is particularly concerned with the effects of communication on working relationships. The method of communication chosen will depend on the situation.

1 *Direct discussion* – face-to-face conversation.
   a With a customer, to find out their requirements and to advise them.
   b Between colleagues, about a particular job.
   c With management and supervisory staff, to talk over procedures and problems.
2 *Writing*
   a Recording customer and vehicle details.
   b Preparation of job sheets. Inter-department memos.
   c Letters. Reports. Manuals. Notices.
3 *Telephone*
   a Quick communication with customers, suppliers and others.
   b Between departments and colleagues.
   c Quick contact with emergency services.

Chapter 1

Clutches

Clutches	2
Clutch plate constructional features	5
Clutch operation and adjustment	6
Multi-plate clutches	10
Checking the clutch pressure plate	11
Diagnostics	13
Torque and power transmitted by friction clutches	14
Investigation	14
Calculations (clutch linkage)	16

CLUTCHES

In a vehicle transmission system the clutch transmits the engine torque to the gearbox. State three other purposes which the clutch fulfils.

....

....

....

....

....

....

The clutch assemblies shown opposite are typical examples of light and heavy vehicle units, which are manually operated by the driver controlling a pedal. Label the illustrations and briefly describe the significant features of the clutch operation.

DRIVE

....

....

....

RELEASE

....

....

....

The main differences between clutch assemblies other than coil or diaphragm spring operation are:

diameter (increased for greater torque capacity)
clamping springs (increased in strength or number for increased torque capacity)
withdrawal mechanism (push or pull)
drive to pressure plate (lug or strap).

DIAPHRAGM SPRING TYPE

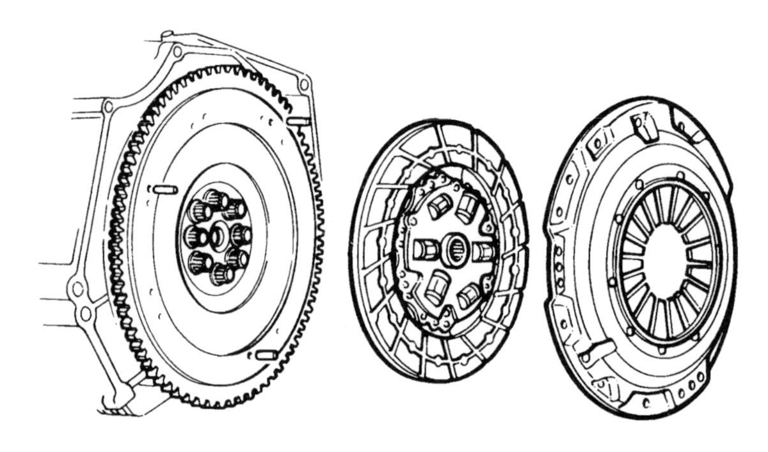

COIL SPRING TYPE (HGV)

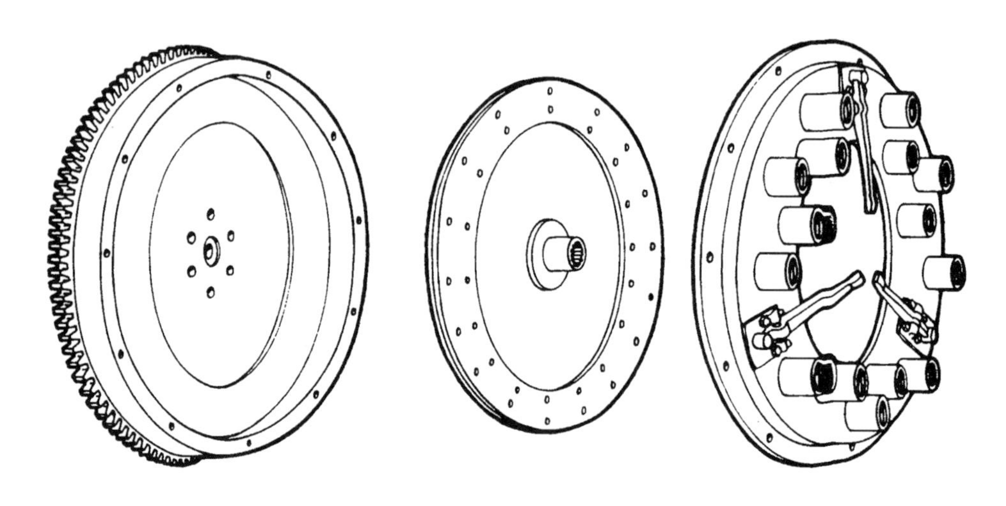

The Multi-spring Clutch

A simple multi-spring type of clutch is illustrated below. Label the drawing and state the function and operation of the parts listed opposite.

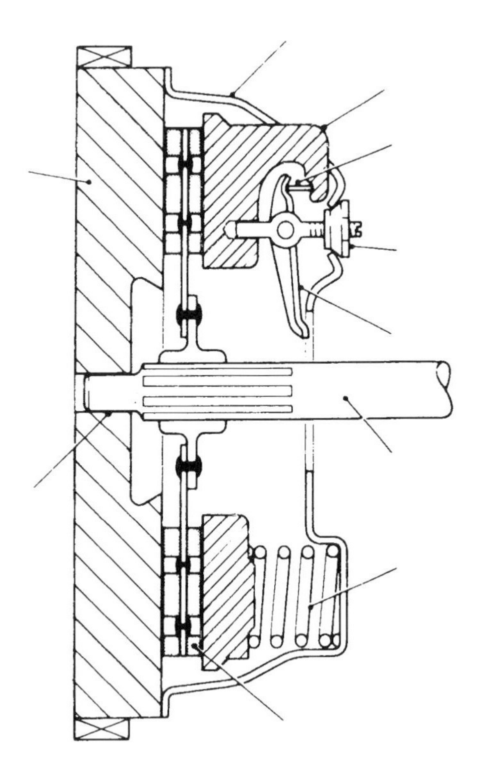

Power flow

In the clutch shown, 'lugs' formed on the pressure plate fit into rectangular slots in the cover to provide drive from cover to pressure plate.

Add arrows to the drawing to indicate the 'power flow' through the clutch from the fly wheel to the primary shaft.

Cover

....

....

....

Coil springs

....

....

....

Pressure plate

....

....

....

Release levers – eye bolts and pins

....

....

....

Adjusting nuts

....

....

....

'Knife-edge' struts

....

....

....

....

The Diaphragm Spring Clutch

A single-plate diaphragm clutch is shown below; label the drawing and add arrows to show the power flow through the clutch from the flywheel to the primary shaft.

A is release bearing movement
B is pressure plate movement

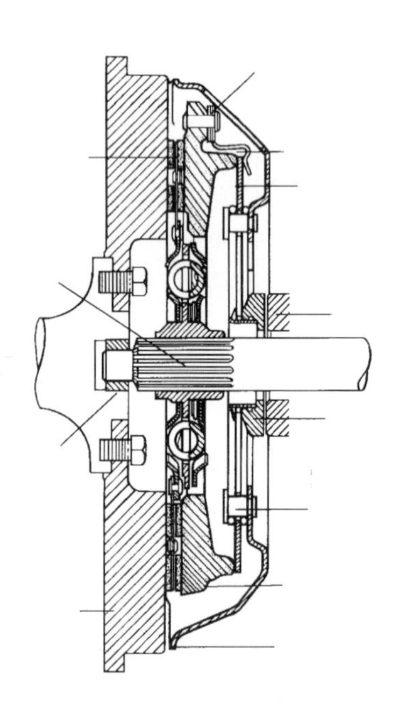

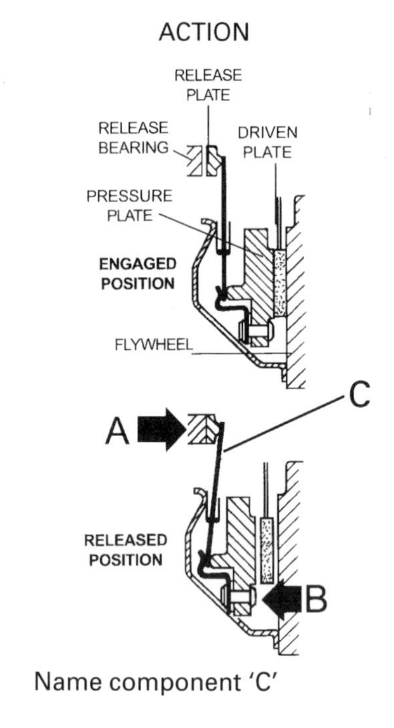

Name component 'C'

....

What particular feature of a clutch pressure plate is illustrated opposite?

....

....

....

....

....

With a coil spring type pressure plate, as the springs are compressed (pedal depressed) the spring force increases and as the springs are extended (owing to lining wear) the spring clamping force reduces. The diaphragm spring has the opposite characteristics: see the graph. Add to the graph the load/displacement curve for a coil spring.

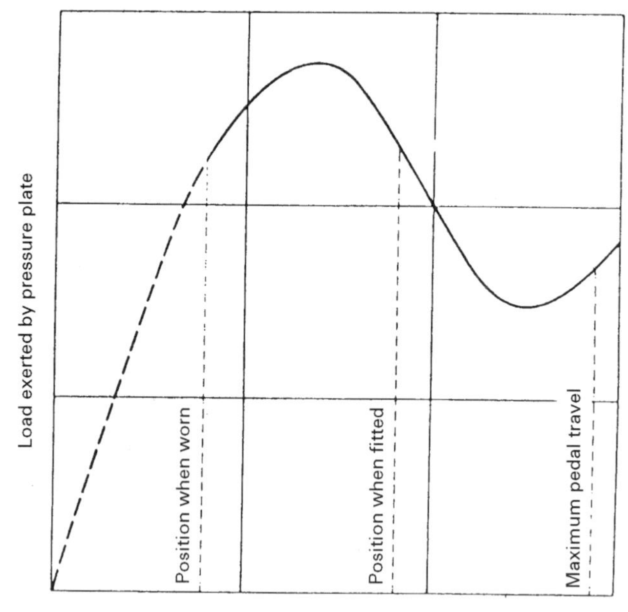

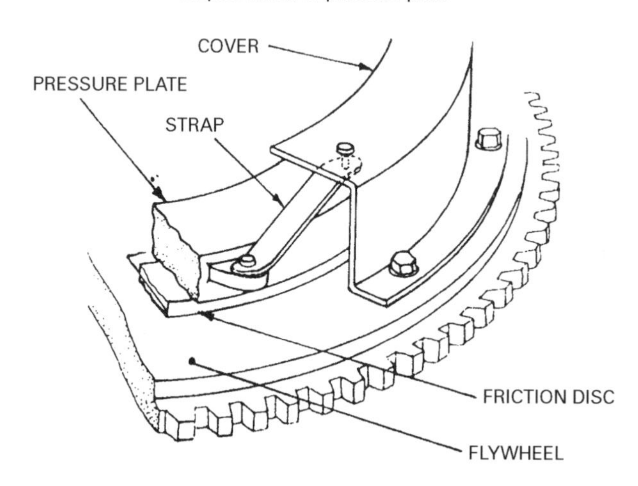

CLUTCH PLATE CONSTRUCTIONAL FEATURES

The two main types of clutch plate in general use are shown below:

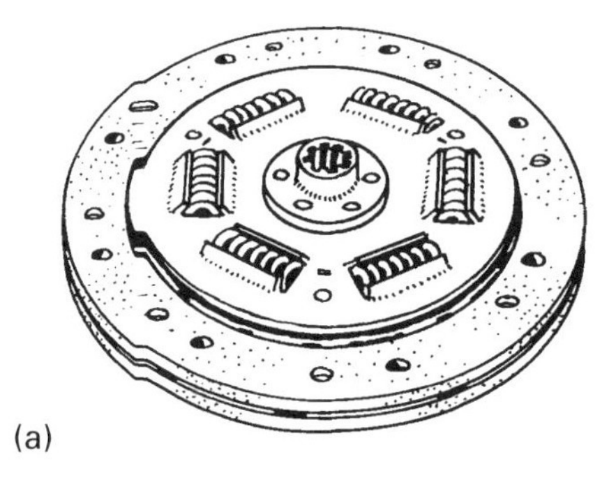
(a)

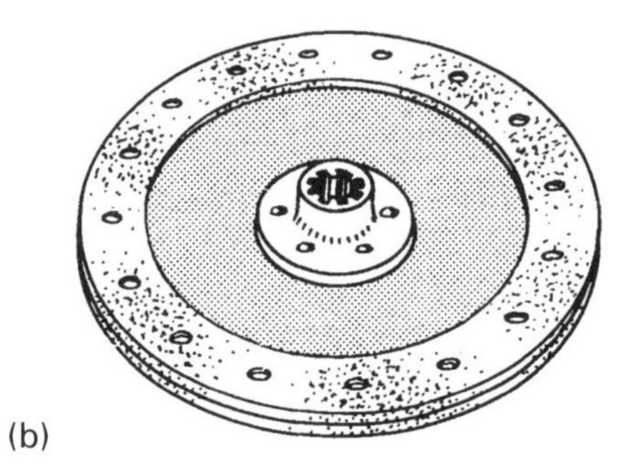
(b)

Type (a) ....

....

Type (b) ....

....

Application
The spring centre type is used on most light vehicles.

The spring hub cushions drive take-up and absorbs torsional vibration due to engine torque fluctuations. Examine a dismantled spring type clutch centre plate and explain how torsional vibration is absorbed.

....

....

....

....

....

....

....

Friction materials – clutch plate facings or linings

The friction material from which the clutch linings are made must: maintain its frictional qualities at high temperatures; be hard wearing; withstand high pressure and centrifugal force; and must also offer some resistance to oil impregnation.

Several materials are currently in use for friction linings. Asbestos bonded with resin has been the most widely adopted material in the past, but many firms now use non-asbestos materials.

Name two other ORGANIC (resin bonded) materials used for clutch plate facings.

....

An alternative to the organic type of lining is used in many heavy duty applications; this is ....

The drawings below illustrate how the friction linings are attached to the clutch centre plate:

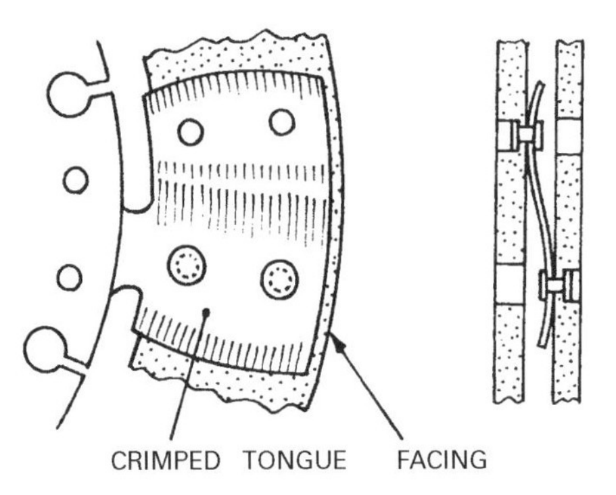

Examine a clutch plate and observe the effect of this method of securing the facings. State the purpose of such an arrangement:

....

....

....

....

....

CLUTCH OPERATION AND ADJUSTMENT

State four methods of clutch operation: ....

....

The clutch-release bearing is carried on a withdrawal lever or fork which pivots in the clutch housing. One arrangement is shown at (a) below; make a similar sketch at (b) to show an alternative.

(a) (b)

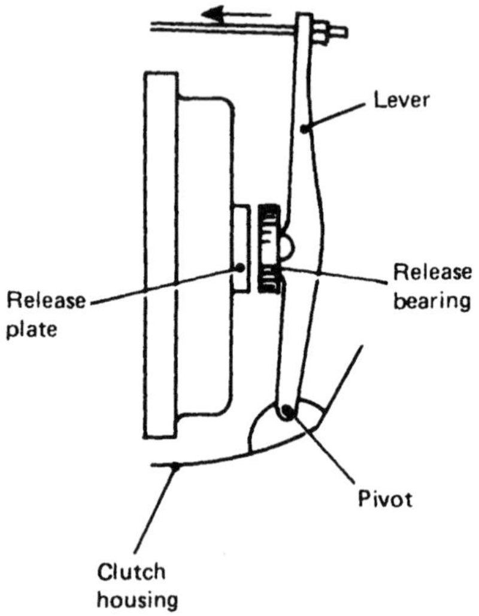

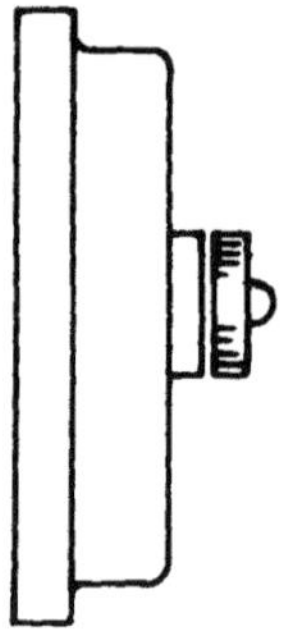

Lever

Release plate

Release bearing

Pivot

Clutch housing

In its released position the release bearing is just clear of the release plate or diaphragm fingers; or, in some systems it may be in very light continuous contact. Why is the disengaged position of the release bearing critical with regard to clutch operation?

....

....

....

....

Label the clutch operating system shown below and describe its action during operation:

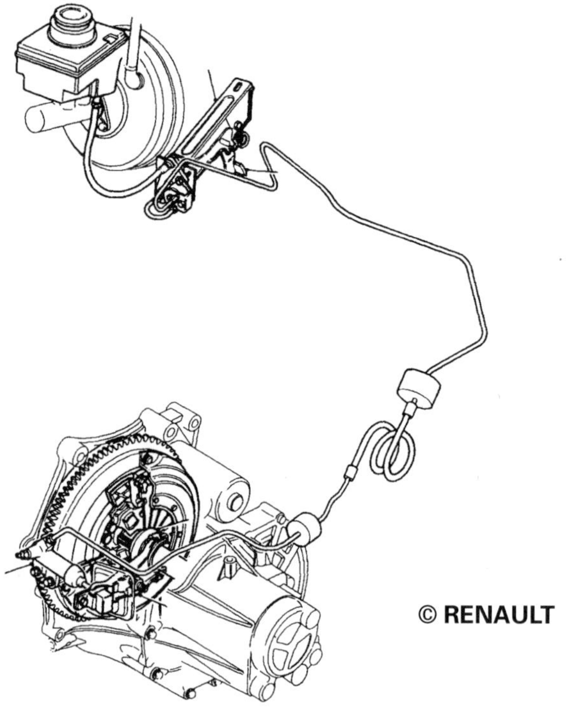

© RENAULT

....

....

....

....

....

A common fault with this system is fluid leakage from the slave cylinder. What are the effects of this on clutch operation?

....

....

....

....

Investigation

Examine a vehicle with a cable operated clutch and describe, with the aid of a sketch, the method of adjustment.

....

....

....

....

....

State clutch faults which would create the need for:

(a) increasing free movement

....

(b) decreasing free movement

....

Self-adjusting Mechanisms

The slave cylinder in a hydraulic clutch operating system can provide automatic adjustment to compensate for friction lining wear. Describe briefly how this is achieved:

....

....

....

....

....

An automatic clutch adjustment system is shown below. Label the drawing and describe briefly how the system operates:

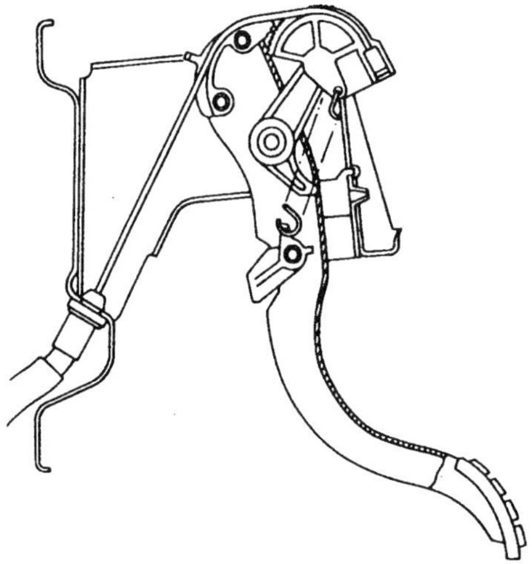

....

....

....

....

....

....

....

....

....

....

State the effects of a broken quadrant tension spring in the mechanism above:

....

....

....

....

Release Bearing

The release bearing transmits the axial thrust to the release levers, plate or diaphragm or the clutch pressure plate during engagement and disengagement.

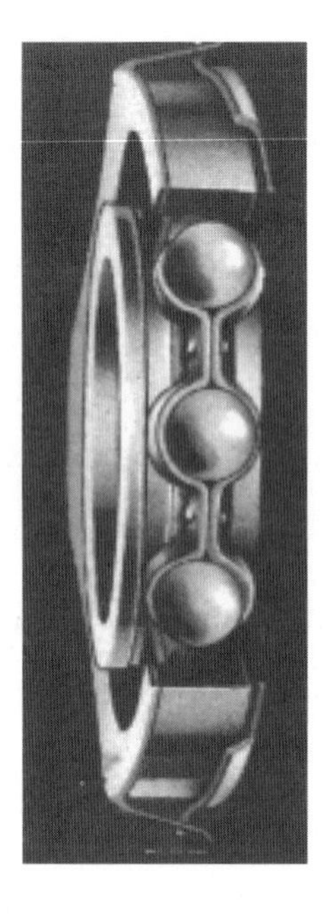

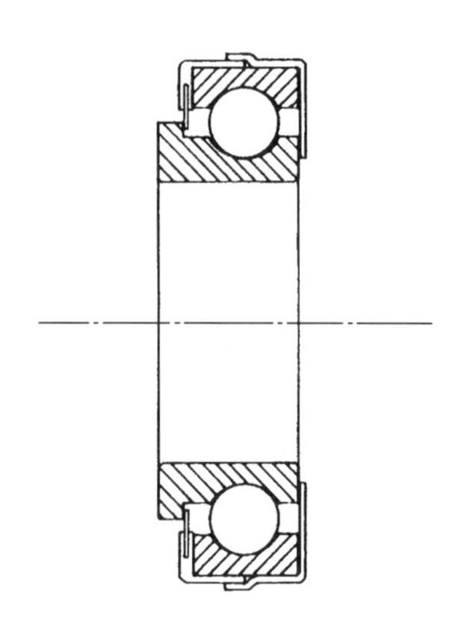

The bearing shown is a single row deep groove ball bearing. This type of bearing will withstand the axial thrust imposed during operation.
How is this type of release bearing lubricated?

....

....

....

State the purpose of the ball bearings:

....

....

....

Release bearing carrier

To achieve correct contact with the pressure plate the ball bearing type release bearing moves parallel to the primary shaft; it may be permanently in contact with the pressure plate or held just clear. The release bearing is very often mounted on a carrier which slides on the primary shaft sleeve to provide linear movement. Label the drawing.

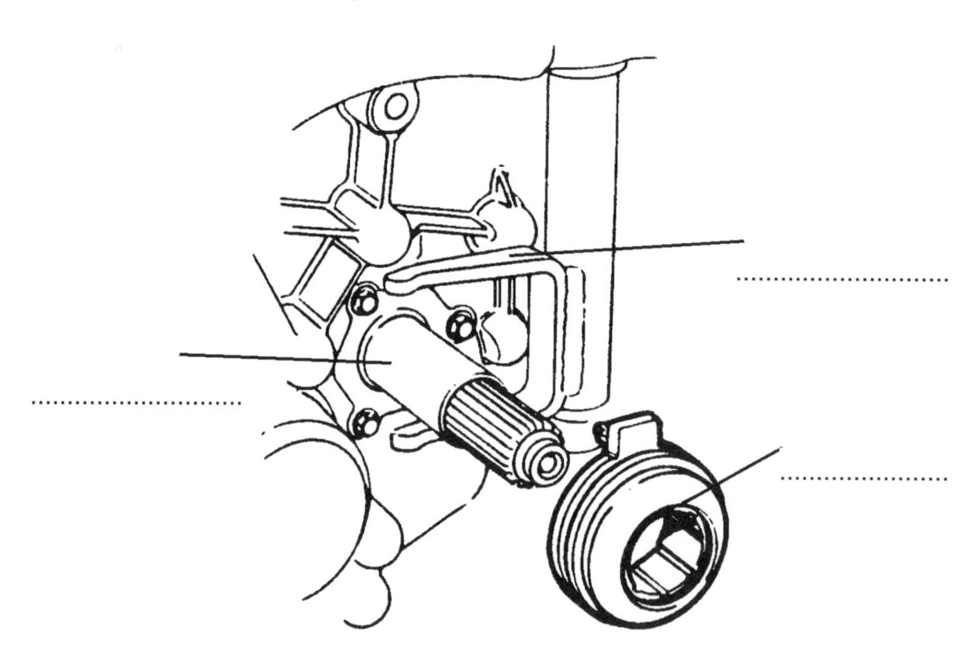

Spigot bearing

The front end of the first motion shaft is supported in a bearing located in the flywheel; name two types of bearing used for this purpose and add one to the drawing above.

....

How is a spigot shaft bush in the flywheel lubricated?

....

....

Air Assisted Clutch Operation (HGV)

Power assisted clutch operation reduces the effort required at the pedal. One arrangement employs a clutch servo, which is an air/hydraulic unit. The servo is located at the clutch housing to actuate the clutch shaft lever. Study the section of a clutch servo below and describe its operation:

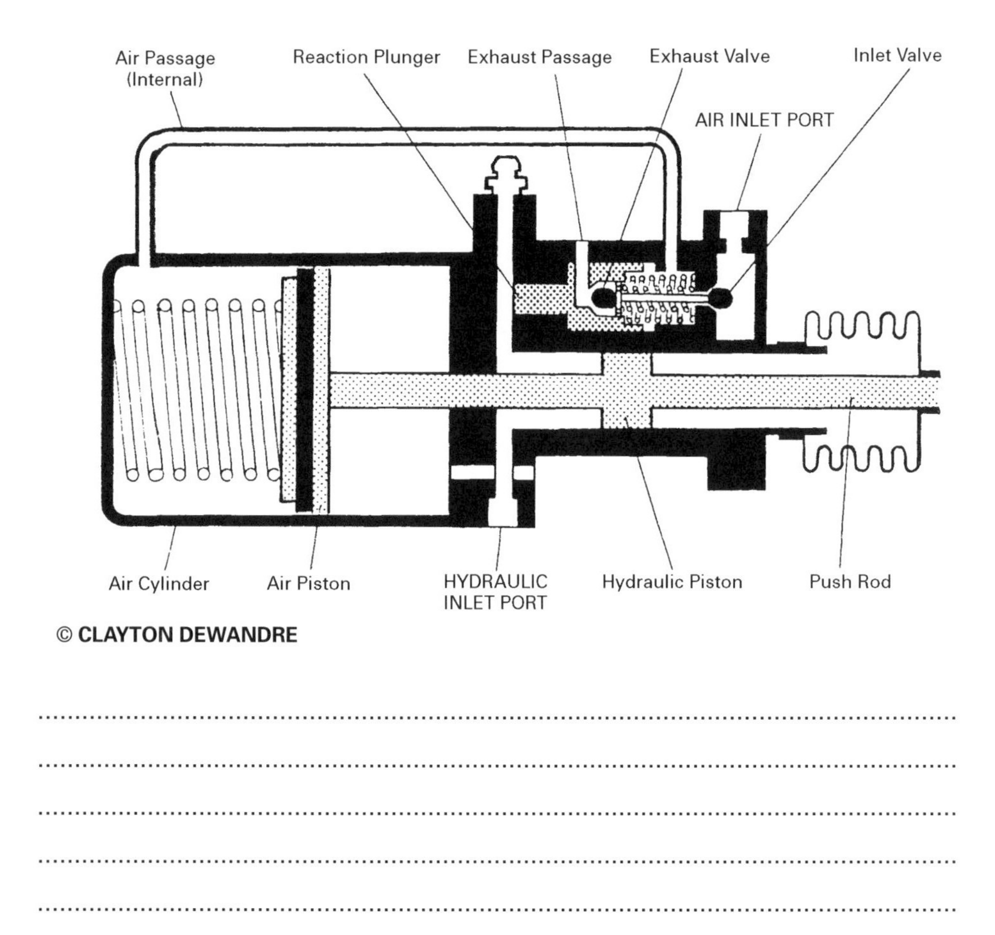

© CLAYTON DEWANDRE

....

....

....

....

....

....

....

....

Clutch Stop

The purpose of a clutch stop, or brake, is to prevent the centre plate and primary shaft assembly from spinning or rotating too long following clutch disengagement.

On which vehicles are clutch stops used and why?

....

....

OPERATION (clutch stop)

Describe the action of the system shown below:

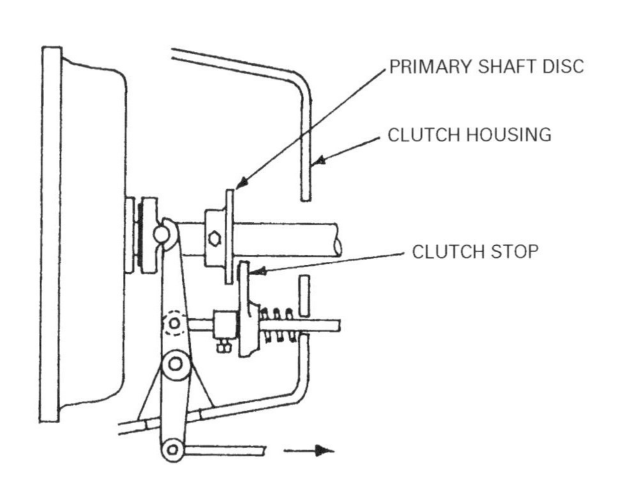

....

....

....

....

....

MULTI-PLATE CLUTCHES

The multi-plate clutch is a clutch assembly which utilises more than one clutch plate.

There are three basic reasons for using multi-plate clutches:

1. *Allows a reduction in clutch diameter* ....
2. ....
3. ....

A typical example of an HGV multi-plate clutch is the twin disc type shown opposite. The use of an intermediate plate and two clutch plates doubles the number of frictional driving faces in contact. Label the drawing and describe the operation of the clutch:

....

....

....

....

....

....

....

....

....

....

Describe an alternative method of transmitting the drive to the intermediate pressure plate:

....

....

....

Twin Disc Pull Type Clutch Assembly

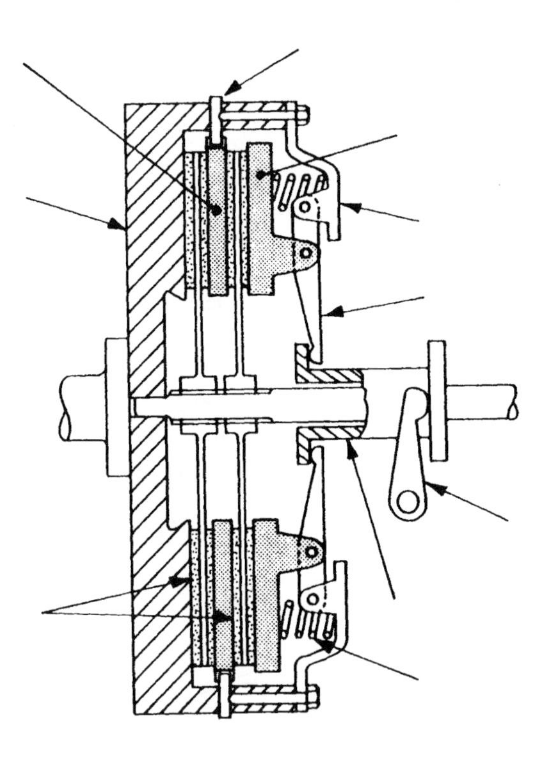

This is an example of a multi-plate, pull type clutch as used on high performance race and rally cars. It is a compact type of clutch which provides increased strength, plus low weight and low inertia.

AP RACING CLUTCH

Show on the drawing below how the flywheel should be checked for run-out.

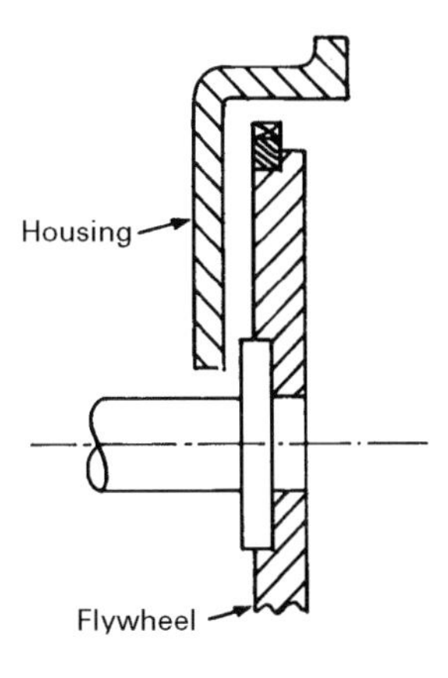

Indicate the mean radius and state the maximum run-out.

Maximum run-out ....

A hydraulically operated clutch needs to be bled on certain occasions. When are these and what equipment is needed?

....

....

....

....

The cover must be located on the 'flywheel dowels' during reassembly otherwise vibration will occur.

State the purpose of the dowels.

....

....

CHECKING THE CLUTCH PRESSURE PLATE

Visual checks on a clutch pressure plate would be for surface cracks, scoring and spring finger breakage. Describe the checks shown below. As an alternative to the procedure shown in the upper diagram a straight edge may be placed across the cover mounting flange.

....

....

....

....

....

....

....

....

Clutch System and Component Protection

How might the system and its components be protected during use or repair against the hazards listed?

(a) Ingress of dirt ....

....

....

(b) Moisture absorption by fluid ....

....

....

(c) Mixing of incompatible fluids ....

....

....

(d) Fluid contamination ....

....

....

(e) Friction face contamination ....

....

....

(f) Excessive facing wear ....

....

....

(g) Shock loading ....

....

....

(h) Undue cable and pivot wear ....

....

....

Clutch Maintenance Adjustments

Preventive routine maintenance is necessary in order to ensure clutch efficiency, prolong its life and minimise the risk of failure. List six important maintenance tasks:

....

....

....

....

....

....

....

Outline clutch tests in respect of the following:

Engagement and disengagement ....

....

....

....

Abnormal noise ....

....

....

Abnormal vibration ....

....

....

....

Clutch brake operation ....

....

....

DIAGNOSTICS: CLUTCH FAULTS

State a likely cause for each symptom/system fault listed below. Each cause will suggest any corrective action required.

SYMPTOMS	FAULTS	PROBABLE CAUSES
Partial or total loss of drive; vehicle speed lower than normal compared with engine speed	Slip	
Difficulty in obtaining gears, particularly first and reverse gears	Drag	
Difficulty in controlling initial take-up drive	Fierceness or snatch	
Shudder and vibration as vehicle moves off from rest	Judder	
Squeak on first contact with clutch pedal; rattle on tickover	Squeak or rattle	
Difficulty in obtaining gears	Spin	

Complete the table below in respect of the general rules for efficiency and precautions to be observed during clutch maintenance and repair:

OPERATION	GENERAL RULES
Lifting and supporting; preventing distortion	
Obtaining correct free play	
Correct fitting of centre plate	
Ensuring component cleanliness	
Avoiding fluid spillage; disposal of waste	
Use of clean fluid	

TORQUE AND POWER TRANSMITTED BY FRICTION CLUTCHES

The factors affecting the torque transmitted by a friction clutch are:

the coefficient of friction between the contact surface	μ
the total force exerted by the pressure plate springs	W
the mean radius of the clutch linings	r
the number of pairs of contact surfaces	n

These values are multiplied together to obtain the torque transmitted by a friction clutch, that is:

torque transmitted = μWrn

What is the mean radius of a clutch plate whose friction rings are 0.25 m outside diameter (r_1) and 0.15 m inside diameter (r_2)?

Mean radius (r) = ―――――――― = ... m

How many pairs of contact surfaces has a single-plate clutch? ....

Having found the torque (T), if the speed of rotation of the clutch (N) is known it is a simple matter to calculate the power transmitted by the clutch, using the formula:

Power = $2\pi NT$

INVESTIGATION

Conduct a coefficient of friction experiment using various sections of clutch lining, and state the coefficient of friction for:

(a) a new clutch lining ....

(b) a polished clutch lining ....

(c) a glazed clutch lining ....

To show the effect of the number of pairs of contact surfaces:

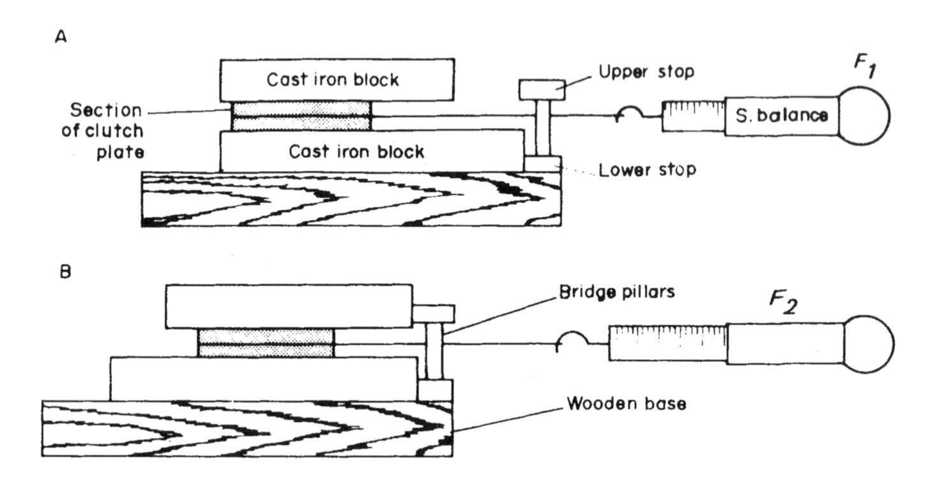

1. Using apparatus similar to that shown above, determine the force required to slide the friction material on the lower CI block while carrying the upper CI block (A above).
2. Note force required to slide friction material between the CI blocks (B above).

   Spring balance reading before contacting stop (F_1) =
   Spring balance reading after contacting stop (F_2) =

What effect have the number of contact surfaces on the force of friction?

....

....

How can this effect be used to advantage in a clutch assembly?

....

....

PROBLEMS

1. A single-plate clutch has a coefficient of friction of 0.3 and the friction linings have a mean radius of 75 mm. If the total force exerted by the pressure plate spring is 2200 N, calculate the torque transmitted by the clutch.

   Torque transmitted = μWrn

   $$\text{torque transmitted} = \frac{0.3 \times 2200 \times 75 \times 2}{1000}$$

   $\therefore$ torque transmitted = 99 N m

2. A single-plate clutch has facings of 225 mm outside diameter, 135 mm inside diameter and a coefficient of friction of 0.24. The nine pressure plate springs each exert a force of 320 N. Calculate:

   (a) the maximum torque that could be transmitted by the clutch.

   (b) the power transmitted by the clutch at a speed of 50 rev/s.

3. Determine the coefficient of friction between the contact surfaces of a single-plate clutch which will transmit a torque of 125 N m, if the total spring force acting on a friction plate of 100 mm mean radius is 2500 N.

4. A single-plate clutch has a mean radius of 80 mm and a coefficient of friction of 0.3. What force must be exerted by each of six springs to enable the clutch to transmit a torque of 105 N m?

5. A multi-plate clutch has three friction plates of 60 mm mean radius running in oil. The coefficient of friction is 0.15 and the total spring force is 1000 N. Calculate the torque and power transmitted by the clutch at 40 rev/s.

CALCULATIONS (CLUTCH LINKAGE)

Two types of lever used in clutch linkages are shown below. In each type of lever the load and effort will each produce a 'clockwise' and 'anti-clockwise' turning moment about the fulcrum or pivot. The turning moment is the product of the *force* and the *perpendicular distance from the fulcrum.*

When a state of balance (equilibrium) is maintained:

Clockwise moments = anti-clockwise moments

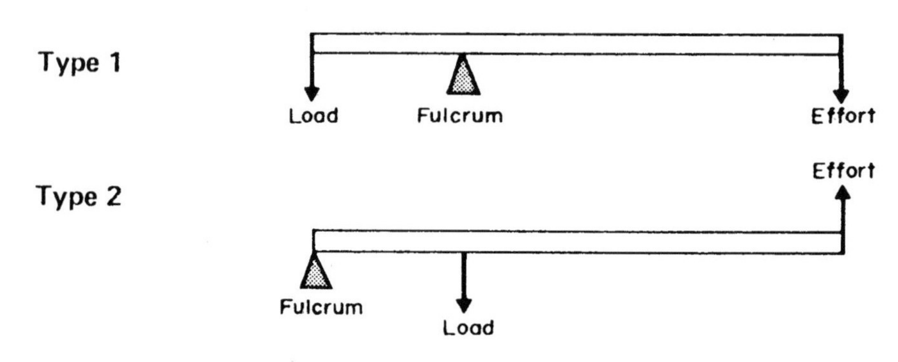

1. Calculate the pull on the cable for the pedal lever shown when the force exerted at the pedal is 90 N.

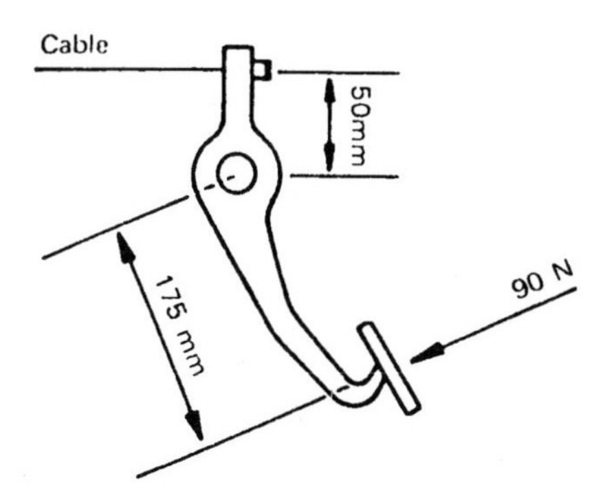

2. Calculate the force exerted by the slave cylinder push rod below when the force at the release bearing is 800 N.

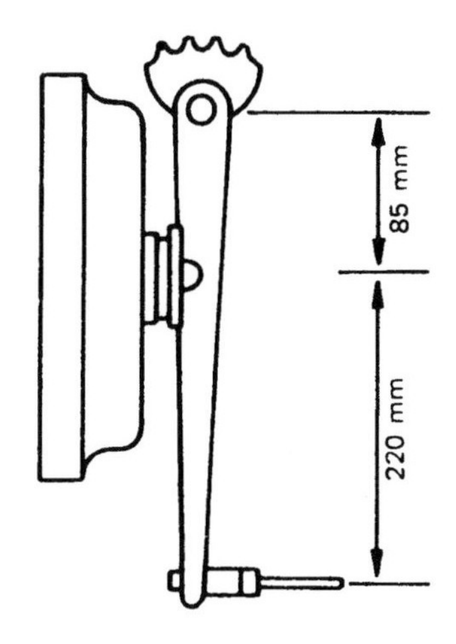

In question (1), owing to the ratio of the lever, the movement at the input (pedal) is $3\frac{1}{2}$ times greater than the movement at the output (cable). This is known as the MOVEMENT RATIO.

MOVEMENT RATIO (MR) = ...

FORCE RATIO (FR) = ... or ....

% EFFICIENCY = ... × 100

3. Calculate the force required at right angles to the clutch pedal lever to operate the clutch with the mechanism shown below:

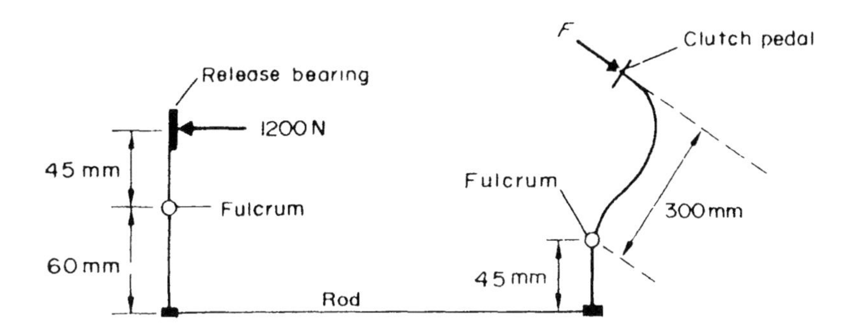

5.

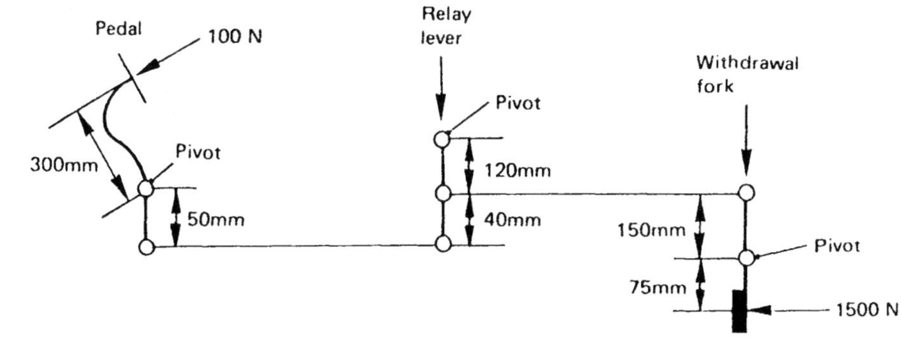

Calculate the MR, FR and EFFICIENCY for the HGV clutch-operating system shown above.

4. In a cable-operated clutch a pedal effort of 90 N is applied through a distance of 150 mm. If the force applied at the release bearing is 540 N, calculate MR, FR and EFFICIENCY for the system if the movement of the release bearing is 20 mm.

6. In an HGV the force required at the release bearing to disengage the clutch is 2400 N. Calculate the pedal effort if the clutch operation system has a movement ratio of 18:1 and an efficiency of 90%.

Chapter 2

Manual Gearboxes

Manual gearboxes 19
Function 20
Constant-mesh gearbox 21
Synchromesh devices 22
Five-speed gearbox 24
Selector mechanisms 25
Investigation 25
Gearbox bearings and mountings 27
Gearbox and linkage routine maintenance and lubrication 28
Auxiliary gearboxes (HGVs) 29
Range change gearbox 30
Twin layshaft gearbox 31
Speedometer/tachograph drive arrangement 32
Diagnostics 33
Testing and test equipment 33
Investigation: gearbox and operating linkage removal and refitting 34
Power take off (PTO) systems 35
Four-wheel drive (4WD or 4 × 4) 36
Two-speed transfer box 37
Differential locks 38
Viscous coupling (VC) 39
Gear ratio, torque ratio, efficiency 42
Calculations: gear ratio 43
Calculations: torque ratio 43
Motion 44
Compound gear trains 45
Investigation 47

MANUAL GEARBOXES

Location

In a vehicle transmission the gearbox is located between the clutch and final drive gears. Its actual location on the vehicle, however, depends on the layout of the main components.

Name the drive arrangements on this page and arrow the gearbox on each layout.

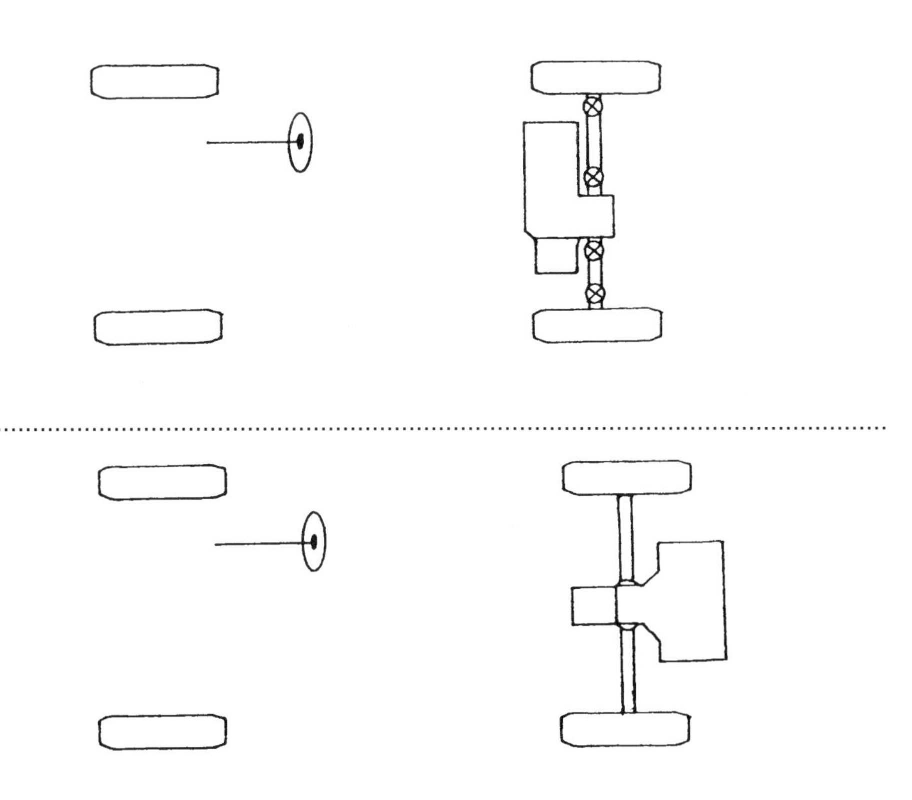

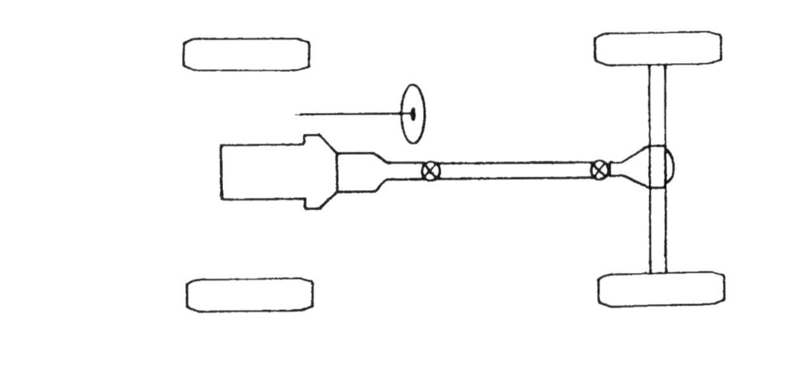

....

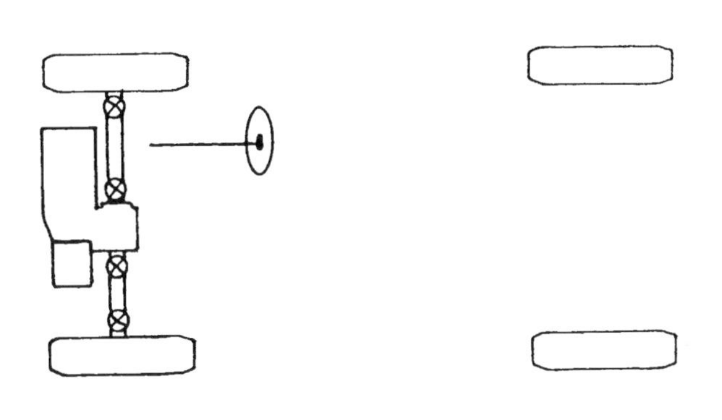

....

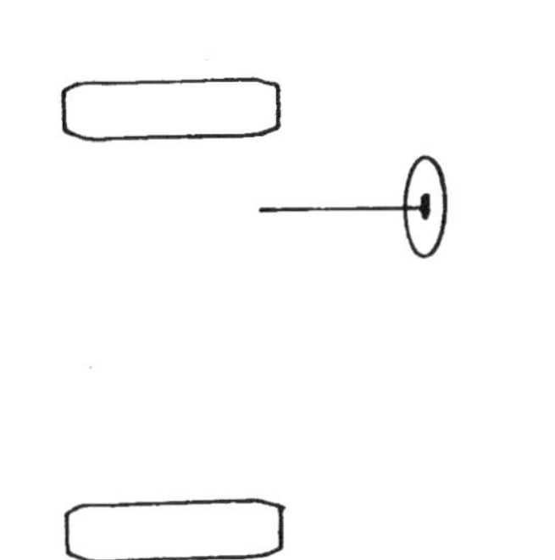

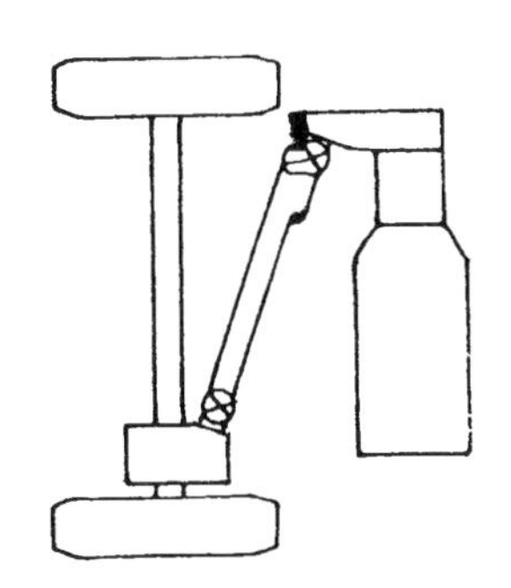

....

FUNCTION

Basically the gearbox serves a number of purposes:

1. *To multiply (or increase) the torque (turning effort) being transmitted* ....

....

....

....

....

When the gearbox multiplies engine torque, a speed reduction occurs between gearbox input and output shafts. When the torque multiplication is high the vehicle speed is .... The gear in which maximum torque multiplication is obtained is ....

The torque output from the gearbox is therefore varied according to the SPEED AND LOAD REQUIREMENT of the vehicle.

The range of GEAR RATIOS provided by the gearbox enables the engine's torque and speed characteristics to be used most effectively.

Give an example of the gear ratios provided by a gearbox employed in a modern car or HGV.

Vehicle's make .... Model ....

....

....

....

Modern HGV gearboxes provide a range of gear ratios far in excess of those available to the car driver. Why is this?

....

....

....

....

Types of gearing

The spur gear, the helical gear and the double helical gear are all types of gears used in gearboxes.

By observation in the workshop, complete the sketches below to show the tooth arrangement for each type:

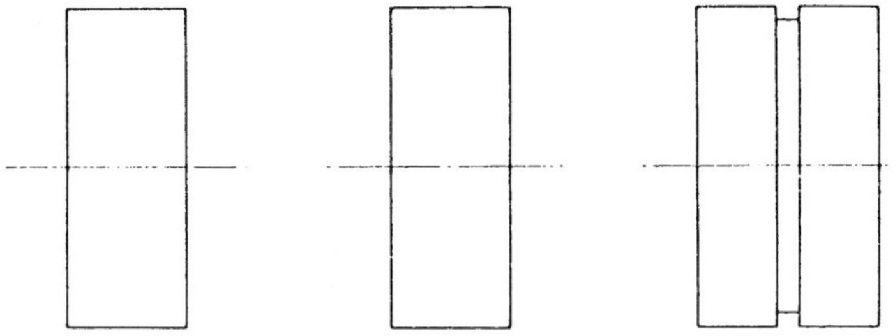

Spur Helical Double helical

Torque multiplication

This can be achieved by using different-sized gearwheels, for example, transmitting the drive from a small-diameter gearwheel to a large-diameter gearwheel.

Consider the pair of gears shown below:

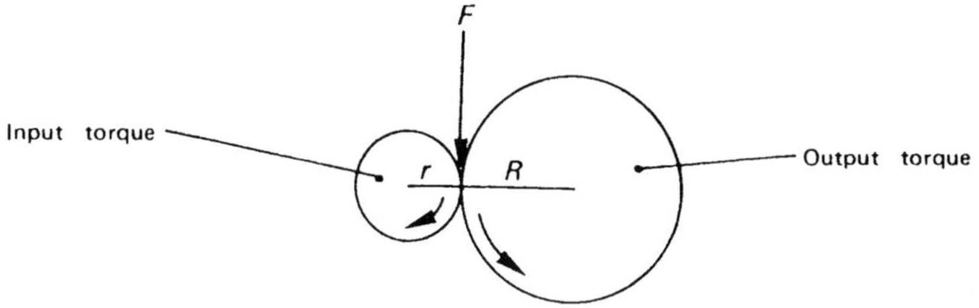

F

Input torque r R Output torque

The input torque produces a force *(F)* which is transmitted from the small gear to the large gear.

Describe how input torque is multiplied:

....

....

....

CONSTANT-MESH GEARBOX

In the constant-mesh gearbox, as the name implies, the gearwheels are permanently in mesh. Constant meshing of gears is achieved by allowing the mainshaft gearwheels to rotate on bushes. Dog clutches, which are normally part of a synchro-mesh hub, splined to the mainshaft, provide a positive connection when required, to allow the drive to be transmitted to the output shaft.

Principle of operation:

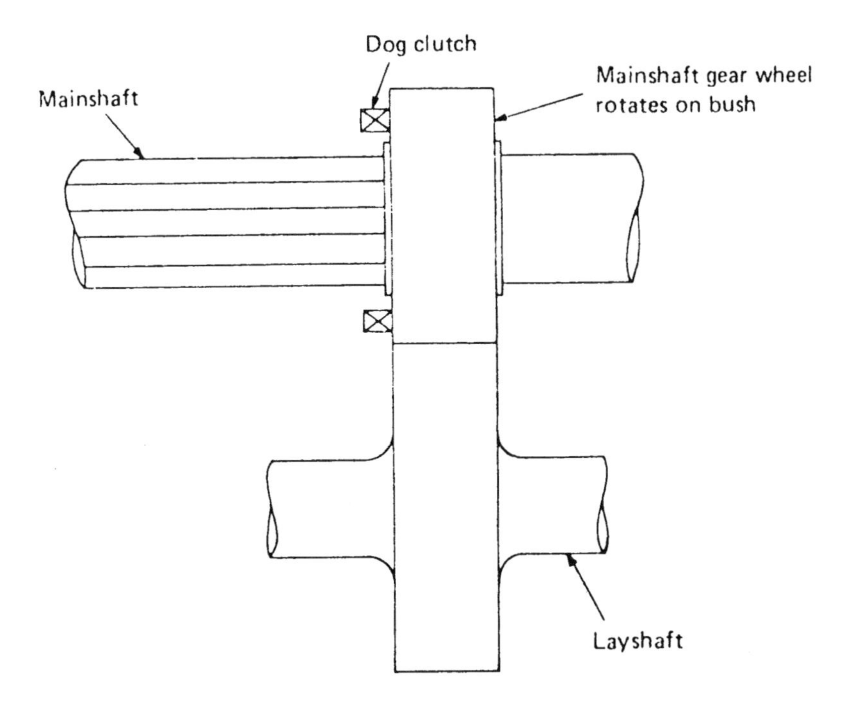

Power flow

(a) Complete the drawing above to show how a dog clutch member is used to connect the mainshaft gearwheel to the mainshaft, and add arrows to indicate the power flow from the layshaft to the mainshaft.

(b) Add arrows to the drawings opposite to trace the power flow through the constant-mesh gearbox in the gear positions shown. Sketch the dog clutches in their correct positions.

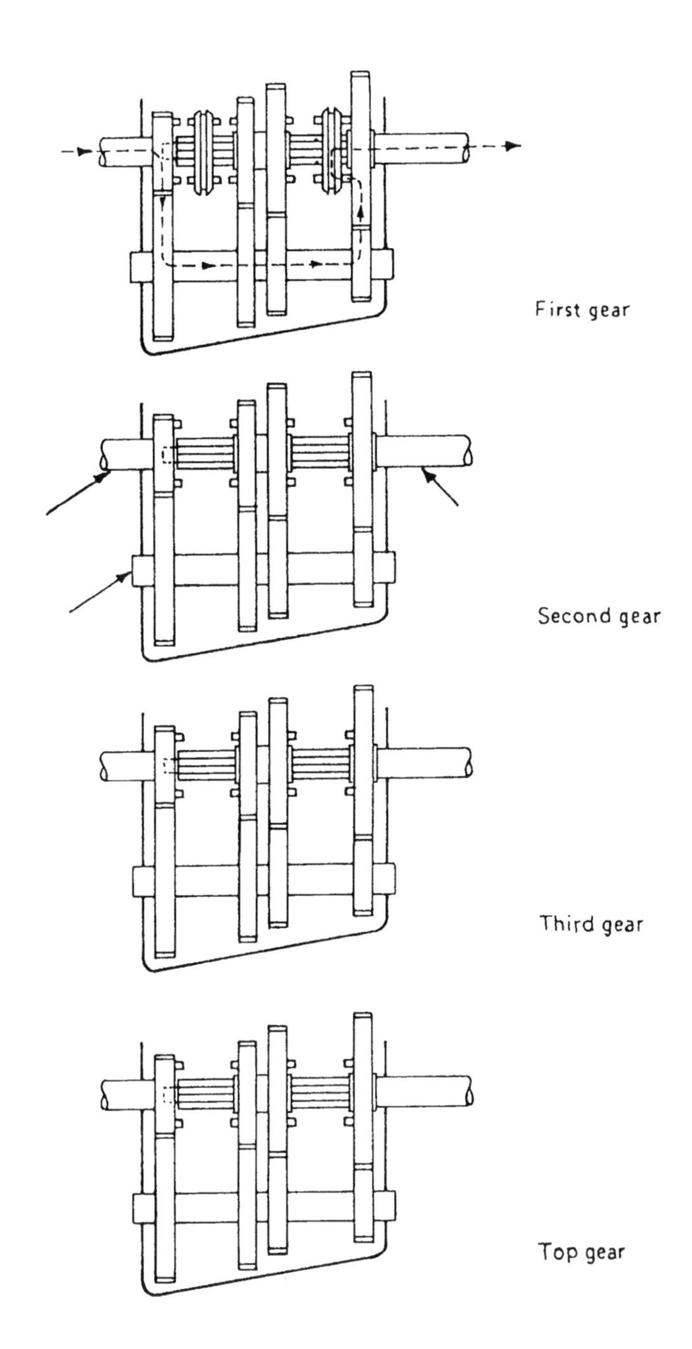

SYNCHROMESH DEVICES

When a gear-change is made, the speeds of the meshing dog clutch teeth must be equalised in order to avoid clashing which would result in noise and wear. As the name implies, the synchromesh device synchronises the speeds of the dog teeth during gear-changing. An early type of synchromesh device (now obsolete) is the CONSTANT LOAD type. This relatively simple arrangement can be used to aid initial understanding of what is quite a complex mechanism in a modern gearbox.

The constant-load synchro-hub is shown in its neutral position at A opposite and engaging third gear at B.

Describe its action during engagement:

....

....

....

....

....

....

....

....

....

....

....

State the factors which affect the frictional force, and hence the speed of synchronisation on the cones:

....

....

....

Constant-load synchromesh device

A

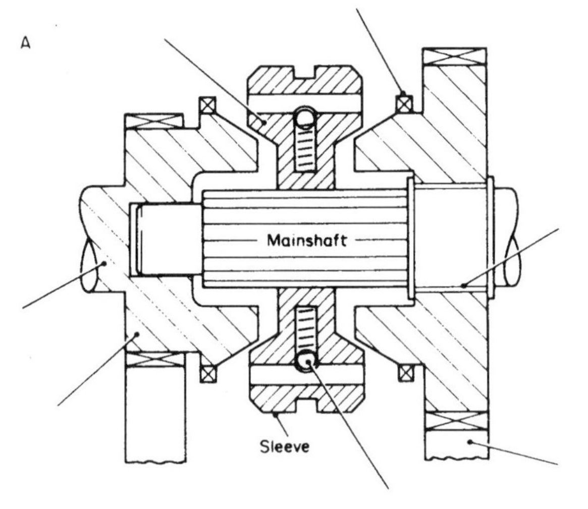

Complete the labelling on this drawing.

B

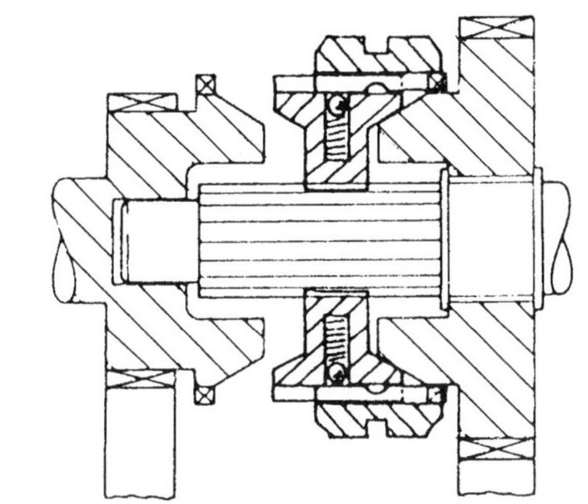

State the reason why this device is known as the 'constant-load' type:

....

....

....

....

Baulk-ring synchromesh

With the constant-load type of synchromesh, if a quick gear-change is made the dog clutches can come into contact before their speeds are properly synchronised. This causes noise and wear, that is, it is possible to 'beat' the synchromesh.

The baulk-ring synchromesh is a development of the constant-load type. The characteristics of the baulk-ring type are that:

(a) the synchronisation of the dog clutches is quicker, thus allowing a quicker gear-change;

(b) the dog clutches cannot be engaged until their speeds are equal.

Name the component parts of the baulk ring synchromesh unit shown below:

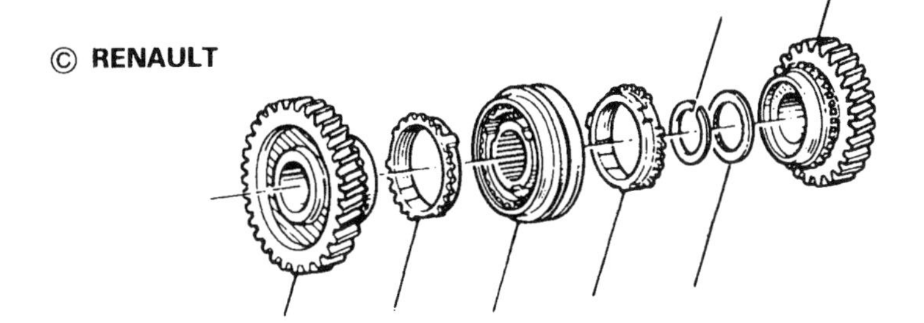

The sketch at A represents a section of a baulk-ring synchromesh unit and the mating dog clutch.

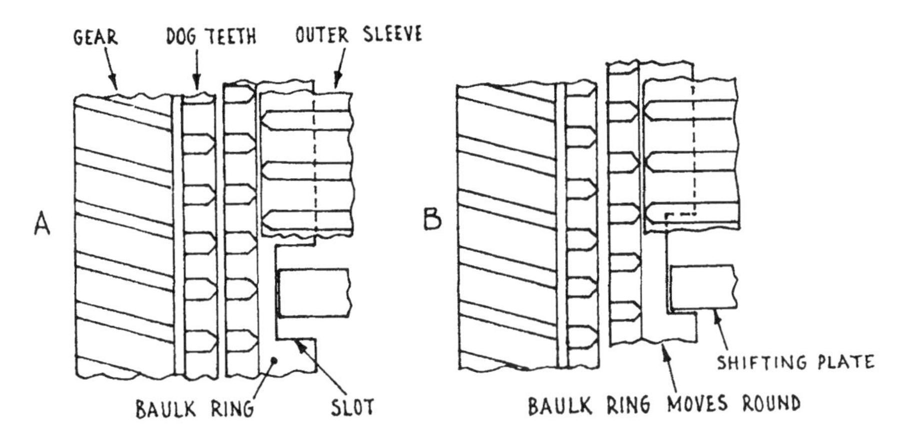

In the position shown at A the outer sleeve (dog clutch) can engage with the dog teeth on the gearwheel because the baulk ring teeth are aligned with the gearwheel dog teeth. At B the baulk ring has rotated to prevent engagement;

Examine a synchromesh unit and observe the limited rotation of the baulk ring.

....

....

....

....

....

....

....

....

....

....

....

....

....

Describe how the baulk ring synchromesh can provide a more rapid gear-change:

....

....

....

....

....

....

....

....

FIVE-SPEED GEARBOX

In most four-speed gearboxes the top ratio is 1:1, that is, the primary shaft and mainshaft are connected to give a direct drive through the gearbox. Many modern vehicles, however, are fitted with a five-speed gearbox in which top gear is an 'overdrive' or 'step-up' gear.

Complete the drawing below to show a five-speed constant-mesh gearbox with top gear as an overdrive:

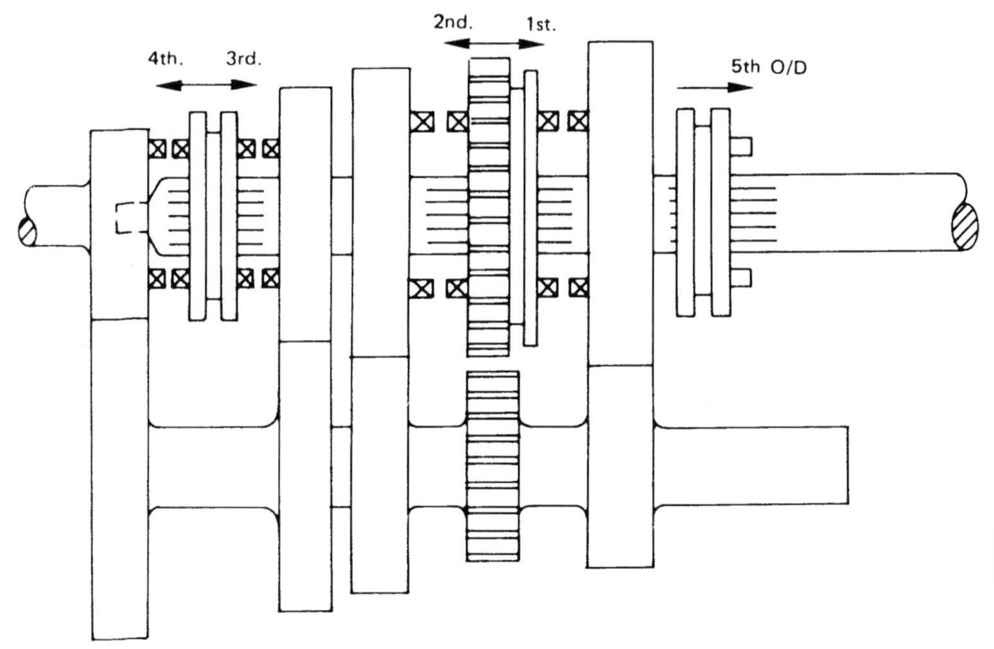

Show, on the drawing above, and explain below how reverse gear is obtained:

....

....

....

(a) Name the drive layout for the transmission shown below:

....

(b) How many speeds has the gearbox? ....

(c) Label the drawing.

(d) Use arrows to indicate power flow through the gearbox in top gear.

(e) State the reason why there is no 'constant-mesh pair' of gears:

....

....

....

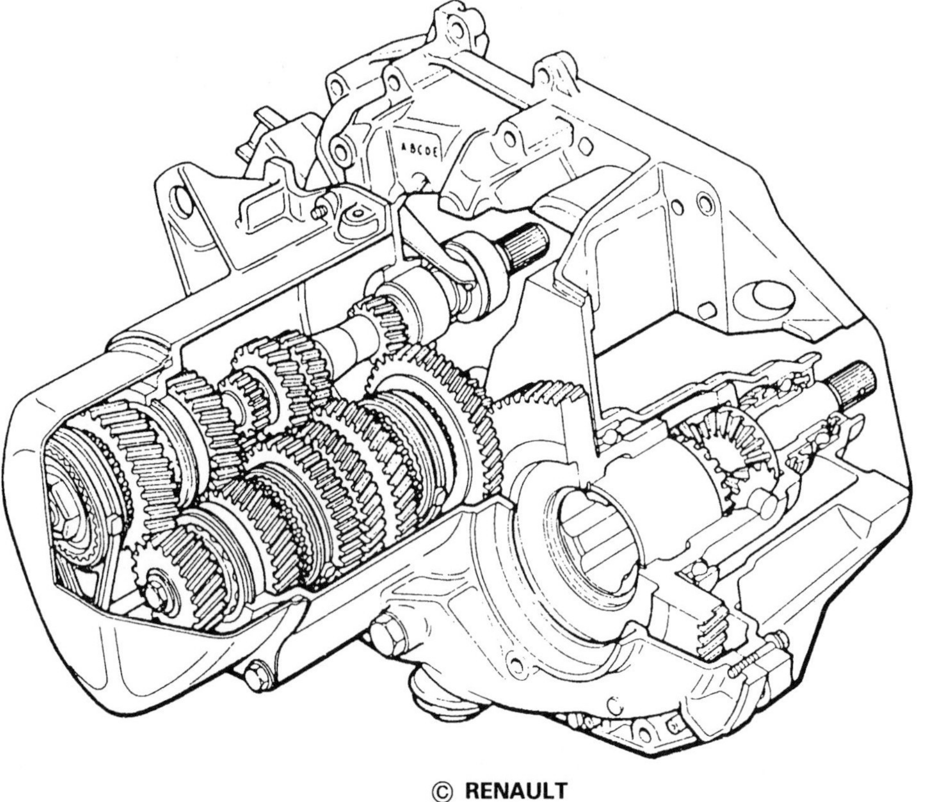

© RENAULT

SELECTOR MECHANISMS

When a gear is selected the movement of the gear lever is transferred to the synchro-hub or gear-wheel through a selector fork. The fork may be fixed to a sliding shaft, or the fork may slide on a fixed shaft. Provision is also made for locking the selector fork in the required gear position and also for preventing movement of more than one selector at a time.

Examine different types of selector mechanism and complete the simplified drawing below to include the selector fork and gear lever.

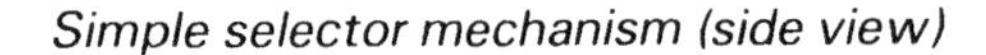

Simple selector mechanism (side view)

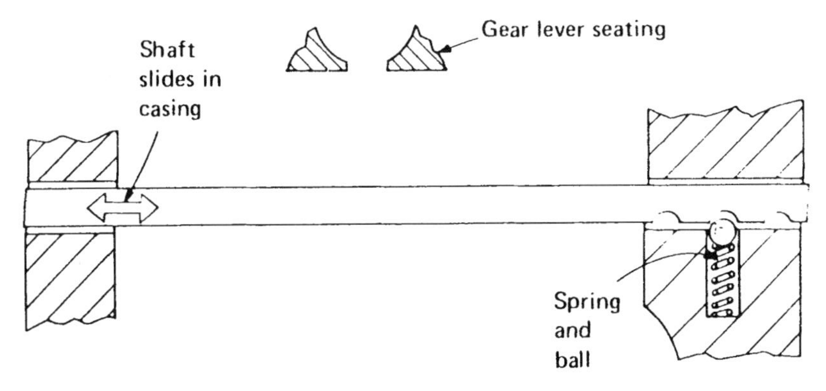

An exploded view of a selector mechanism is shown below; label the drawing.

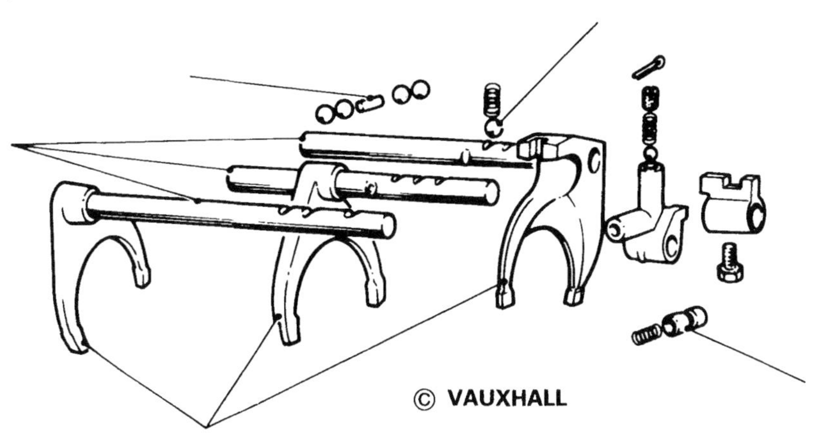

INVESTIGATION

Examine a 'sectioned' gearbox which has a DIRECT ACTING gear shift.

State:

1. the number of gears ....

2. the number of selector rails and forks ....

Show by sketching below

(a) how accidental selection of reverse gear is prevented:

(b) how the REVERSING LIGHT switch is operated:

Interlock mechanisms

The interlock mechanism prevents the engagement of two gears at once. The drawing at A below shows a .... interlock mechanism.

Examine various gearboxes and show by sketching at B one other form of interlock mechanism.

A

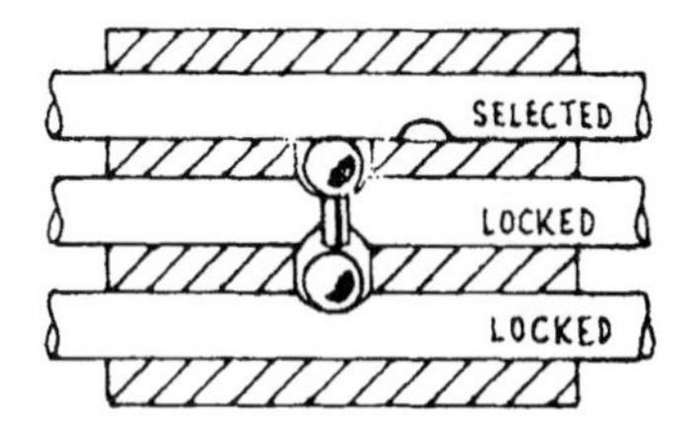

B

Remote control gearshift systems

Give reasons for using remote control systems:

....

....

....

....

....

The mechanical remote control shown below allows the use of a short gear lever.

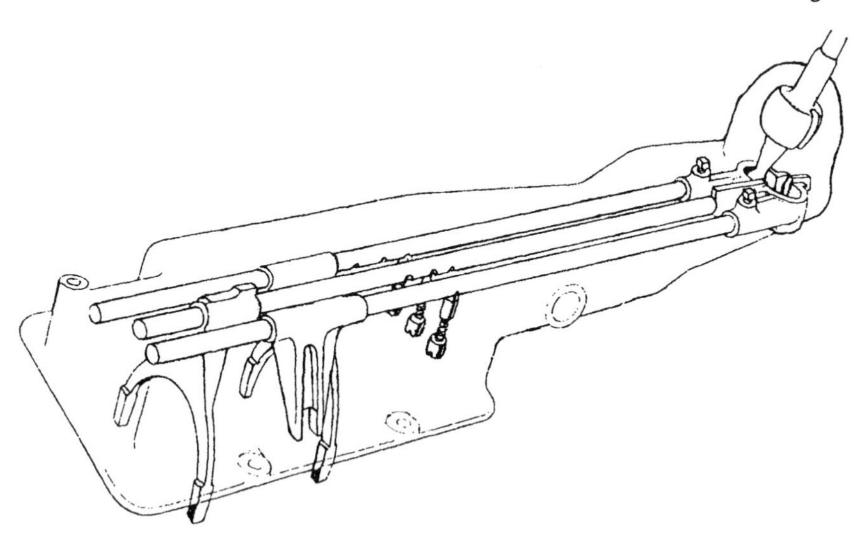

Study an alternative remote control system, such as electrical, pneumatic etc., on a vehicle or by use of manuals, trade journals etc. and with the aid of a simple diagram describe the system:

GEARBOX BEARINGS AND MOUNTINGS

By means of arrows on the drawing opposite indicate the location of all the gearbox bearings in this HGV gearbox.

Complete the labelling by naming the bearings and, below the drawing, give the main reasons for using the particular bearings in each location.

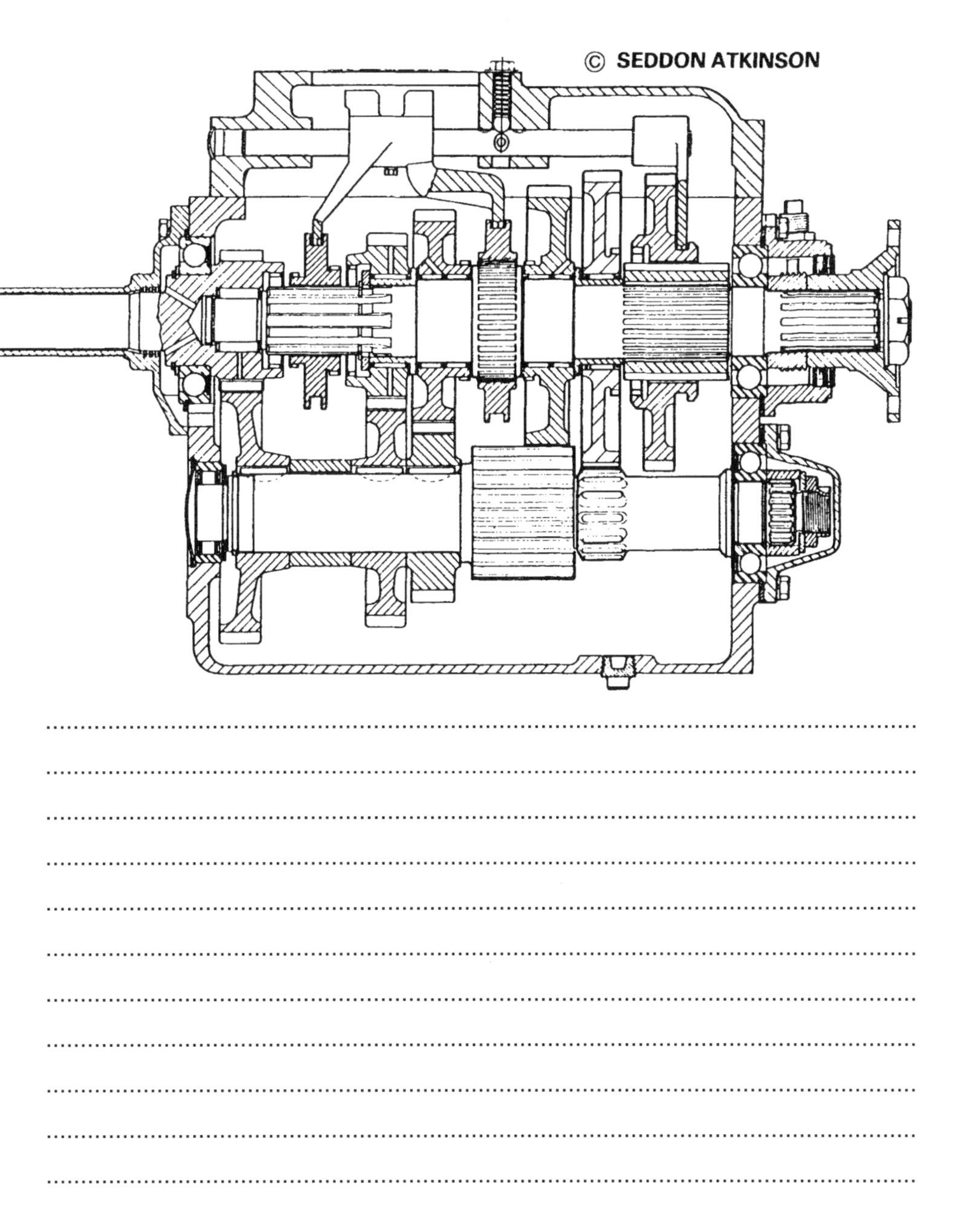

© SEDDON ATKINSON

....

....

....

....

....

....

....

....

....

....

....

....

....

The flexible mountings through which the gearbox is attached to the vehicle frame reduce noise and prevent vibration being transmitted to the vehicle structure. Typical mountings for a transverse front wheel drive engine and transmission are shown below:

a

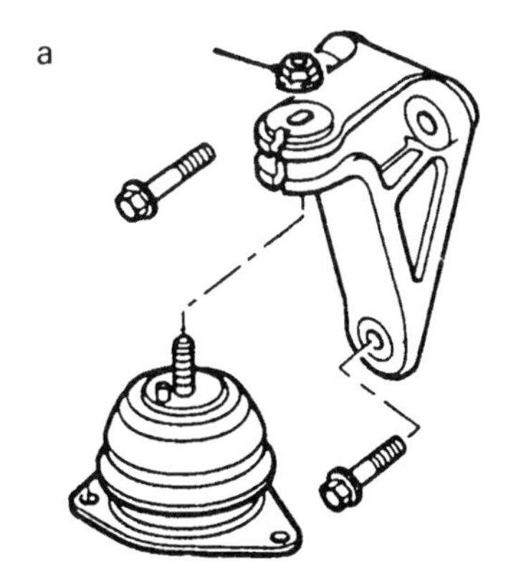

b

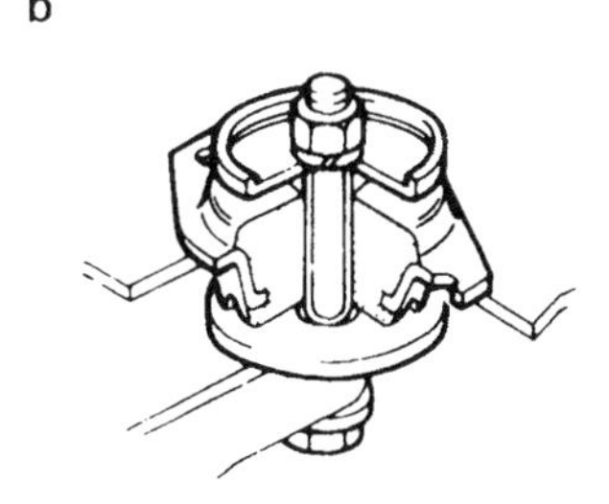

© ROVER

The cut-away drawing at (b) shows how the rubber is shaped and housed within the mounting assembly so as not to be in direct compression when loaded. Why is this?

....

....

....

Oil sealing

Name the two oil seals shown below and give examples of their uses in manual and automatic gearboxes:

....

....

....

....

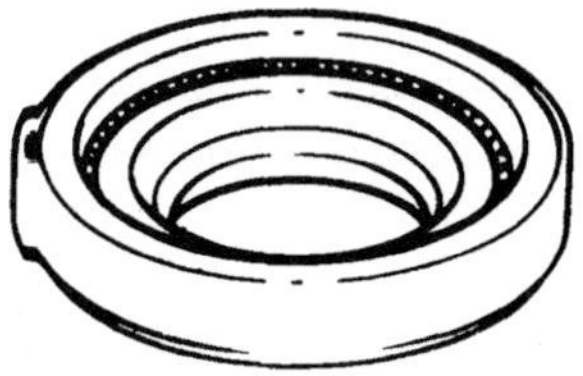

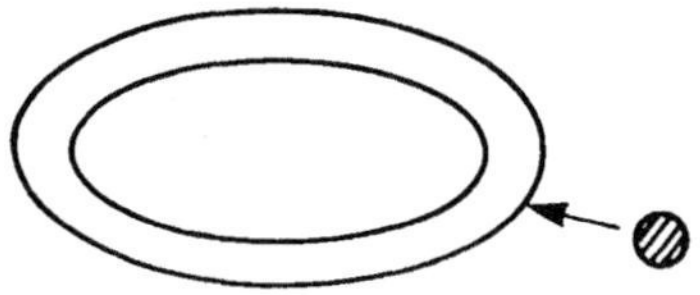

Add an oil return scroll and slinger washer to the gearbox primary shaft shown below and describe the action of these oil-retaining devices:

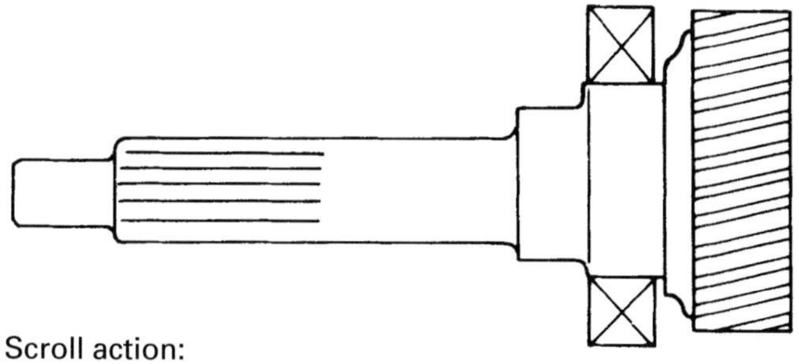

Scroll action:

....

....

....

Slinger washer:

....

....

....

GEARBOX AND LINKAGE ROUTINE MAINTENANCE AND LUBRICATION

Regular maintenance and proper lubrication are essential to extend gearbox life, improve efficiency and minimise failure risk of components.

List SIX maintenance items applying to gearbox and linkage:

....

....

....

....

....

....

Transmission lubricants

The main functions of a gearbox lubricating oil are:

....

....

....

....

State the reason why different lubricants are used in manual and automatic gearboxes:

....

....

....

....

....

....

....

AUXILIARY GEARBOXES (HGVs)

It is now common practice for heavy goods vehicles to employ ten or more gear ratios. One method of increasing the number of gear ratios is to operate an auxiliary gearbox in conjunction with a five- or six-speed gearbox.

Two types of auxiliary gear arrangements in use are:

SPLITTER GEARBOXES and ....

State the reason why one gearbox is referred to as the 'splitter' type:

....

....

....

....

....

....

....

....

A splitter-gear arrangement is shown opposite in tandem with a five-speed gearbox. Describe, below the drawing, the operation of the gearbox, including gear-selection procedure.

Describe how the splitter gear is usually controlled from the driving position:

....

....

....

....

....

....

The splitter gearbox

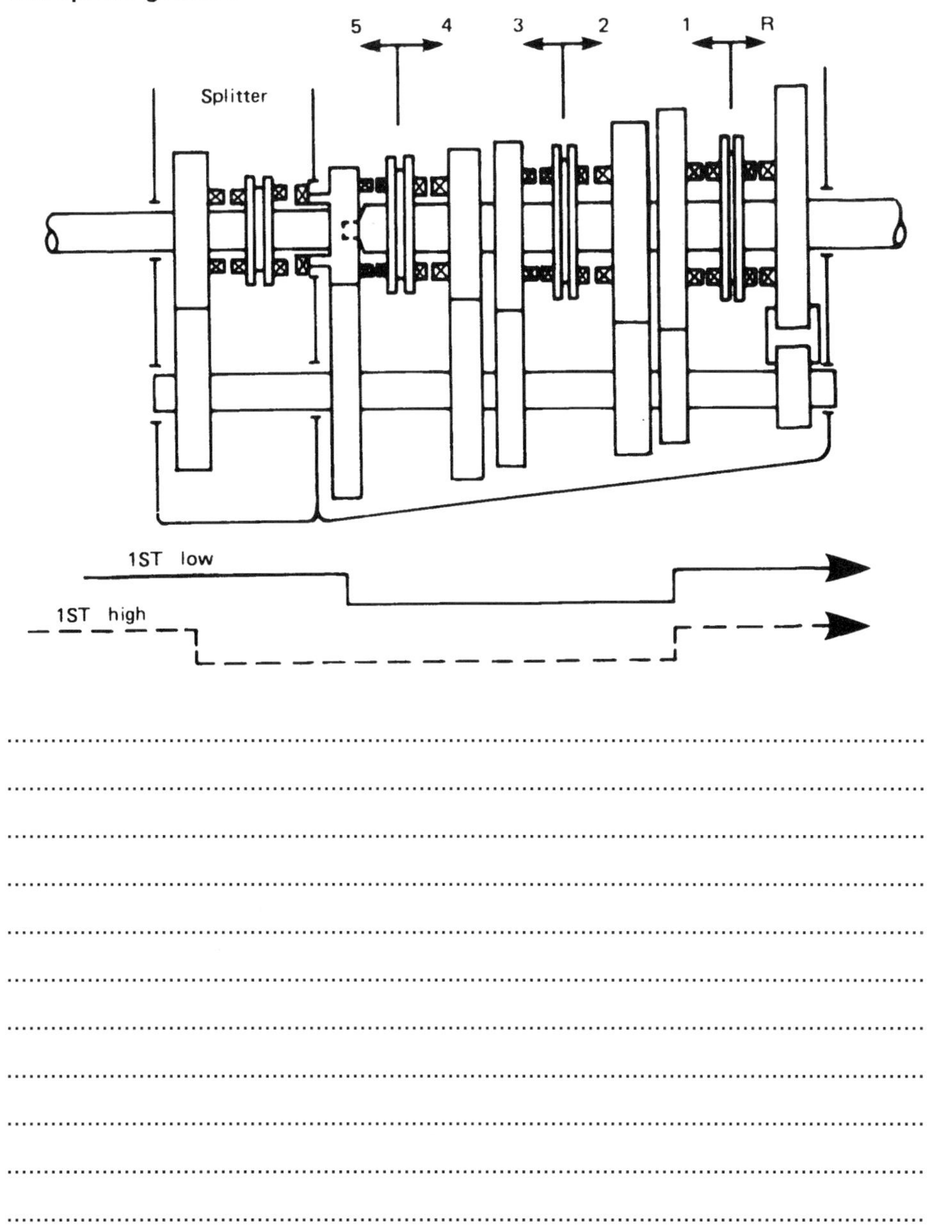

....

....

....

....

....

....

....

....

....

....

RANGE CHANGE GEARBOX

The range change is an auxiliary gearbox, usually attached to the rear of the main gearbox, which provides a low range of gear ratios and a high range. With this gearbox the driver would change as normal through the gears from first to fifth, with the auxiliary gear in low. He would then engage first gear again, but this time in high ratio, that is, 'straight through' the auxiliary gearbox, and gear-change as normal up to fifth gear, thus obtaining ten speeds.

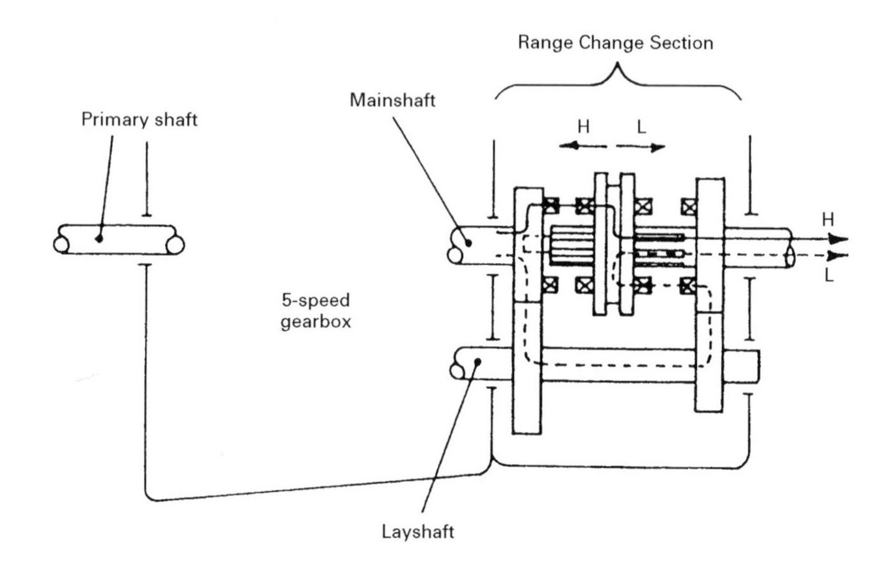

Briefly describe powerflow through the auxiliary gearbox:

....

....

....

....

....

The shift pattern, with typical gear ratios, for a range change gearbox is shown below. From the information given on the drawing, state the auxiliary reduction gear ratio.

Auxiliary reduction gear ratio: ....

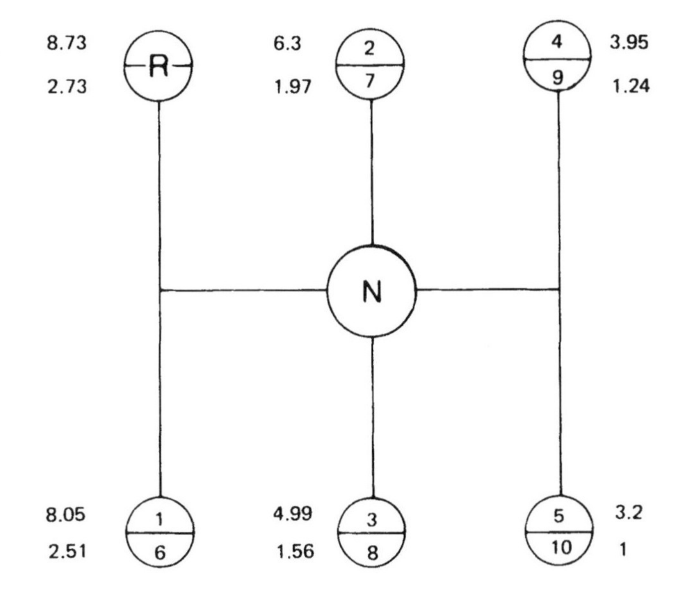

Although ten gear ratios are available to the driver, he would not necessarily select every ratio up and down through the gears. It is usual procedure to 'skip' gears and select the gear appropriate to the speed and load of the vehicle.

TWIN LAYSHAFT GEARBOX

An extremely popular HGV gearbox is the twin layshaft or twin counter-shaft type. In this design the drive from the primary gear is transmitted to the mainshaft gears via two layshafts which are positioned on either side of the mainshaft.

The simplified drawing below shows the layout of a ten-speed transmission, which incorporates a 'floating' mainshaft. The leading edges of the dog clutch teeth are chamfered to provide a synchronising action as the teeth engage.

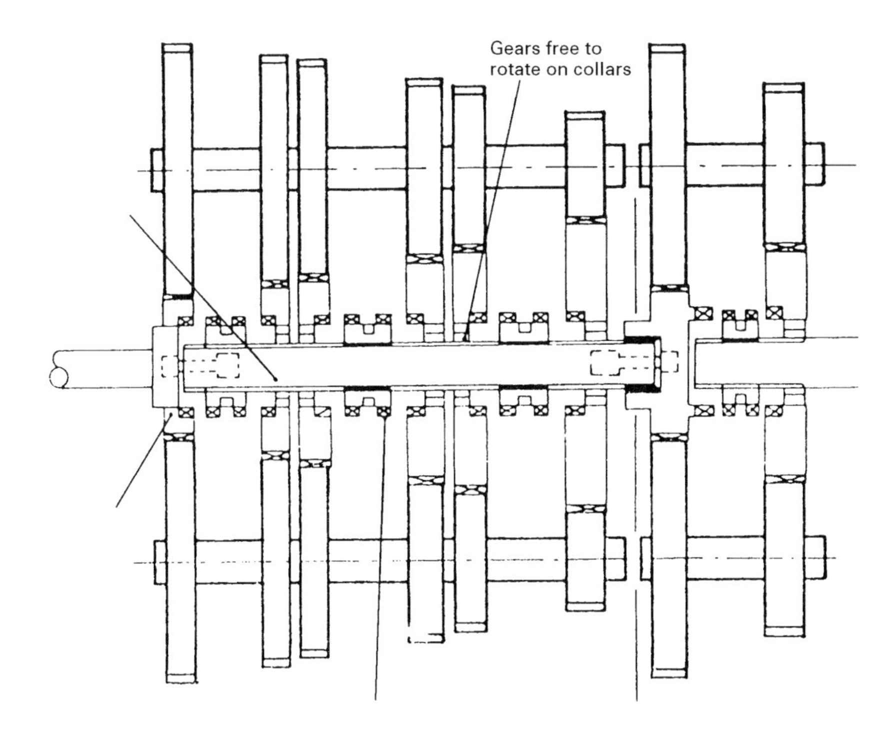

Label the drawing above and name the gears.

Torque balance

The upward-tooth load on one side of the mainshaft gears is balanced by the downward-tooth load on the opposite side; the mainshaft gears do therefore 'float' between the layshaft gears, that is, the mainshaft gears are located radially by the layshaft gears.

Describe how the tooth-load balance or torque balance is achieved:

....

....

....

....

....

....

State the advantages of the twin layshaft type of gearbox compared with the single layshaft type:

....

....

....

....

....

....

....

....

....

....

....

....

....

SPEEDOMETER/TACHOGRAPH DRIVE ARRANGEMENT

The speedometer or tachograph is usually driven by a gear arrangement (shown below) at the rear end of the gearbox mainshaft. The driving gear shown is a push fit on the rear of the mainshaft and is clamped between the mainshaft rear bearing and the U/J flange.

(a) (b)

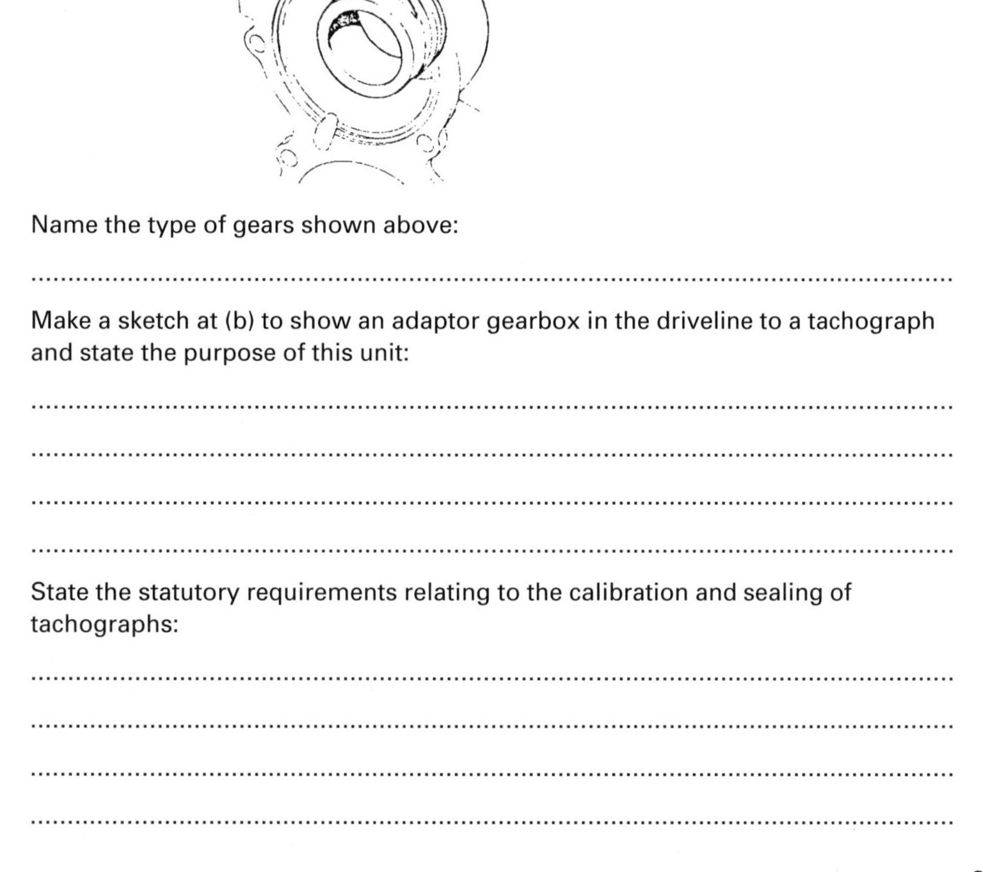

Name the type of gears shown above:

....

Make a sketch at (b) to show an adaptor gearbox in the driveline to a tachograph and state the purpose of this unit:

....

....

....

....

State the statutory requirements relating to the calibration and sealing of tachographs:

....

....

....

....

Indicate on the gearbox shown below the location of oil seals, gaskets, level plug and vent:

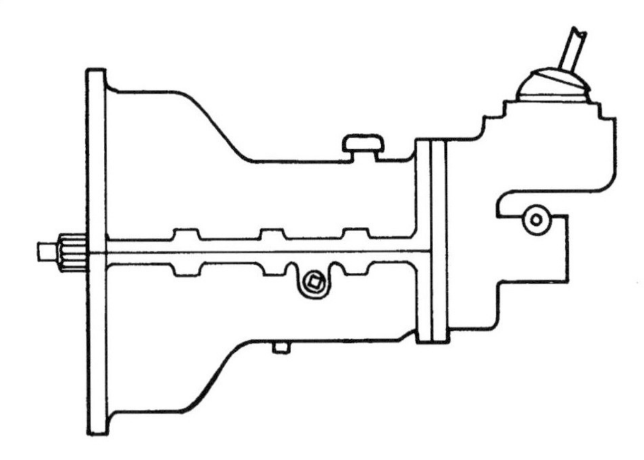

State the purpose of:

Gaskets

....

....

Level plug

....

....

Vent

....

....

State the quantity and type of oil used to lubricate a typical car and HGV gearbox:

Car make .... Model ....

Oil quantity .... Type ....

HGV make .... Model ....

Oil quantity .... Type ....

DIAGNOSTICS: GEARBOX – FAULTS

State a likely fault for each symptom listed below.
Each fault will suggest any corrective action required.

Symptoms	Faults
Noisy operation	
Difficulty in obtaining a certain gear	
Jumping out of gear	
'Sloppy' gear lever action	
Oil leakage from rear of gearbox	
Regular ticking or knocking noise	
Gearbox 'locked up solid', that is shafts will not rotate	
Inoperative speedo/tacho	

TESTING AND TEST EQUIPMENT

Problems associated with the gearbox become evident during testing and checking procedures which include:

Road testing
Operating the vehicle on a rolling road dynamometer
Visual external examination of the gearbox, linkage and mountings
Examination of the oil.

Carry out diagnostic tests on a gearbox either by operation on road, or dynamometer, or on a partially dismantled gearbox. List the equipment used and faults found.

Vehicle make .... Model ....

EQUIPMENT

....

....

....

FAULTS

....

....

....

....

Describe the operation and maintenance of ONE item of test equipment:

....

....

....

....

....

INVESTIGATION: GEARBOX AND OPERATING LINKAGE REMOVAL AND REFITTING

Gearbox

As a result of workshop experience, or by reference to a workshop manual, describe the procedure for removing a typical gearbox (for linkage, see opposite).

Vehicle make .... Model ....

Gearbox type ....

Linkage

Additional to the gearbox removal, describe how to remove, replace and adjust the gear/change linkage:

While carrying out gearbox work, how might the box and linkage be protected against the following hazards?

Oil leakage ....

Ingress of dirt and moisture ....

Linkage seizure ....

Pressure build up ....

POWER TAKE OFF (PTO) SYSTEMS

A PTO enables a wide range of installed equipment, towed and standing machinery to be driven by the vehicle's engine, for example, generators, pumps, winches etc. Dual purpose vehicles used by farmers and construction workers and specialist vehicles (Volvos shown) use PTOs.

Name four power take off points for the PTO:

1. *Gearbox mainshaft* .... 2. ....

3. .... 4. ....

Complete the simple line drawing below to show how a PTO can be geared to the layshaft.

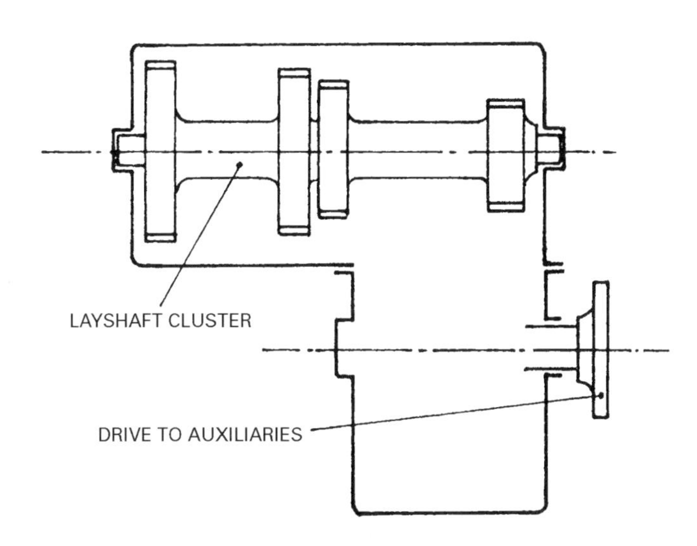

In what type of transmission is the centre PTO used and how is it operated by the driver?

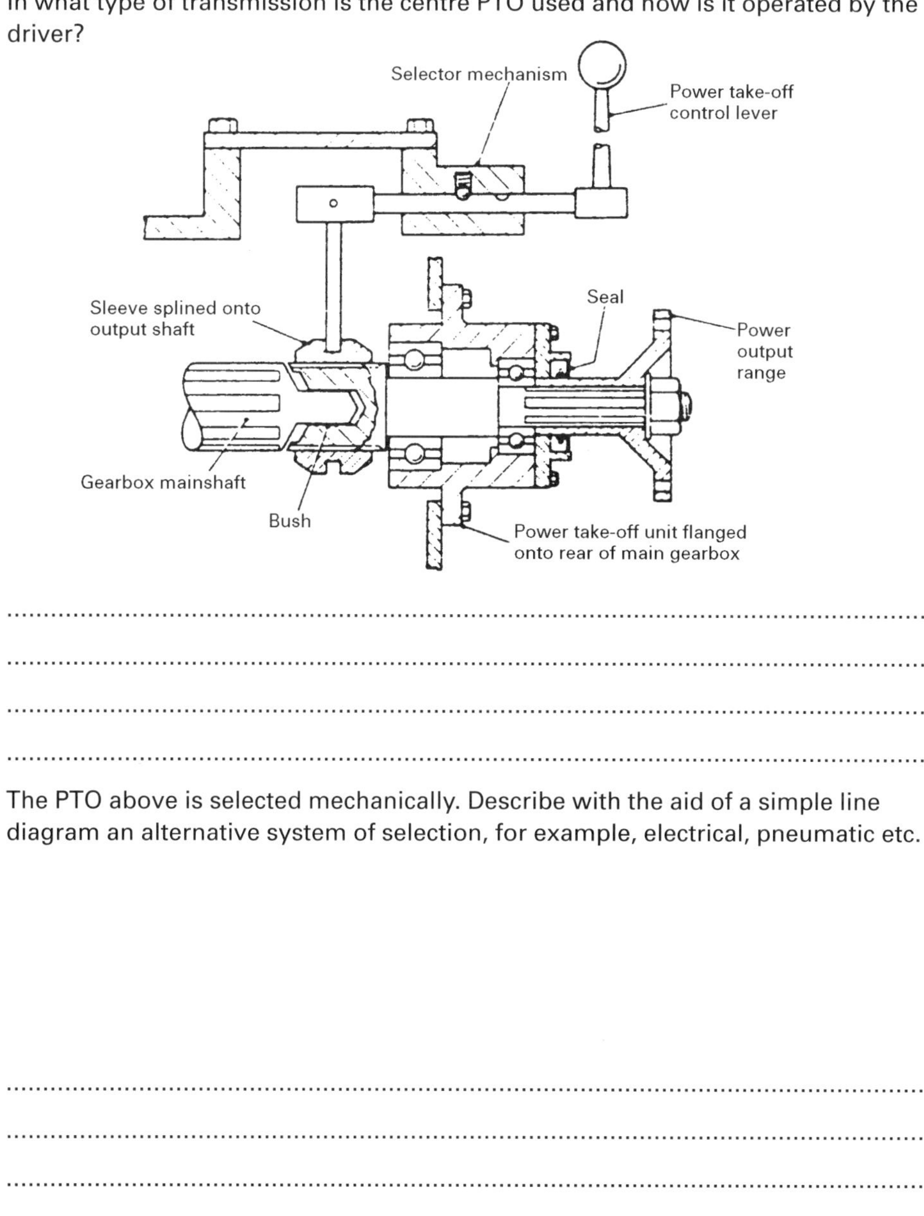

....

....

....

....

The PTO above is selected mechanically. Describe with the aid of a simple line diagram an alternative system of selection, for example, electrical, pneumatic etc.

....

....

....

....

FOUR-WHEEL DRIVE (4WD or 4 × 4)

Four-wheel drive vehicles are either:

(a) *Heavy truck type vehicles operating for part of the time off the normal road on uneven, soft or slippery surfaces; or*

(b) ....

(c) ....

With the simple system shown, four-wheel drive is engaged for use on uneven, soft or slippery surfaces. It should be disengaged for normal road use. Why is disengagement necessary?

....

....

....

Label the drawing below and state purpose of the arrangement shown:

....

....

Layout

Complete the drawing below to show the front and rear final drive gears and show how the drive to the front and rear axle can be disconnected:

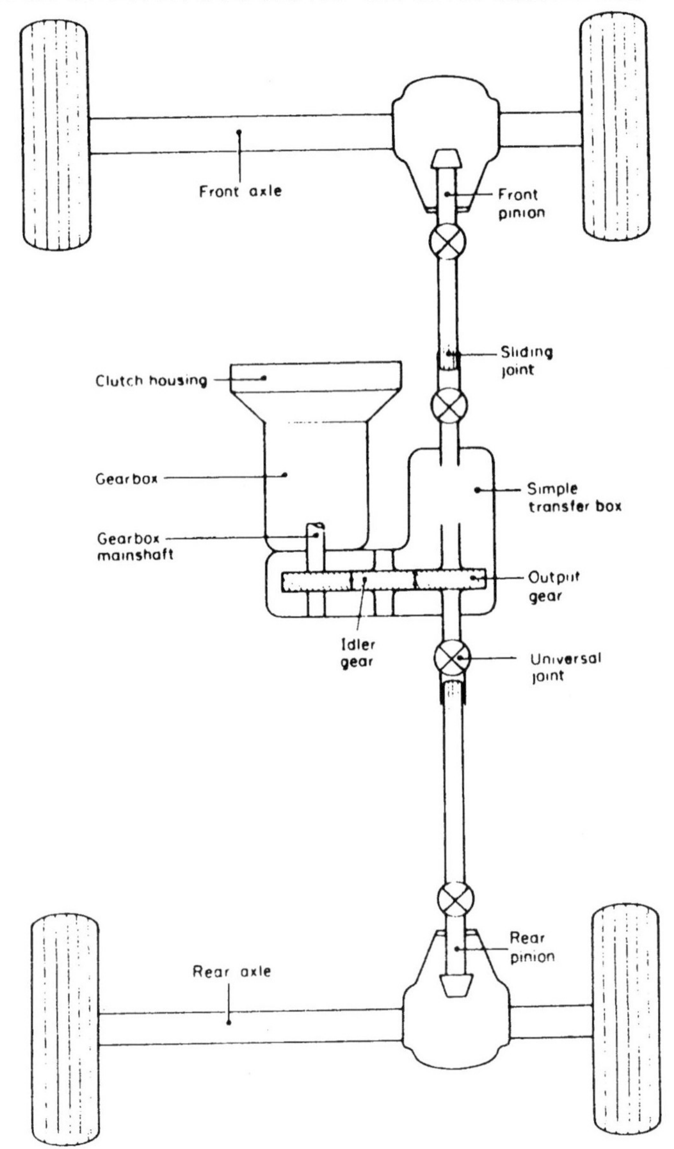

TWO-SPEED TRANSFER BOX

Complete the drawing and labelling of the two-speed transfer box below and describe its operation in low and high ratio:

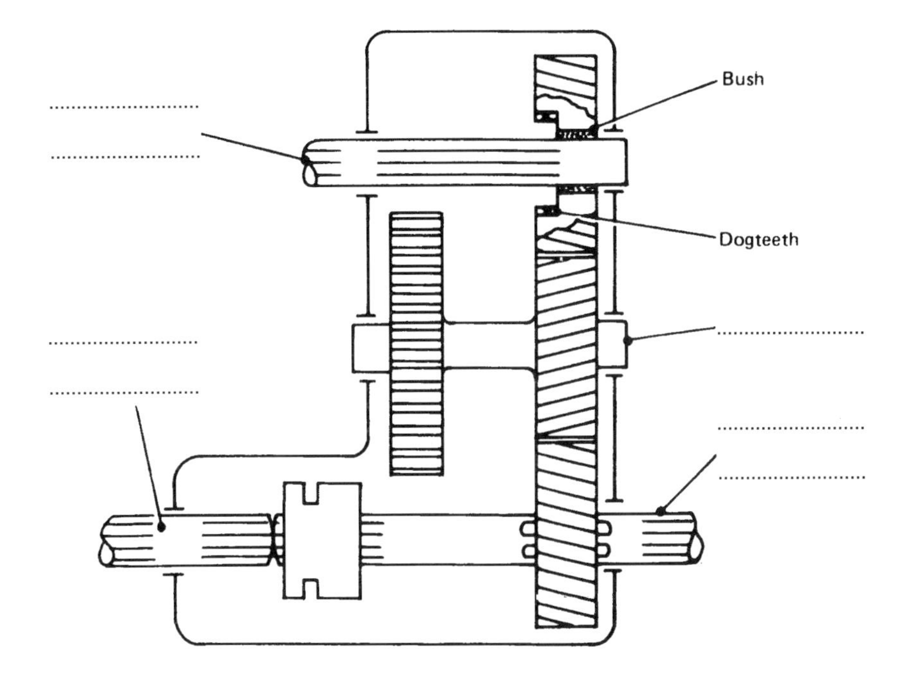

OPERATION (two-speed transfer gearbox)

....

....

....

....

....

How does a two-speed transfer box affect the transmission gearing as a whole?

....

....

....

Complete the drawing below to show how the front and rear propeller shafts can be driven through a differential:

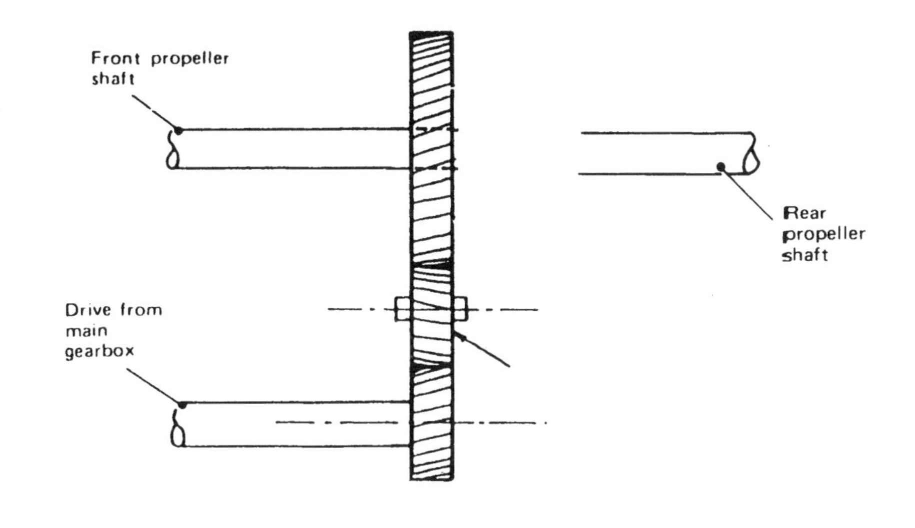

State the reason for using a third differential:

....

....

....

To obtain full benefit of four-wheel drive when operating off the road it is usual practice to use a differential lock on the third differential.

DIFFERENTIAL LOCKS

If one driving wheel of a vehicle encounters a soft or slippery surface, the wheel will spin and the torque required to drive it will be negligible. Because of the action of the differential the other non-slipping wheel will receive the same negligible torque. The vehicle will therefore be immobilised.

The differential lock, as the name implies, locks the differential and allows maximum tractive effort allowed by the road surface to be utilised at each wheel.

Suggest three possible applications of the differential lock:

1. *Farm tractors* ....

2. ....

3. ....

What safety precautions should be taken when working on the transmission of a vehicle fitted with a differential lock?

....

....

....

....

....

....

....

....

....

....

....

If two elements of a differential are locked together, the entire assembly is locked. The device shown at the top opposite operates on this principle.

Complete the sketch at 'B' below to show the differential lock engaged, and explain how the action locks the differential:

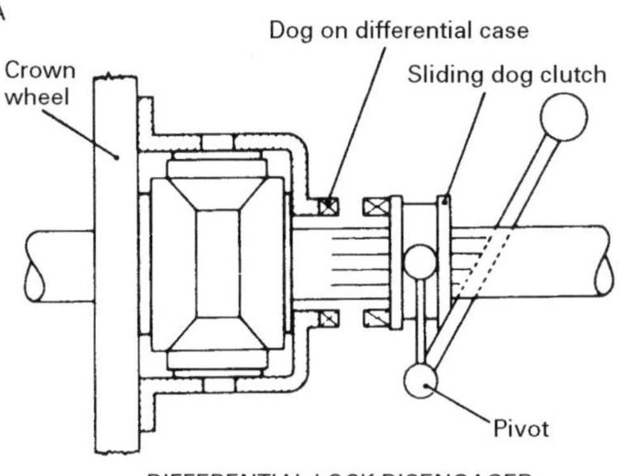

DIFFERENTIAL LOCK DISENGAGED

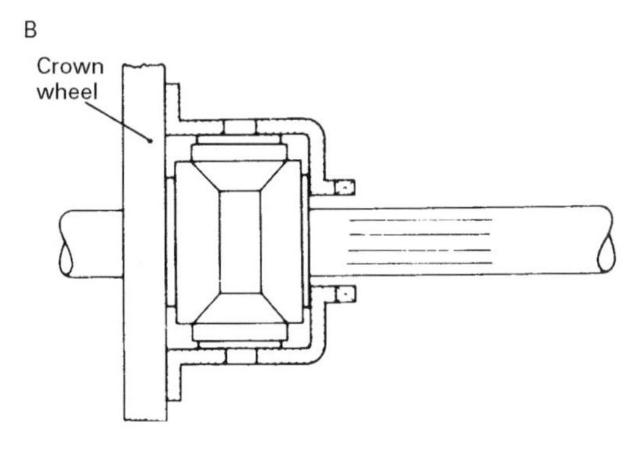

OPERATION

....

....

....

....

....

....

....

VISCOUS COUPLING (VC)

Four-wheel drive operation under all conditions can be achieved by the use of LIMITED SLIP DEVICES. The system shown opposite employs viscous couplings in the centre (third) differential and in the rear differential. These couplings control wheelspin and greatly improve traction and roadholding in all drive conditions without the need for the engagement of manual differential locks by the driver. Further information on limited slip differentials on page 88.

On a car application, such as the one shown opposite, it is usual to divide the driving torque UNEQUALLY between front and rear wheels.

Give a typical percentage TORQUE SPLIT and state the reasons for this.

TORQUE SPLIT: FRONT .... REAR ....

....

....

....

....

....

....

Where in the transmission is the torque split achieved?

....

....

How is the drive transmitted to the front propeller shaft?

....

Complete the labelling on the layout shown below:

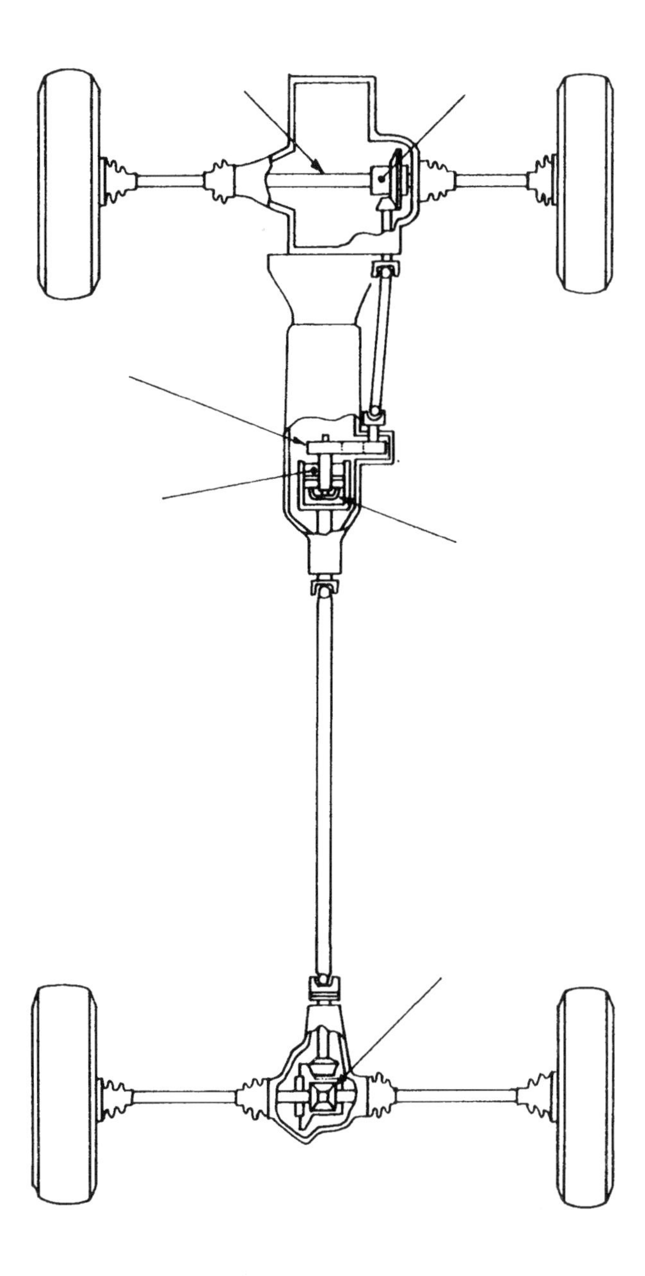

Viscous Coupling – Operation

The structure of a viscous coupling is similar to that of a multi-plate clutch. The coupling consists of a number of INNER and OUTER discs.

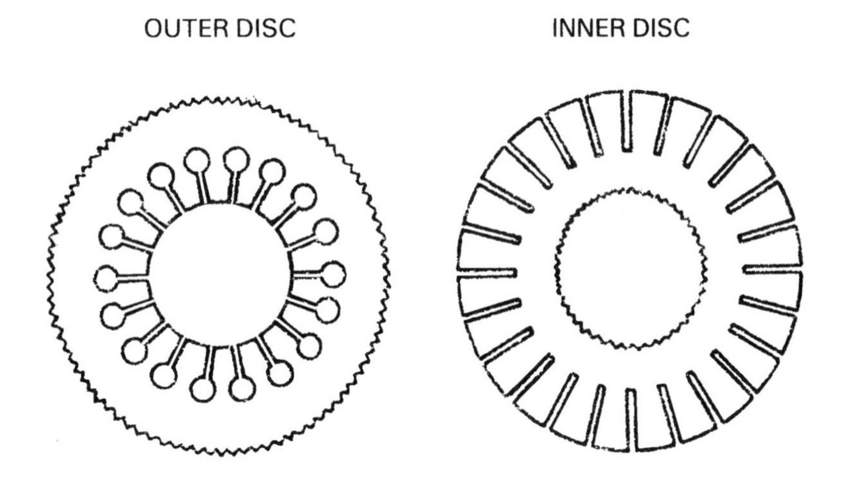

The inner discs are splined on to an inner carrier shaft and the outer discs are splined on to the inside of an outer housing. A small clearance or gap between the discs is maintained by interposed spacer rings. The gap between the discs is filled with a high viscosity silicone fluid.

Torque transmission through a VC is based on the transmission of shearing forces in the fluid.

Viscous couplings, as already stated, will control differential spin. In what other capacity are they employed in vehicle transmissions?

....

....

....

....

Complete the drawing below to show the inner and outer discs in the simplified layout:

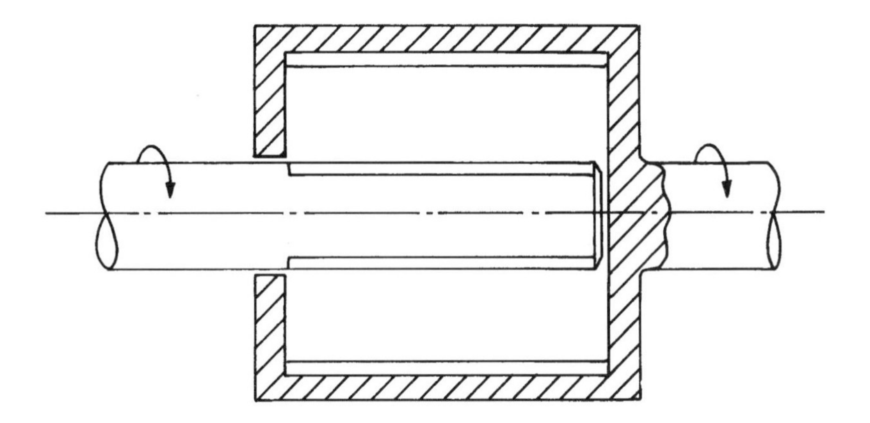

How is torque transmitted from the inner shaft to the outer housing?

....

....

....

....

....

....

....

....

Viscous Coupling and Differential

The viscous coupling, or viscous control, when used in conjunction with a differential will limit or control spin on shafts being driven via the differential, for example, front/rear propeller shafts or drive shafts.

The simplified drawing below shows a VC incorporated into a final drive/differential unit.

Label the drawing and explain how the VC limits drive shaft spin.

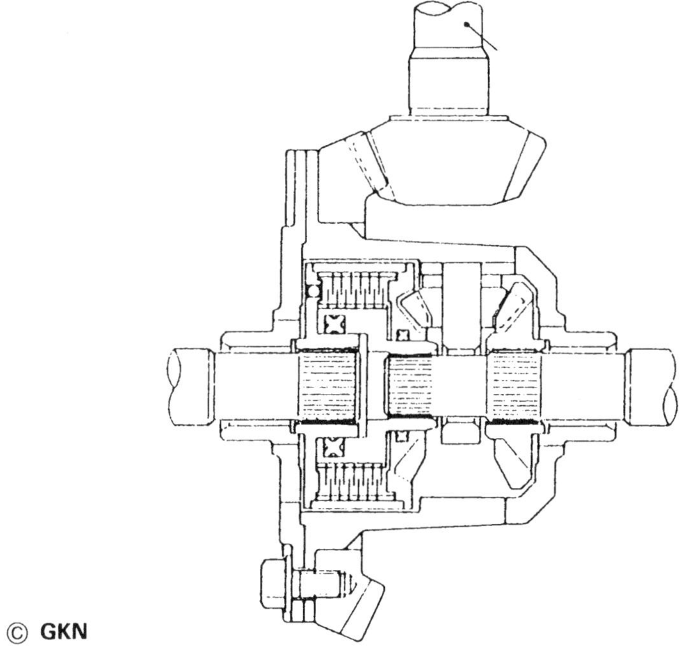
© GKN

....

....

....

....

....

....

Centre Differential

The centre differential shown below is an epicyclic gear set. Complete the drawing by adding a VC which would control spin between front and rear propeller shafts:

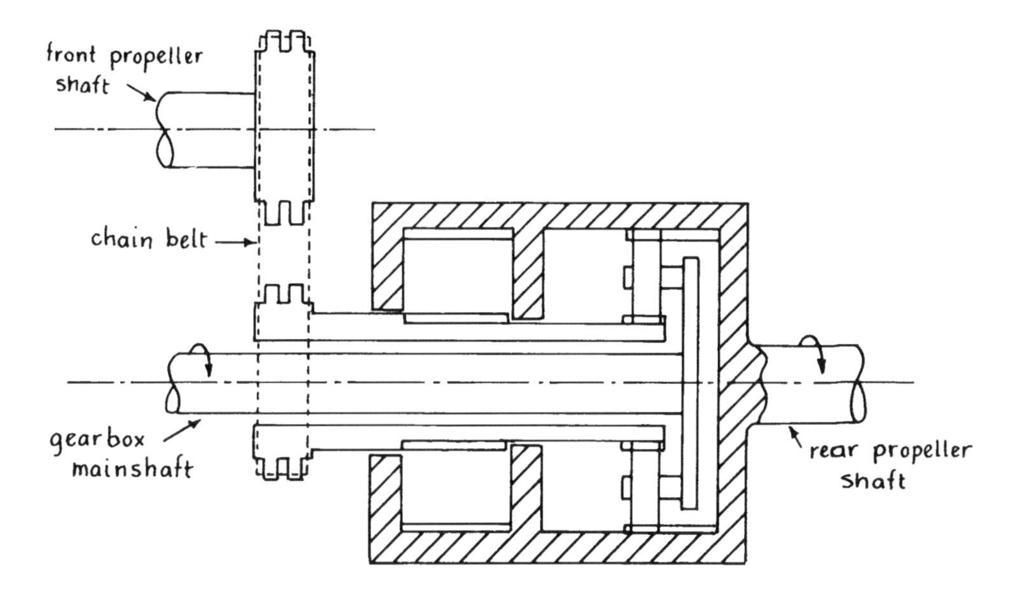

How does this differential provide the 2:1 torque split?

....

....

....

....

....

GEAR RATIO, TORQUE RATIO, EFFICIENCY

Gear ratio

When two gearwheels are meshed to form a 'simple' gear train, the *gear ratio* can be expressed as a ratio of gear teeth or a ratio of gearwheel speeds.

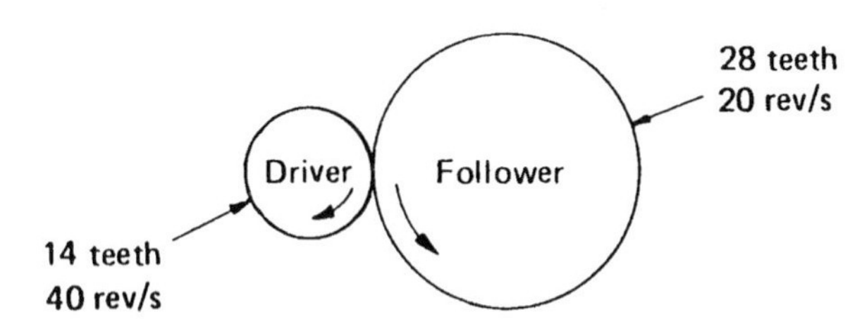

For the pair of gears represented above:

$$\text{Gear ratio} = \frac{\text{No. of teeth on follower}}{\text{No. of teeth on driver}} =$$

or using gear wheel speeds,

Gear ratio =

When speeds are used the ratio is sometimes expressed as

....

(Note: see associated problems on next page.)

Torque ratio

It has already been stated that the gearbox multiplies engine torque; thus the relationship between gearbox input torque and output torque can be expressed as a ratio:

Torque ratio =

It does not follow that the torque ratio will be the same as the gear ratio, this is because some of the input torque is used to overcome friction in the gear assembly.

Efficiency

The efficiency of a gear train is an indication of the friction present. This must therefore be taken into account when calculating the torque ratio.

(a) What would be the efficiency of a gear train if the gear ratio and torque ratio were the same?

....

(b) How is efficiency calculated?

....

(c) State the effects on the following of heavy oil in a gearbox compared with a thin oil:

(1) efficiency ....

(2) torque ratio ....

(3) gear ratio ....

(d) In second gear the gearbox input torque is 250 N m and the output torque is 400 N m. Calculate the efficiency of the gearbox if the second gear rato is 2 : 1.

Torque ratio =

Efficiency =

(e) State the gear ratios for a modern vehicle which is fitted with a four-speed and reverse gearbox:

Vehicle make and model..................................... Year ....

Ratios

1st gear .... 2nd gear .... 3rd gear ....

4th gear .... reverse ....

CALCULATIONS: GEAR RATIO

(a) The input gear of a pair of gearwheels has 12 teeth and the output gear has 30 teeth; the gear ratio is:

(a) 2.5 : 1 (b) 2.75 : 1 (c) 3 : 1 (d) 18 : 1

Answer ....

(b) The constant-mesh gearwheels in a sliding-mesh gearbox have 14 and 28 teeth respectively. Calculate the gear ratio for this pair of gears and the speed of the layshaft when the primary shaft speed is 60 rev/s.

....

....

....

....

(c) With a vehicle in second gear and the engine speed at 70 rev/s, calculate the second gear ratio if the propeller shaft speed is 28 rev/s.

....

....

....

(d) A gearbox has a ratio of 3 : 1 in second gear. When the primary shaft speed is 50 rev/s calculate the speed of the mainshaft.

....

....

....

(e) Complete the table below:

Gear ratio	**4.5 : 1**	**3 : 1**	
Input shaft speed (rev/s)	54		49.5
Output shaft speed (rev/s)		15	18

CALCULATIONS: TORQUE RATIO

(a) Calculate the efficiency of a gearbox given the following data: third gear ratio 1.5 : 1, engine torque 250 N m, propeller shaft torque 350 N m.

....

....

....

....

(b) If the gear ratio of a simple gear train is 3 : 1 and the efficiency is 90%, calculate the output torque and torque ratio when the input torque is 200 N m.

....

....

....

....

(c) In first gear the gear ratio is 4 : 1 and the efficiency of a gearbox is 85%, calculate the torque ratio when the torque transmitted by the engine is 350 N m.

....

....

....

....

....

(d) Complete the table below:

Gear ratio	**2.5 : 1**	**4 : 1**	
Torque ratio	2.5 : 1		6 : 1
Efficiency		95%	93.75%

MOTION

(a) ROLLER OR GEAR OPERATION

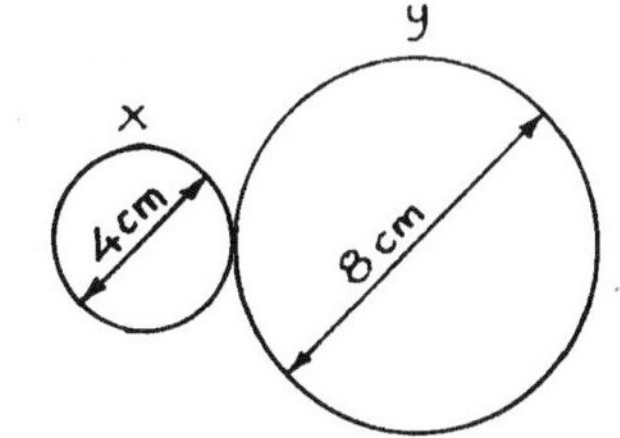

(b) BELT OR CHAIN OPERATION

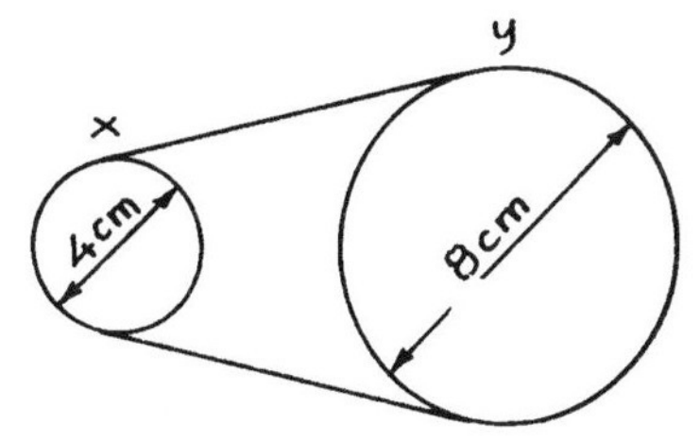

The method of transmission shown above is ....

Indicate the direction of rotation of the larger wheel in each case.

Rotational speed is expressed in ....

Circumferential speed is expressed in ....

Calculate the rotational speed of wheel *y* when the speed of wheel *x* is 50 revs/s.

Let: d_1 = dia. wheel *x* $\quad$ d_2 = dia. wheel *y*

n_1 = speed wheel *x* $\quad$ n_2 = speed wheel *y*

Rational speed wheel *y* =

―――――――――― =

Circumferential speed = πdn

Circumferential speed wheel *x* =

Circumferential speed wheel *y* =

$\therefore d_1 n_1 = d_2 n_2$

What does this prove about the relationship between the wheels in drive trains such as those shown above?

....

....

....

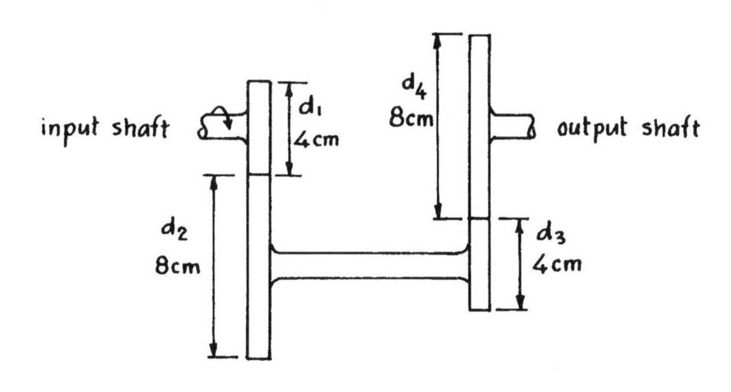

The two pairs of gearwheels shown above provide a gear reduction between input and output shafts. This method of transmission (as employed in a motor vehicle gearbox) is .... It is a COMPOUND gear train.

How does the direction of rotation of the input shaft compare with that of the output shaft?

....

The speed of the output shaft relative to that of the input shaft depends on the overall gear ratio. To determine overall ratio, the ratio of the first pair of gears is multiplied by ....

....

Use the diameters of the gears shown above to calculate overall ratio and output shaft speed when input speed is 1000 revs/s.

By what other methods can the ratio be obtained?

....

....

....

COMPOUND GEAR TRAINS

In the motor-vehicle gearbox, to obtain a gear reduction and to facilitate gear-changing, we use two pairs of gears arranged as shown below. This is known as a 'compound' gear train.

To calculate the gear ratio of such an arrangement, we multiply the ratio of one pair of gears by the ratio of the other pair. For example:

$$\text{Gear ratio} = \frac{\text{number of teeth on follower}}{\text{number of teeth on driver}} \times \frac{\text{number of teeth on follower}}{\text{number of teeth on driver}}$$

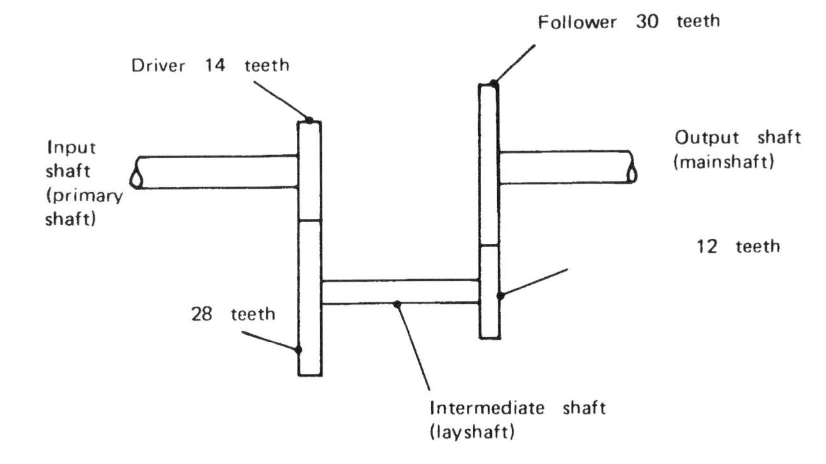

Gear ratio = ―――――― ―――――― =

Complete the table below:

Input driver	Input follower	Output driver	Output follower	Gear ratio
15	25		30	
	30	15	35	

State the relationship between each pair of gearwheels in a compound gear train:

....

....

PROBLEMS

1. In a certain gearbox the constant-mesh wheels have 14 and 30 teeth respectively. The first gearwheel on the mainshaft has 28 teeth and the first gearwheel on the layshaft has 16 teeth. Calculate the first gear ratio.

....

....

2. In a three-speed gearbox the constant-mesh gears have 24 and 36 teeth respectively. The first gearwheel on the mainshaft has 45 teeth and the meshing pinion on the layshaft has 15 teeth. Calculate:

   (a) gear ratio in first gear

   (b) rev/s of propeller shaft, when engine speed is 50 rev/s.

....

....

Overall transmission ratio

To obtain the overall transmission ratio for a vehicle the gearbox ratio is multiplied by the final-drive ratio, the final drive being simply another pair of gears.

Complete the following table to demonstrate this fact:

Gearbox ratio	Final-drive ratio	Overall ratio
2.5 : 1	5 : 1	
7.1 : 1	4 : 1	
1.0 : 1	4.35 : 1	

PROBLEMS

1. The final-drive crown wheel on a vehicle has 30 teeth and the pinion has 6 teeth. If the gearbox ratio in first gear is 4 to 1, the overall transmission ratio in first gear would be:

   (a) 20 : 1 (c) 1 : 1

   (b) 9 : 1 (d) 10 : 1

   Ans. (   )

2. In a four-speed gearbox the constant-mesh pinions have 20 and 35 teeth respectively. The second gear on the mainshaft has 30 teeth, and the meshing layshaft gear has 25 teeth. If the rear-axle ratio is 5.5 to 1, calculate:

   (a) the overall gear ratio in second gear

   (b) the propeller shaft speed when the engine speed is 80 rev/s.

....

....

....

....

....

3. The constant-mesh gears in a gearbox have 16 and 28 teeth respectively. If the second-gear layshaft wheel has 18 teeth, calculate the number of teeth on the second-gear mainshaft wheel.

....

....

....

....

....

4. Complete the table below, given that efficiency =

$$\frac{\text{Torque ratio}}{\text{Gear ratio}} = \frac{100}{1}$$

Gear ratio	3 : 1	2.5 : 1	
Torque ratio	2.8 : 1		5.4 : 1
Efficiency		97%	91.5%

5. In a four-speed gearbox the constant-mesh gears have 22 and 40 teeth respectively and the first gearwheel on the mainshaft has 42 teeth. Calculate the torque ratio for the gearbox if its efficiency in first gear is 95%.

....

....

....

....

....

....

....

6. When accelerating in second gear the engine torque for a vehicle is 300 N m and the propeller-shaft torque is 648 N m. If the efficiency of the gearbox in second gear is 96%, calculate the gear and torque ratios.

....

....

....

....

....

INVESTIGATION

To determine the gearbox and final-drive ratios of a vehicle.

Vehicle make .... Model ....

1. Jack up *one* driving wheel just clear of the ground and remove the engine sparking plugs.
2. Put a chalk mark on the tyre and one on the floor to line up with it. Similarly, mark the crankshaft pulley and timing case (or use ignition timing marks).
3. Engage top gear and count the number of rotations of the engine for two revolutions of the jacked-up wheel.
4. Assuming a top gear ratio of 1 to 1, the rear-axle ratio will be:

   axle ratio = Number of engine rotations

   ∴ **axle ratio** =

5. Engage first gear and repeat the operation.

   Overall transmission ratio in first gear = number of engine rotations

   ∴ overall transmission ratio in first gear =

   $$\textbf{1st gear ratio} = \frac{\text{overall ratio}}{\text{axle ratio}}$$

   ∴ **1st gear ratio** =

6. Repeat for second gear:

....

....

....

....

....

7. Repeat for third gear:

....

....

....

....

....

8. Repeat for reverse gear:

....

....

....

....

....

RESULTS

Gear	1st gear	2nd gear	3rd gear	Reverse
Ratio				

State the reason why only one rear wheel is jacked up:

....

....

....

....

State the reason why it is necessary for the jacked-up wheel to rotate twice:

....

....

....

Chapter 3

Automatic Transmission Systems

Automatic transmission	49
Fluid flywheel	50
Torque converter	51
Epicyclic (planetary) gear trains	53
Investigation	53
Automatic gearbox mechanical system	55
Automatic gearbox hydraulic system	56
Oil cooler	57
Gear selection (automatic transmission)	58
Electronic control	59
Continuously variable automatic transmission	60
Semi-automatic transmission	64
Band operated automatic gearbox	65
Semi-automatic hydracyclic gearbox	66
Computer aided gear changing (CAG)	67
Testing and fault diagnosis	68
Diagnostics	70

AUTOMATIC TRANSMISSION

An automatic transmission system fulfils exactly the same requirements as a manual transmission in that it:

(a) *Multiplies engine torque to suit varying load and speed requirements*

(b) ....

(c) ....

In addition to these functions the automatic transmission provides automatic gear changing, that is, the gearbox ratios are selected automatically to meet the speed and load requirement of the vehicle.

What is the difference between FULLY automatic transmission and SEMI-automatic transmission?

....

....

....

....

Most fully automatic transmissions, however, have a manual override facility with which the driver can dictate when gear selections are made. Another feature of modern automatic transmission is a 'mode' selector – for example, sport or urban; Economy or Power drive programmes. These settings affect gear selection according to the way in which the vehicle is being operated. Explain the difference between FIXED RATIO and STEPLESS transmission:

....

....

....

....

....

Show on the drawing opposite (by shading) the location of the automatic transmission unit and state a make and model using it.

Layouts

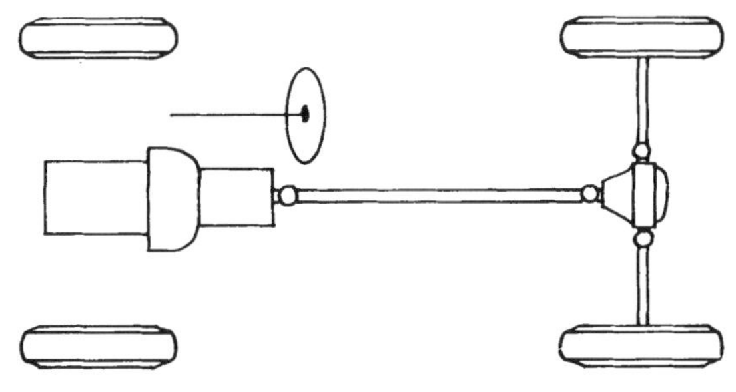

FRONT ENGINE REAR WHEEL DRIVE

Typical make .... Model ....

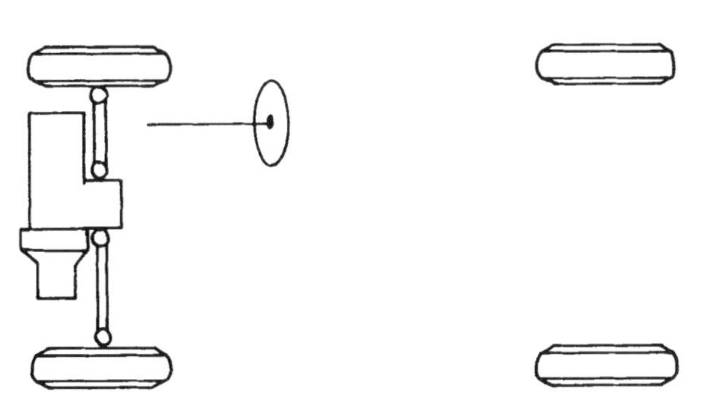

FRONT ENGINE FRONT WHEEL DRIVE

Typical make .... Model ....

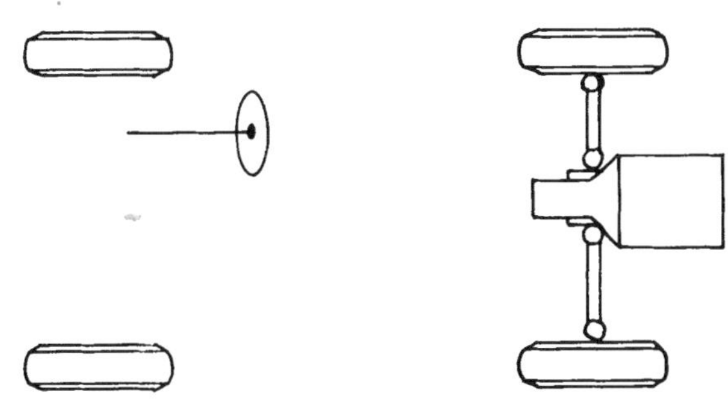

REAR ENGINE REAR WHEEL DRIVE

Typical make .... Model ....

FLUID FLYWHEEL

A fluid flywheel is a hydraulic coupling which is used as an automatic clutch in the transmission system of a vehicle. The unit consists of two main elements, an impeller or driving member and a rotor or driven member. The unit is almost completely filled with fluid. As the engine, and hence the impeller, rotate, the fluid begins to circulate; this transfers torque to the rotor and consequently the gearbox input shaft.

Name the type of gearbox used with this type of coupling:

....

....

Add arrows to the drawing opposite to show the direction of fluid circulation.

State the reason why this type of coupling is generally not used with a normal gearbox:

....

....

Name a modern vehicle using a fluid flywheel:

....

In most modern fluid flywheel applications the fluid flywheel is used in conjunction with a lock-up clutch. State the reason for this:

....

....

....

A fluid flywheel engages automatically, thus simplifying driving technique, and it serves as an effective transmission damper.

Fluid flywheel assembly

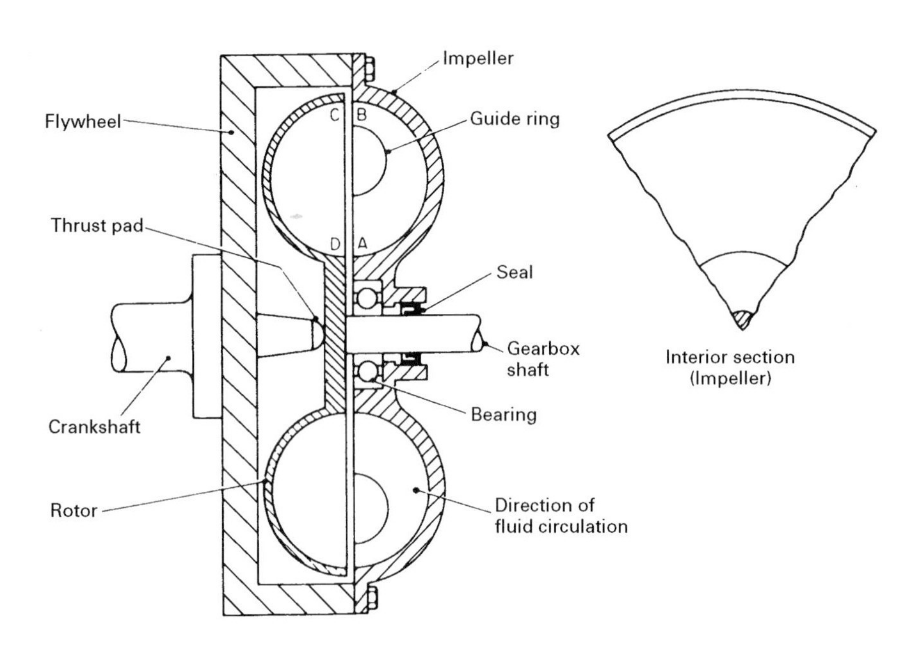

Examine the interior of a fluid flywheel assembly and complete the section of the impeller, shown on the right above, by showing the arrangement of the vanes.

Why does the fluid flow from:

1. A to B? ....

....

2. C to D? ....

....

In what other direction is the fluid moving?

....

Describe how the fluid forces the rotor round:

....

TORQUE CONVERTER

The function and action of a torque converter are somewhat similar to those of a fluid flywheel, but with (in its simplest form) the addition of another fixed, bladed member. The advantage of this arrangement is that when 'slip' is taking place a torque multiplication is obtained.

State the type of gearbox normally used in conjunction with a torque converter:

....

A simple line-diagram representing a single-stage three-element torque converter coupling is shown opposite. Complete the labelling on the drawing and add arrows to indicate the path of oil flow. One significant constructional feature of a torque converter is the shape of the vanes in both the impeller and turbine. The greater the change in fluid direction after striking the turbine vanes, the greater will be the force on the turbine. This change in fluid direction is achieved by curving the vanes.

State the reason why it is necessary for the fluid to pass through a reaction member before going back into the impeller:

....

....

....

....

....

....

....

....

Examine a torque converter and complete the sketch above right by showing the shape of the vanes.

Torque converter

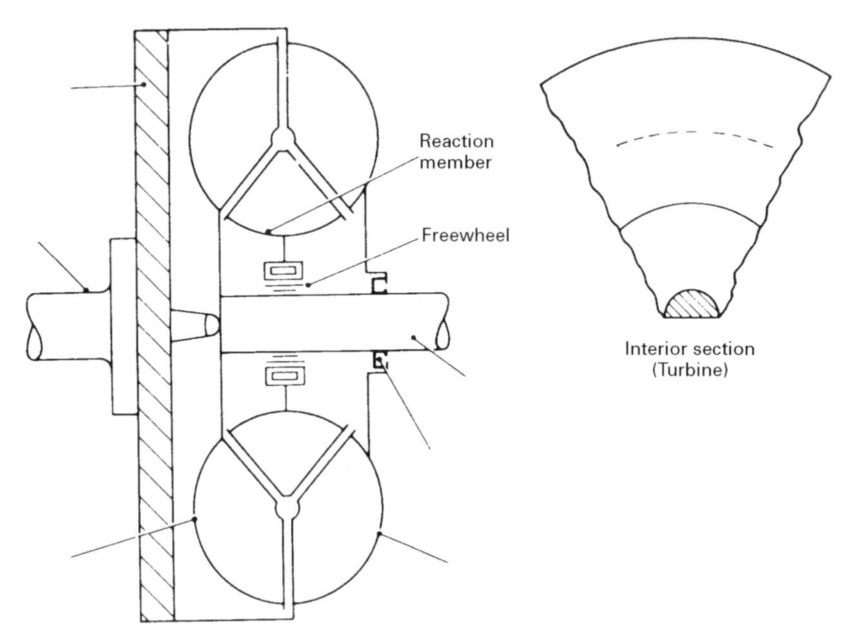

State the reason why the reaction member is mounted on a freewheel:

....

....

....

....

....

....

The converter shown above is a 'three-element' or .... unit,

the maximum torque multiplication for such a unit is ....

State when maximum torque multiplication occurs:

....

....

Name the component parts on the torque converter assembly shown below:

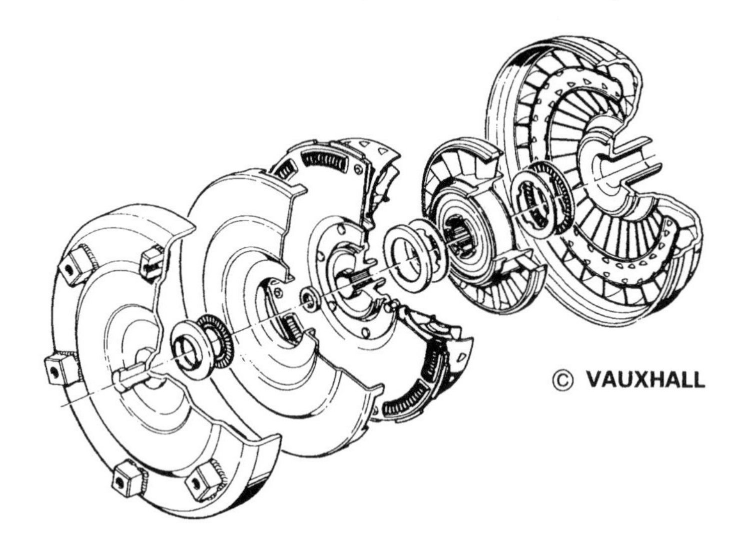

Coupling Characteristics

The graph at (A) shows the efficiency output curve for a constant input speed. For the fluid flywheel, complete the graph at (B) to show the curve for a torque converter:

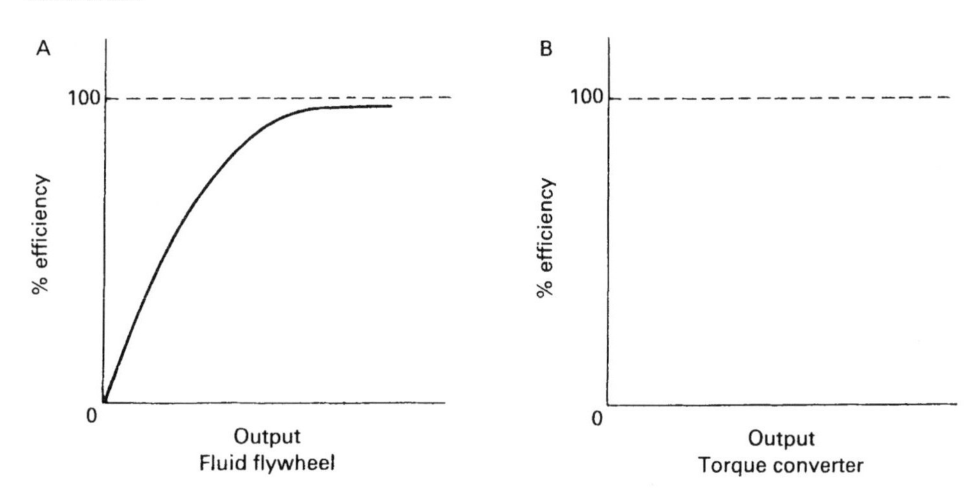

Considering the formula KINETIC ENERGY = $^1/_2\, mv^2$, state the relationship between fluid speed and kinetic energy:

....

....

....

....

The velocity and hence the kinetic energy of the fluid are increased as it moves to the outer radius of the impeller. State how the kinetic energy is converted to a force which produces a driving torque in the turbine or output shaft:

....

....

....

....

....

....

Some energy is lost because of conversion to heat energy, that is, the fluid is heated as it is made to work on the turbine. Under which operating condition is maximum heat generated?

....

....

....

....

....

....

....

....

....

EPICYCLIC (PLANETARY) GEAR TRAINS

Epicyclic gear trains provide, in a very compact manner, various ratios and directions of rotation. This is achieved by holding certain members and applying power to one of the other members. The members are held for gear engagement purposes by brake bands or multi-plate clutches that are normally actuated hydraulically.

List some common motor-vehicle applications of epicyclic gearing:

....

....

....

The compactness of the epicyclic gear train is an advantage. State one other advantage of epicyclic gearing:

....

....

....

....

....

Complete the labelling for the simple epicyclic great train shown below:

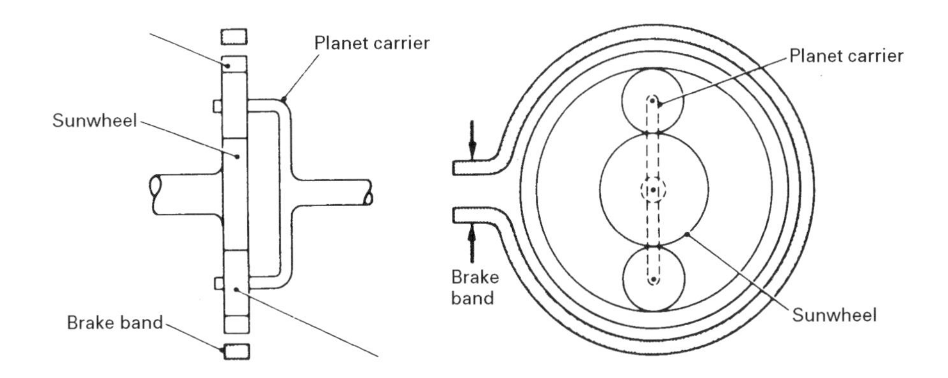

INVESTIGATION

Examine a simple epicyclic gear train and study the three combinations shown below.

Describe alongside each drawing the relationship between input and output in each case. Add arrows to the drawings to indicate the direction of rotation of the wheels.

BRAKE ON

Input sunwheel–output planet carrier:

....

....

....

....

BRAKE OFF

Input sunwheel–output annulus–planet carrier locked:

....

....

....

....

BRAKE ON

Input planet carrier–output sunwheel:

....

....

....

....

....

Describe how it is possible to increase the number of gear ratios available using epicyclic gearing:

....

....

....

Compound epicyclic gear trains

This is a more complex and sophisticated type of epicyclic arrangement than the simple epicyclic gearing mentioned on the previous page. It has the advantages of being able to obtain a greater number of forward ratios, and reverse.

A 'two-element' epicyclic gear train, as used in some automatic gearboxes, is shown opposite; this gear set will provide three forward gears and one reverse gear. One important feature to appreciate is the fact that the planet carrier is mounted on a freewheel or one-way clutch; this will only permit the carrier to revolve clockwise (as seen from the drawing). Examine such a gear set and, in the spaces provided describe its operation in each gear.

1st gear, input to small sun gear

2nd gear, input to small sun gear – large sun gear locked

Top gear, small sun gear and large sun gear locked together – input through both

Reverse gear, input to large sun gear – small sun gear free – freewheel locked

AUTOMATIC GEARBOX MECHANICAL SYSTEM

The drawing opposite represents the complete mechanical system in an automatic gearbox. The various ratios are obtained through a compound epicylic gear set such as outlined on the previous page.

Although the automatic gearbox arrangement shown opposite is an early model, it is a relatively simple example and thus ideal for the purpose of gaining an understanding of automatic gearbox operation.

Study the control system for the gear set and indicate, on the table below right and the drawing opposite, which clutches and brake bands are operative in each gear.

State:

1. The conditions under which the freewheel is operating, and
2. The effect of this action.

....

....

....

State the effect of the application of the rear brake band in first and second gears:

....

....

....

State the reason why the rear brake band is applied in reverse gear:

....

....

....

....

....

....

....

....

Mechanical system fully automatic gearbox

Complete the labelling on the drawing:

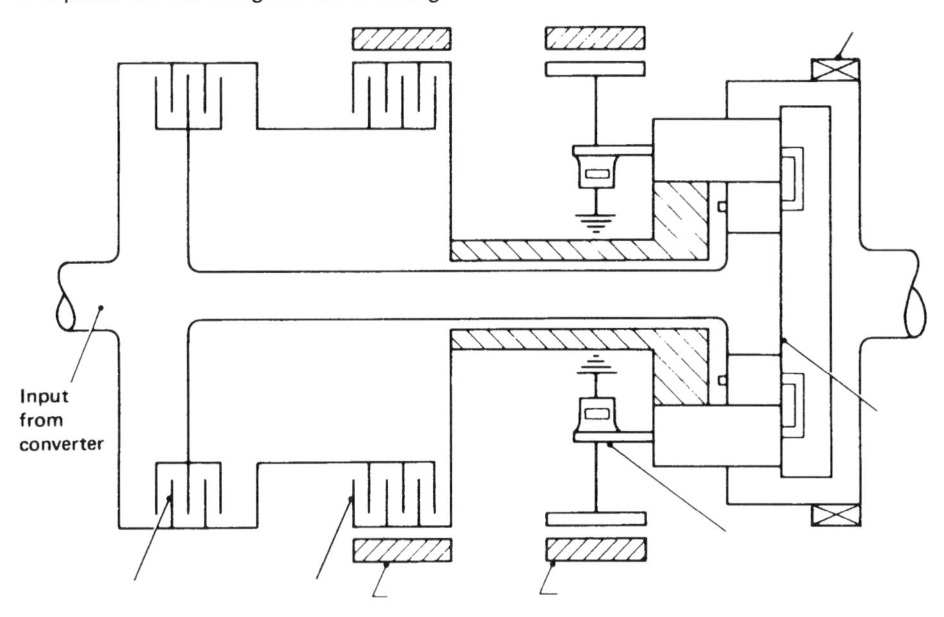

	1st gear	2nd gear	Top gear	Reverse	Neutral
Front clutch					
Rear clutch					
Front brake band					
Rear brake band					

AUTOMATIC GEARBOX HYDRAULIC SYSTEM

The hydraulic system in the automatic gearbox serves three basic purposes. It:

1. maintains a pressurised supply of fluid to the torque converter;
2. provides gearbox lubrication;
3. actuates hydraulic servos during multi-plate clutch and brake band operation.

Name, state the purpose and describe the operation of the component shown below:

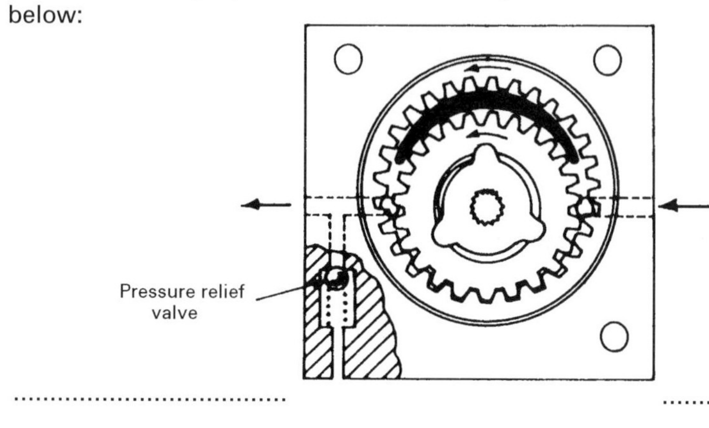

....

....

....

....

....

....

Name the types of fluid seal used and their particular application in an automatic transmission:

....

....

State the reason why the fluid level changes when the selector lever is moved:

....

Hydraulic servo arrangements for brake band and multi-plate clutch operation are shown below. Complete the labelling on the drawings.

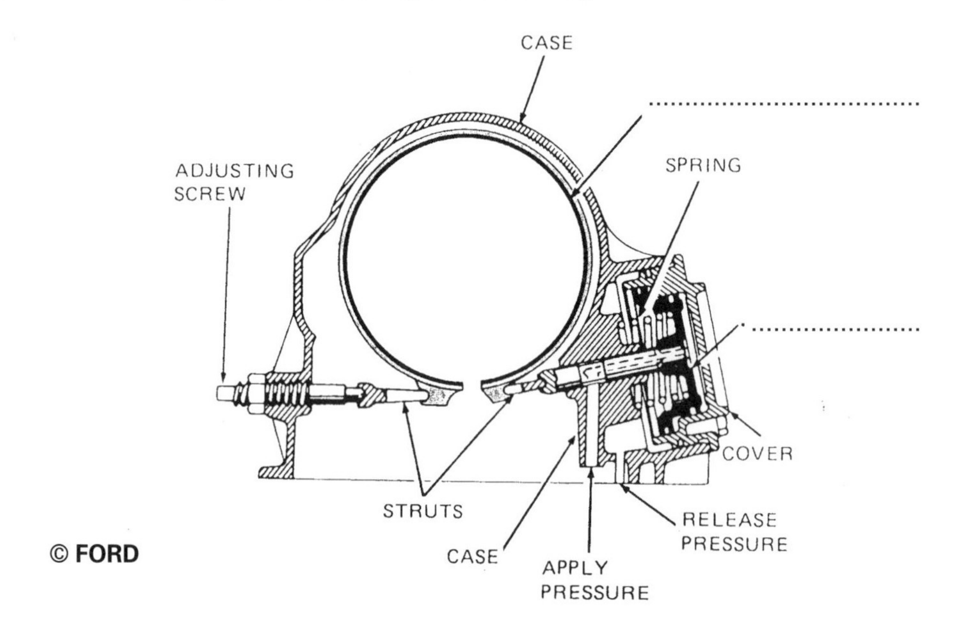

© FORD

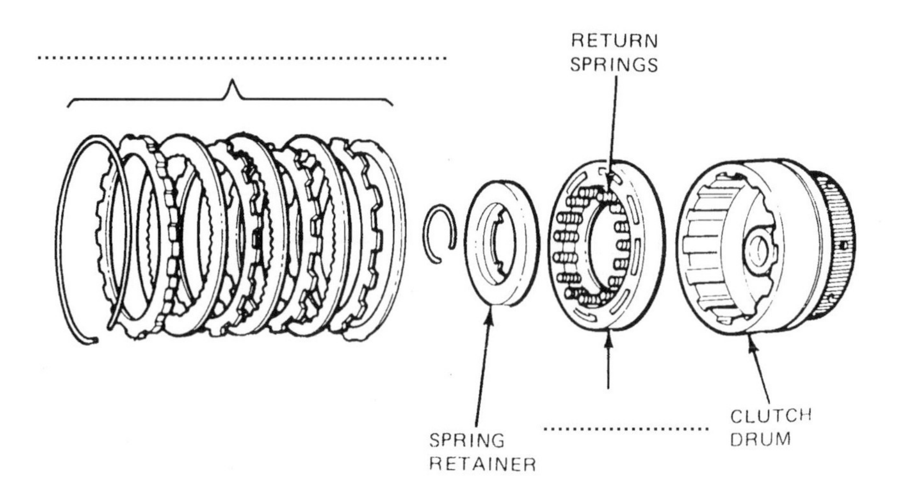

© FORD

A complete automatic transmission system for a front wheel drive vehicle (HONDA) is shown below. Name the numbered parts.

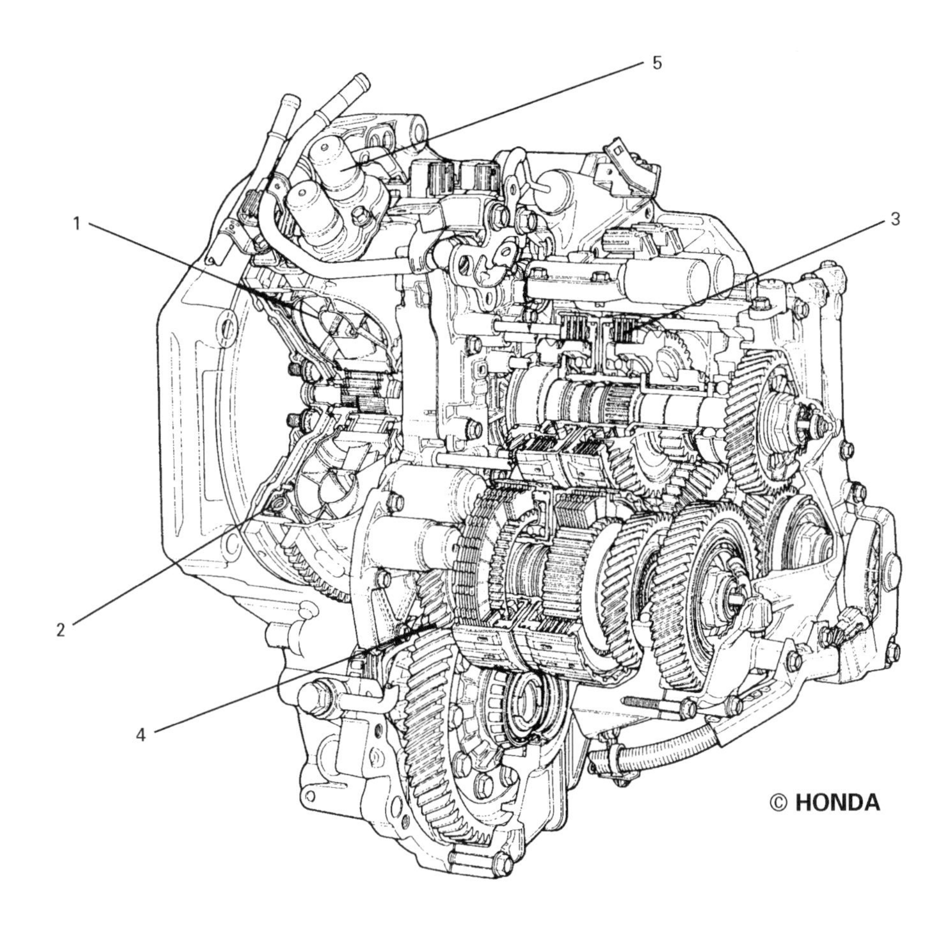

1 .... 2 ....

3 .... 4 ....

5 ....

State the function of the:

A ACCUMULATOR ....

....

....

....

B LOCK UP CLUTCH ....

....

....

....

OIL COOLER

A typical oil cooler system is shown.

1. Label the drawing (add arrows to indicate flow).
2. State the purpose of the system.
3. Describe briefly how it operates:

....

....

....

....

....

....

....

....

An oil cooler is very often a standard fitment on many automatic transmissions, it can however be fitted as a modification. Give reasons why an oil cooler may need to be added to an automatic transmission:

....

....

GEAR SELECTION (AUTOMATIC TRANSMISSION)

A selector lever and quadrant showing the various operating positions for a four-speed automatic transmission are shown below.

State the purpose of each gear-selector position and button 'S'.

P. ....

....

....

....

....

R. ....

N. ....

D. ....

....

....

....

....

3/2/1 ....

....

....

....

....

....

....

S. ....

....

....

Many modern automatic transmissions have a 'SEQUENTIAL' operating mode (see next page). Briefly describe what this means.

....

....

....

....

....

Starter inhibitor

For safety reasons the starter can only be operated in selector positions P and N. This is achieved by an inhibitor switch on the gearbox which is operated by the manual selector mechanism. The four-terminal switch is operative in two positions for starter, or reverse-light operation.

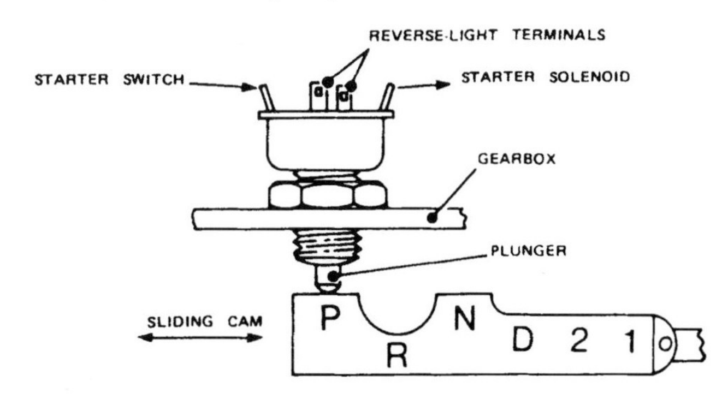

With an electrical shift, a multi-function switch informs the ECU of the selector lever position, this prevents the starter motor operating when the transmission is in gear and also controls the reversing lights.

State the precautions needed when towing a vehicle with automatic transmission when:

(a) engine is defective ....

....

(b) gearbox is defective ....

....

ELECTRONIC CONTROL

An electronically controlled automatic gearbox is basically the same as a hydraulically controlled automatic gearbox in that the main components are torque converter, epicyclic gear set, multi-plate clutches and hydraulic servos.

In addition to these main components, electronic shifting entails the use of:

1. *ECU* .... 2. .... 3. ....

Complete the simple block diagram for such a system and describe briefly how automatic gear changing is achieved.

SENSORS

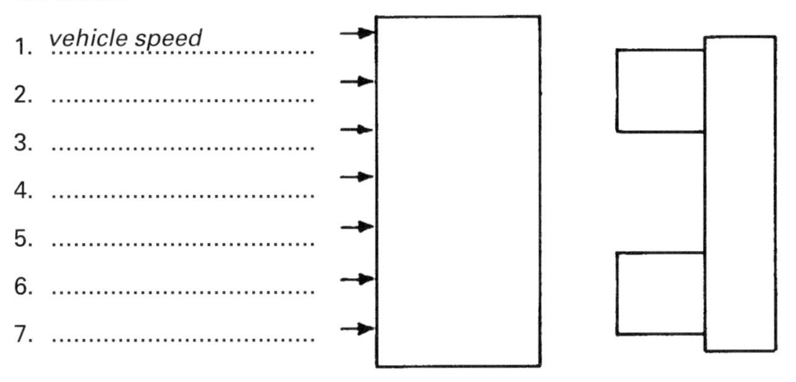

OPERATION

....

....

....

....

....

....

....

Electronic control makes mode selection and operation very much easier and effective.

The layout of an electronic control system for an automatic transmission is shown below.

Name the lettered items.

a ....

b ....

c ....

d ....

e ....

f ....

g ....

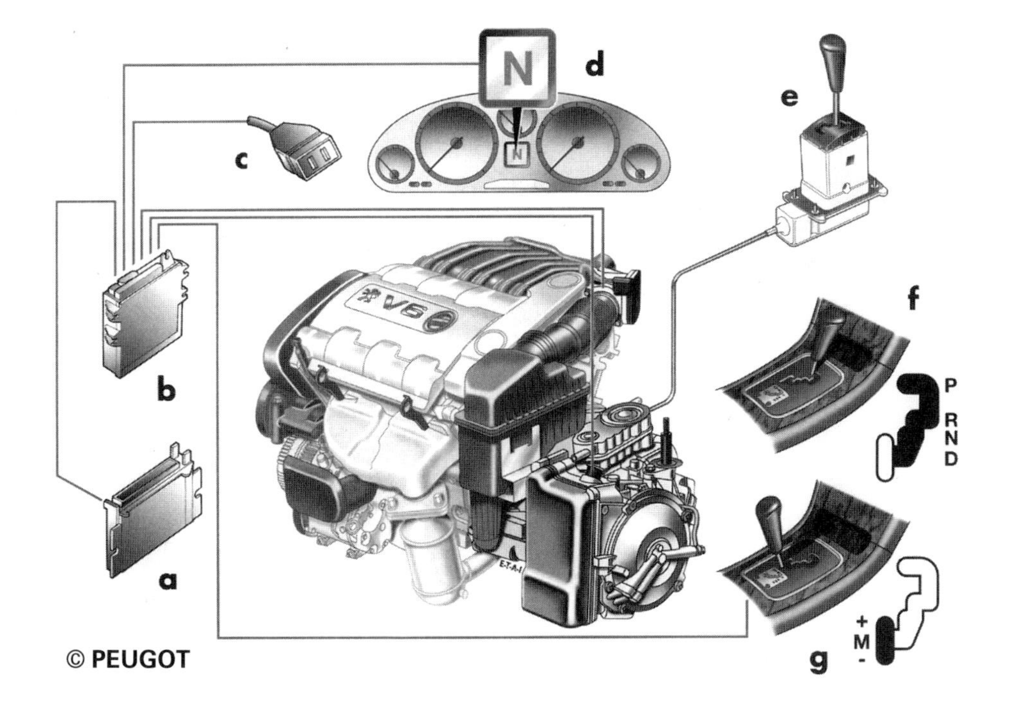

© PEUGOT

CONTINUOUSLY VARIABLE AUTOMATIC TRANSMISSION

Continuously variable or STEPLESS transmission is an alternative to the FIXED RATIO transmission. Unlike conventional automatic transmission, the gear ratios are varied in a smooth, stepless progression to suit driving condition (speed and load).

One transmission arrangement contains two basic elements:

(a) a PLANETARY GEARSET integrated with two wet multi-plate clutches;

(b) a BELT and PULLEY system.

The components named at (a) provide take up from rest and drive to the pulley system.

State the function of (b):

....

....

....

....

A front engine front wheel drive transaxle (FORD) is shown opposite. Complete the labelling on the drawing.

This system does not require a torque converter or a conventional friction clutch. Why is this?

....

....

....

....

List two vehicles using this (or similar) types of transmission:

Make .... Model .... Engine size ....

Make .... Model .... Engine size ....

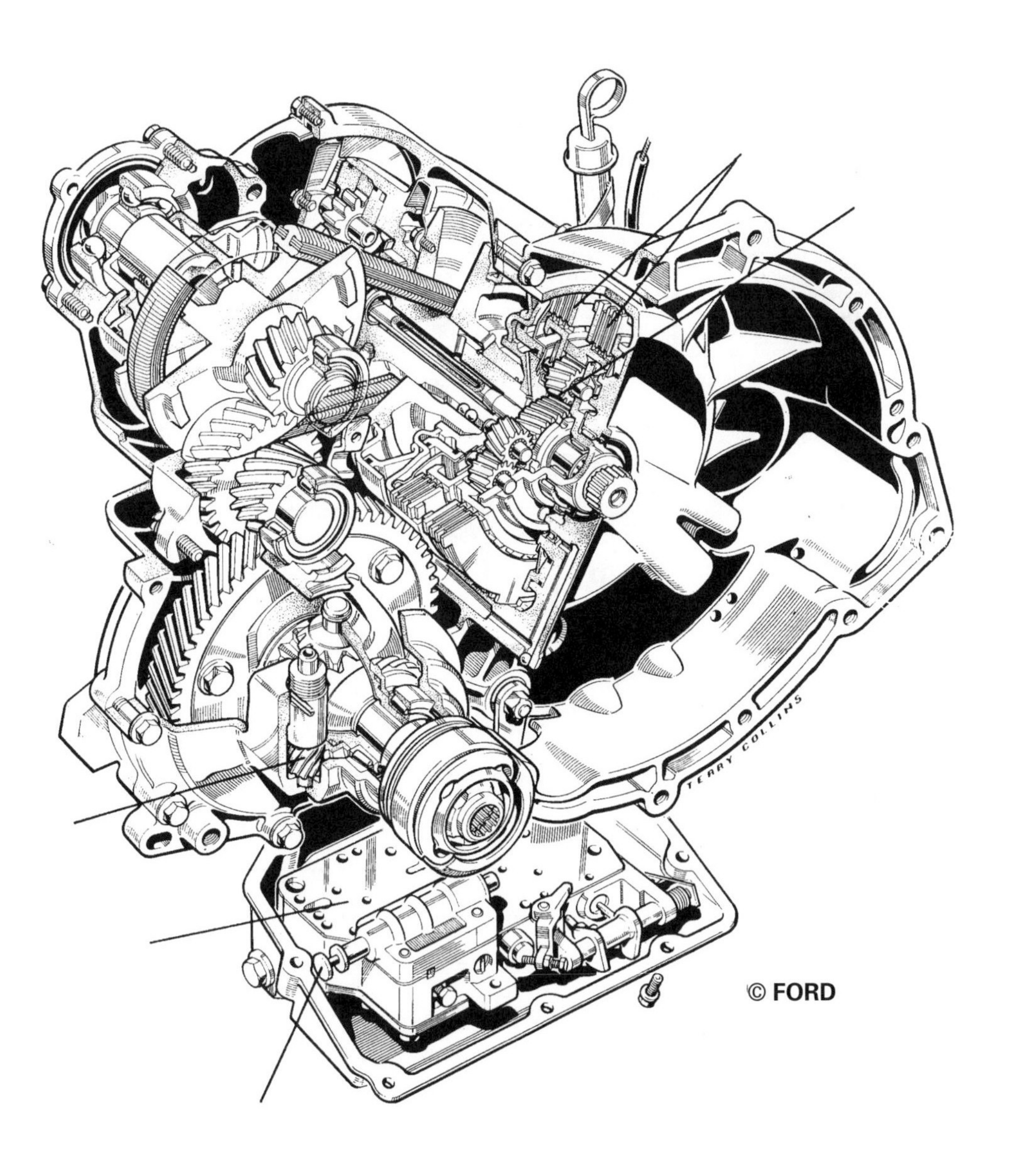

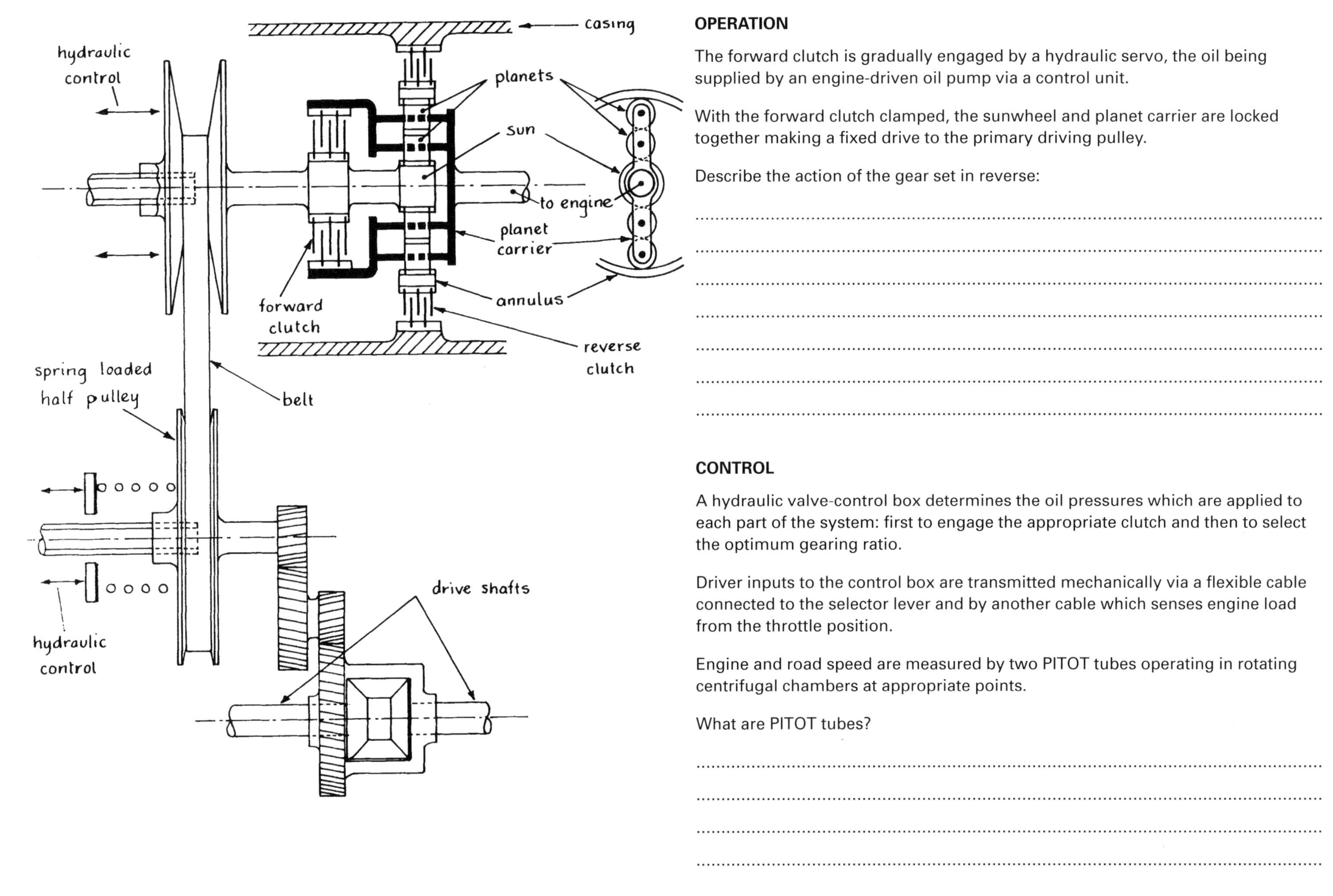

OPERATION

The forward clutch is gradually engaged by a hydraulic servo, the oil being supplied by an engine-driven oil pump via a control unit.

With the forward clutch clamped, the sunwheel and planet carrier are locked together making a fixed drive to the primary driving pulley.

Describe the action of the gear set in reverse:

....

....

....

....

....

....

....

CONTROL

A hydraulic valve-control box determines the oil pressures which are applied to each part of the system: first to engage the appropriate clutch and then to select the optimum gearing ratio.

Driver inputs to the control box are transmitted mechanically via a flexible cable connected to the selector lever and by another cable which senses engine load from the throttle position.

Engine and road speed are measured by two PITOT tubes operating in rotating centrifugal chambers at appropriate points.

What are PITOT tubes?

....

....

....

....

Describe, with the aid of diagrams, the action of the belt and pulley system:

....

....

....

....

....

....

....

....

....

....

Driving pulley width is controlled by: ....

....

Secondary or driven pulley width is controlled by:

....

....

....

....

From what material is the belt made?

....

....

....

LOW GEAR

HIGH GEAR

State the advantages of the continuously variable automatic transmissions over the fixed ratio automatic transmission:

....

....

....

....

The basic layout and power flow for a STEPLESS transmission currently used by HONDA is shown below.

TRANSFER DRIVE GEAR

(OPTION) 4 X 4 TRANSFER

1. FLYWHEEL
2. DRIVE PULLEY
3. STEEL BELT
4. FORWARD CLUTCH
5. REVERSE BRAKE
6. INPUT SHAFT
7. SUN GEAR
8. DRIVE PULLEY SHAFT
9. DRIVEN PULLEY
10. SECONDARY DRIVE GEAR
11. PARKING GEAR
12. FINAL DRIVEN GEAR
13. FINAL DRIVE GEAR
14. START CLUTCH

Use arrows to indicate on the drawing the output for the front wheel drive shafts.

OPERATION

Basically with the FORWARD CLUTCH engaged (pressurised), drive from the engine is transmitted to the drive pulley. The belt transmits drive to the driven pulley and with the START CLUTCH engaged drive is transmitted to the secondary drive gear and final drive.

The effective pulley ratio changes automatically with engine speed due to hydraulic pressure variation acting on the movable faces of the pulleys.

Describe the procedure for checking the fluid level in automatic transmission systems:

....

....

....

....

....

....

....

....

....

....

In addition to checking the fluid level, the condition of the fluid should be examined. Why is this?

....

....

....

....

SEMI-AUTOMATIC TRANSMISSION

An alternative to a fully automatic transmission is a manual gearbox with ROBOTISED gearshift and clutch operation. This is a semi-automatic transmission with two pedal (no clutch pedal) control.

The layout below shows the main control features of this system.

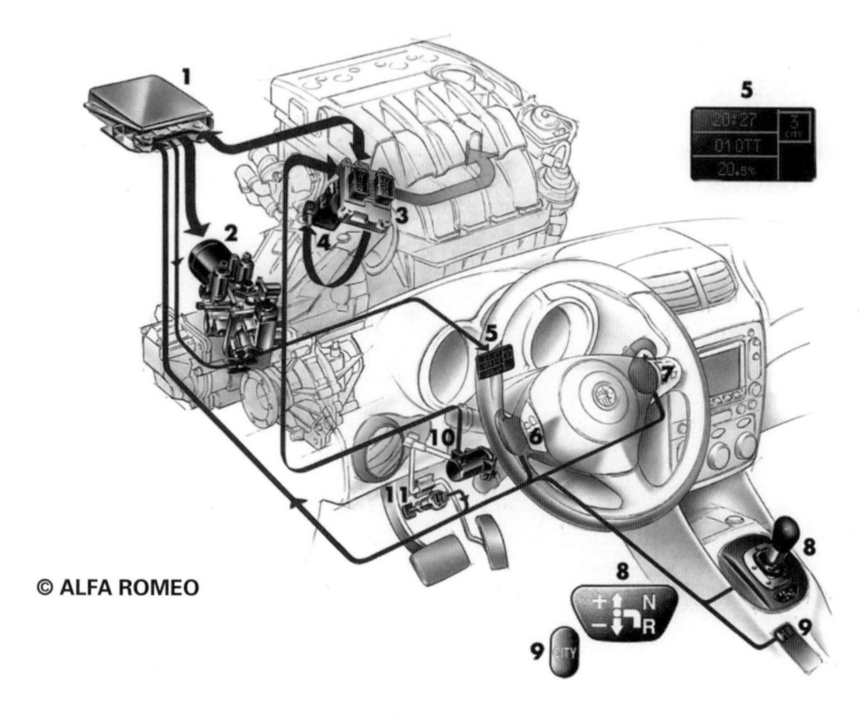

Selespeed gearbox

1) electronic gearbox management unit
2) clutch/gear shift actuator unit
3) electronically controlled throttle
4) electronic engine management unit
5) display on fascia
6) 'Down' control for shifting down through gears
7) 'Up' control for shifting up through gears
8) lever for sequential gear selection and display
9) 'City' control for automatically activating gear changes
10) electronic accelerator potentiometer
11) switch on brake pedal

Study the illustration opposite and complete the electrical circuitry to the schematic layout below.

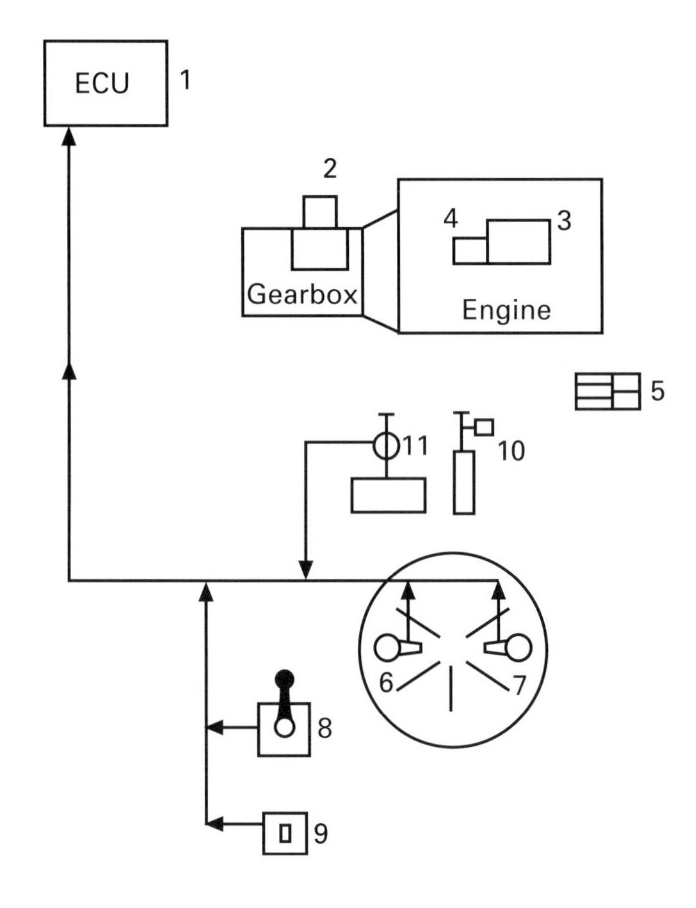

On receiving a signal from the gear lever (or steering wheel button) the ECU will energise solenoid valves in the clutch/gear shift actuator unit. This controls pressure in hydraulic servos to move the gear engagement, release and selection levers; and to operate the clutch. In 'CITY MODE' fully automatic gear shifting is in operation.

BAND OPERATED AUTOMATIC GEARBOX

This system consists of a fluid flywheel in series with a four- or five-speed epicyclic gear set. Brake bands, operated by hydraulic servos, are applied and released to control the power flow and ratio selection in the gear set.

Brake bank operation

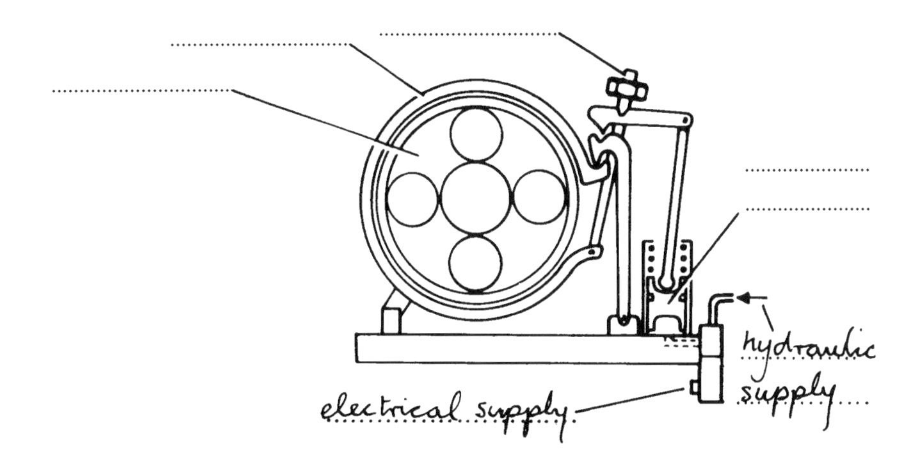

Complete the labelling on the drawing and briefly describe the action of the system during brake band application.

....

....

....

....

....

....

The drawing below illustrates the layout of the control system for a fully automatic (HYDRACYCLIC) transmission popular on PSVs. Describe the operation of the system.

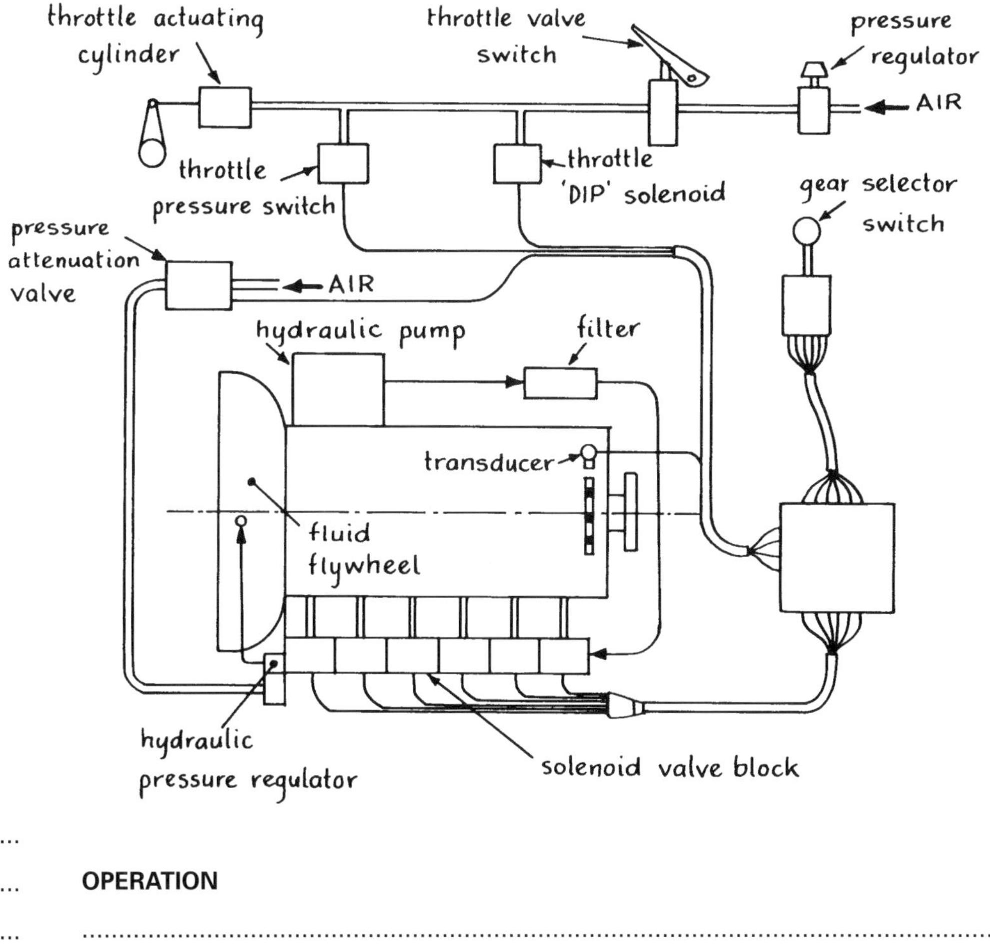

OPERATION

....

....

....

....

OPERATION (continued)

The system usually incorporates a 'friction retarder', which comes into operation when the brakes are applied, and a 'lock up' clutch. What is the purpose of a lock up clutch?

An alternative to the fully automatic system is shown opposite. This is a SEMI-AUTOMATIC system which utilises the same gearbox. The brake band servos are applied hydraulically as in the fully automatic system; in earlier gearboxes compressed air was used.

SEMI-AUTOMATIC HYDRACYCLIC GEARBOX

AIR

throttle pressure switch

hydraulic pump

ECU

OPERATION

As the driver selects the gears, the electrical solenoids are energised to actuate the hydraulic valves. What is the purpose of the ECU (TRANSLATOR), Throttle Pressure Switch and Transducer in this system?

COMPUTER AIDED GEAR CHANGING (CAG)

An optional transmission system for HGVs (SCANIA) is CAG. The physical action of gear changing is performed by air-powered actuating cylinders. It is a form of PRE-SELECTOR transmission.

Main components

Add appropriate numbers to the components indicated on the drawing and state the function of each component:

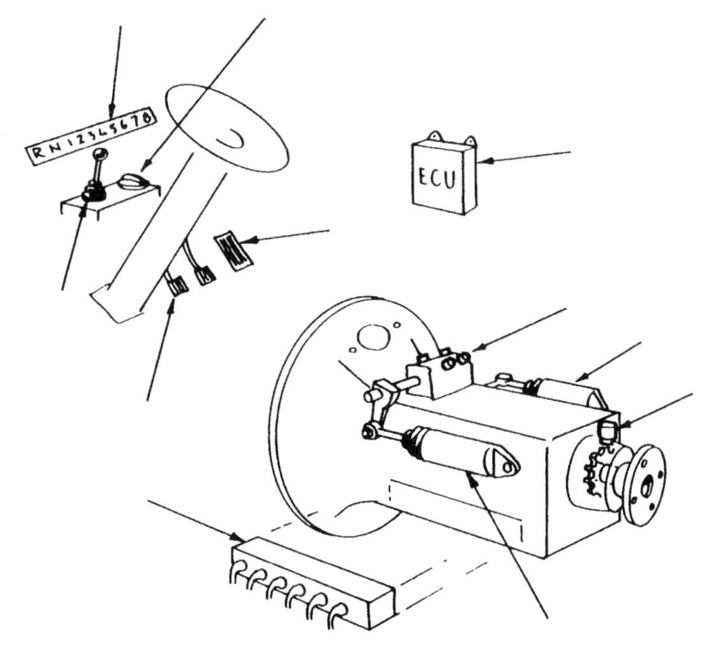

1. SHIFT MODE SELECTOR

....

....

....

2. SHIFT STALK

....

....

3. COMPUTER

....

....

....

....

4. SOLENOID VALVES

....

....

....

....

5. SENSORS

....

....

....

....

....

6. ACTUATING CYLINDERS

....

....

....

7. CLUTCH PEDAL

....

....

....

8. GEAR DISPLAY

....

....

....

....

TESTING AND FAULT DIAGNOSIS

One method of testing a torque converter coupling is to carry out a STALL TEST on the vehicle. Outline this procedure, including any safety measures adopted during the test:

....

....

....

....

....

....

....

....

....

....

For satisfactory operation the engine stall speed should be approximately .... revs/min. Indicate the possible faults for the results given below.

Engine speed 300 revs below specified stall speed:

....

Engine speed 600 revs below specified stall speed:

....

Stall speed too high:

....

....

....

....

....

A road or chassis dynamometer test is another way in which torque converter faults are diagnosed; the road test will also confirm faults suspected during a stall test.

State how the road test would indicate:

1. Slipping stator ....

....

....

....

2. Seized or locked stator ....

....

....

....

If the converter pressure is low 'cavitation' will occur, resulting in slippage, noise and vibration. What are the effects if converter pressure is too high?

....

....

....

With the faults at 1 and 2 above it is necessary to replace the converter as it is normally a sealed unit. State the safety precautions to be taken when draining the fluid:

....

....

....

....

The fluid coupling requires very little maintenance apart from topping up with fluid; however, leakage could result from a worn seal and worn bearings could allow the faces of the impeller and rotor to come into contact, giving noisy operation.

How can the system be protected during use or repair against the following hazards?

1. Overheating

....

....

....

2. Aeration and foaming

....

....

....

3. Contamination of fluid

....

....

....

4. Ingress of dirt and moisture

....

....

....

5. Fluid leakage

....

....

....

6. Mechanical damage

....

....

....

Routine preventive maintenance will improve reliability and efficiency, and maximise the life of a transmission system. List the preventive maintenance checks and tasks associated with automatic transmission:

Check: ....

....

....

....

....

....

....

....

....

....

....

....

....

....

....

List the general rules for efficiency and any special precautions to be observed when carrying out maintenance on the transmission:

....

....

....

....

....

....

....

....

....

....

DIAGNOSTICS: AUTOMATIC TRANSMISSION – SYMPTOMS, FAULTS AND CAUSES

State a likely cause for each symptom/system fault listed below. Each cause will suggest any corrective action required.

SYMPTOMS	FAULTS	PROBABLE CAUSES
Loss of drive; flare up on changes	Worn multi-plate clutch	....
Incorrect gear selection; starter operates in other than P or N	Worn or incorrectly adjusted selector linkage	....
Incorrect gear-change interval	Broken throttle signalling device (cable or vac pipe)	....
Incorrect gear-change interval	Incorrectly adjusted kickdown cable	....
Harsh engagement of gears, excessive creeping	High engine idling speed	....
Slipping clutches	Low line pressure	....
Bumpy gear shifts	High line pressure	....
Slip in one gear	Oil leakage	....
Park will not hold car	Broken parking pawl	....

Chapter 4

Drive Line

Drive line shafts and hubs	72
Propeller shafts	73
Universal joints	73
Constant velocity joints	75
Hubs	77
Steered front axle	78
Dead axle	78
Diagnostics	79
Drive line shafts and hubs: protection during use	80
Maintenance	80

DRIVE LINE SHAFTS AND HUBS

Drive line

Identify the drive line arrangements on this page. Name the major parts and give a brief description of each.

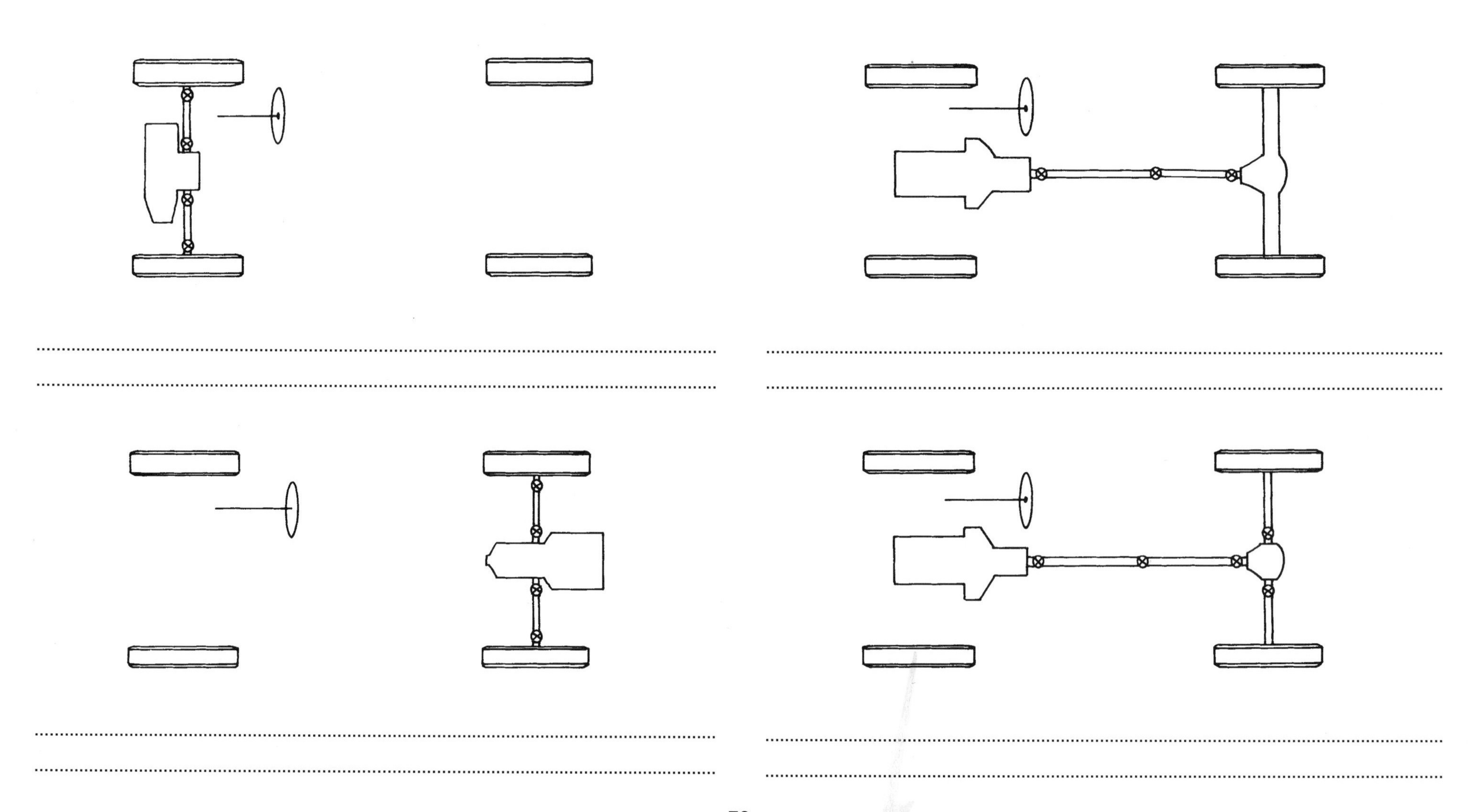

PROPELLER SHAFTS

The propeller shaft transmits the drive from the gearbox to the final drive gear. Name the major parts in the drawing below.

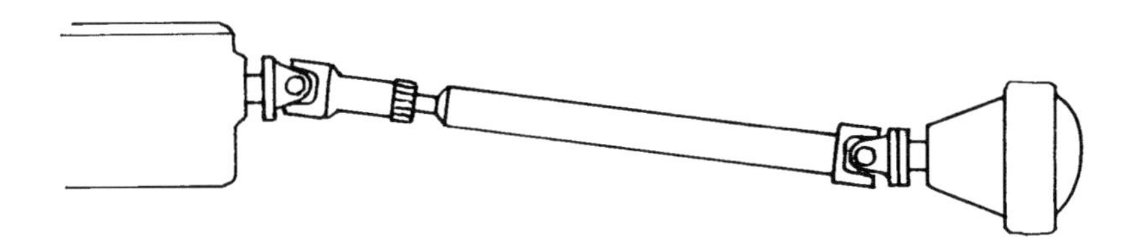

State the purpose of:

(a) the universal joints (UJs)

....

....

....

(b) the slip joint

....

....

....

Give reasons for using a tubular propeller shaft rather than a solid shaft:

....

....

....

What alignment feature is vital during service work on the above shaft?

....

....

....

How does the arrangement below primarily differ from the one opposite?

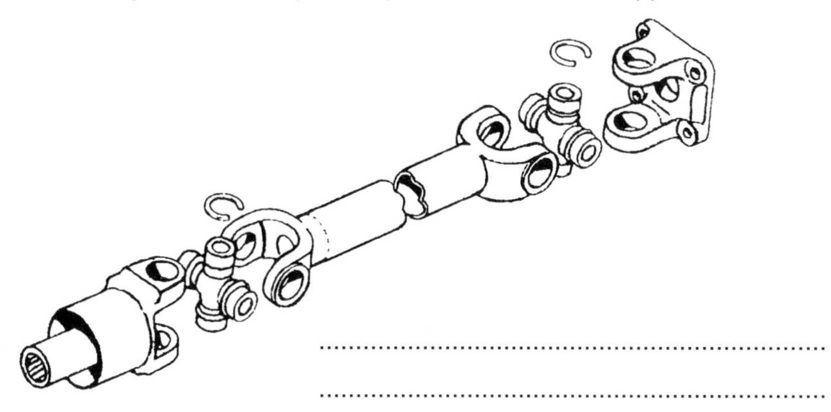

....

....

UNIVERSAL JOINTS

A very popular type of universal joint is shown above. The joint is efficient, compact, easy to balance and it will transmit the drive through quite large angles.

Name this type of joint ....

Examine a dismantled joint and complete the drawing below to show a section through a trunnion:

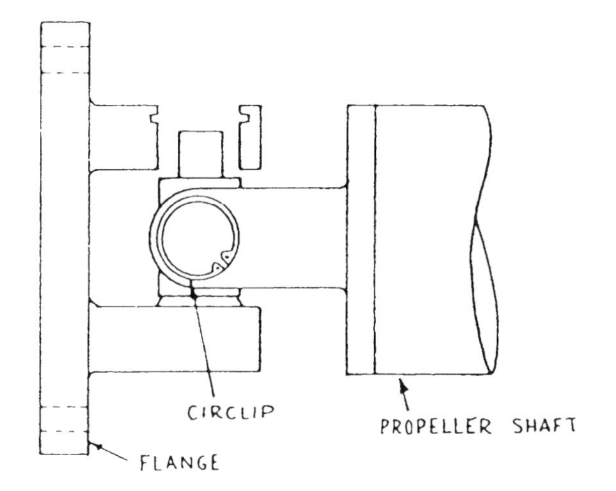

The Hookes joint is an efficient and compact form of UJ and can accommodate a relatively high angular deflection (20°).

This rubber hexagonal joint (Doughnut type) is very often used in the application shown opposite.

Label the drawing.

....

....

Why is it used in the driveline shown?

....

....

....

....

When installed, the rubber is pre-stressed. How is this achieved and what is the reason for it?

....

....

....

....

The 'composite disc' joint shown here is strong and compact and can operate with angular deflection of up to 2.5°.

GKN Automotive

Two-piece propeller shaft

Vehicle application: Make .... Model ....

Give reasons for using this arrangement and label the illustration:

....

....

....

Label the flexibility mounted centre bearing shown below:

State TWO vehicles which use a propeller shaft centre bearing:

MAKE .... MODEL ....

MAKE .... MODEL ....

Universal joint characteristics

When drive is being transmitted from one shaft to another through a Hookes-type joint, if one shaft swings through an angle its velocity will vary, that is, the shaft will accelerate and decelerate slightly during every 180° of revolution. This is more pronounced if the angle through which the shaft swings is great, for example, the drive to a steered wheel.

State the conditions necessary if a universal joint is to transmit 'constant velocity':

....

....

....

Hookes-type joint

The drawing below illustrates the problem with a Hookes-type joint; the line B is bisecting the angle between the two shafts. Draw a centre line through the joint driving contacts.

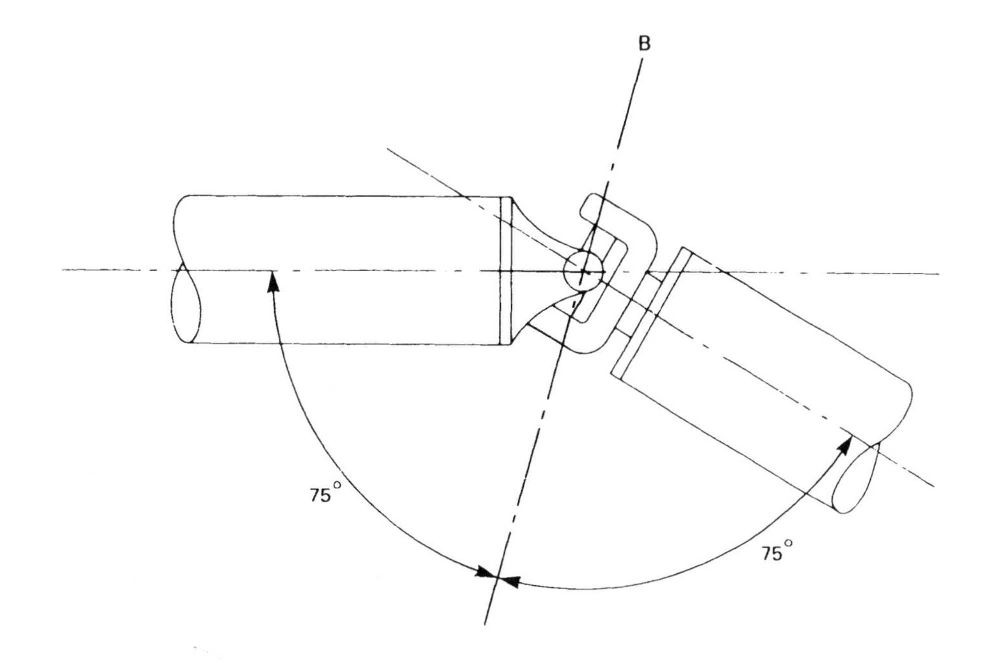

CONSTANT VELOCITY JOINTS

Constant velocity can be transmitted from one shaft to another through varying angles by a special type of joint known as a 'constant velocity joint'. Alternatively, the problem can be overcome by using an intermediate shaft with a Hookes joint at each end to connect the input and output shafts. The irregularity introduced by one joint is then cancelled out by the equal and opposite irregularity introduced by the second joint.

Constant velocity joint

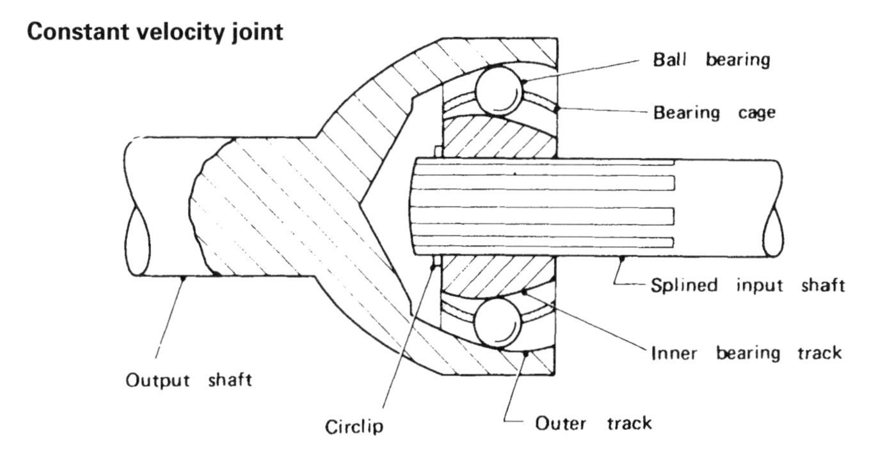

The joint shown above will transmit constant velocity through varying angles. The driving contacts always lie on a plane bisecting the angle between the input and output shaft. Describe how this is achieved:

....

....

....

....

....

....

Plunge type CV joint

In which part of the drive line would the joint shown below be employed, and how does its action compare with the normal type CV joint?

© HARDY SPICER

State the purpose of the gaiter:

If constant angular velocity is to be transmitted from the gearbox mainshaft to the final drive pinion via the drive line shown below, certain conditions have to be met. State these and indicate them on the illustration:

Drive arrangements

The tractive effort produced as a result of driving torque at the tyre/road contact point results in a driving thrust on the vehicle in order to propel it. A 'reactive thrust' and indeed a 'reactive torque' are therefore inherent in the tractive process.

Study the drive arrangements, comment briefly on the action of each during vehicle operation and label the drawings:

HUBS

The number, type and location of bearings employed in a hub assembly depend, to a large extent, on the weight of the vehicle.

Three main classifications of hubs are:

1. *Semi- or non-floating*
2. ....
3. ....

The three hub types are shown opposite. Explain briefly the constructional and operational characteristics of each; label the drawings and state the type of vehicles which use them.

Semi-floating ....

....

....

....

Three-quarter floating ....

....

....

....

Fully floating ....

....

....

Describe how the hub bearings are adjusted during assembly of the fully floating hub:

....

....

....

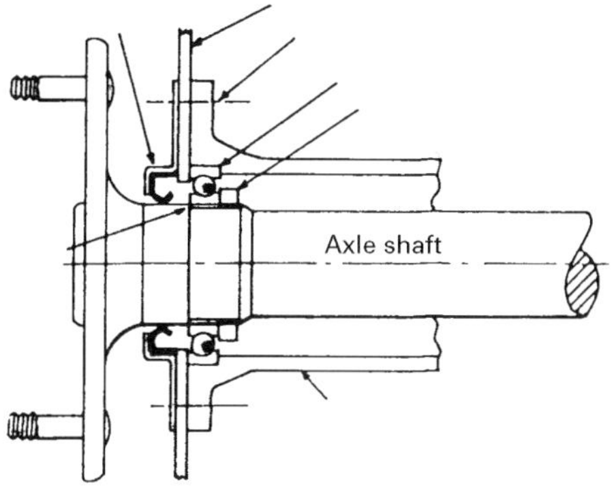

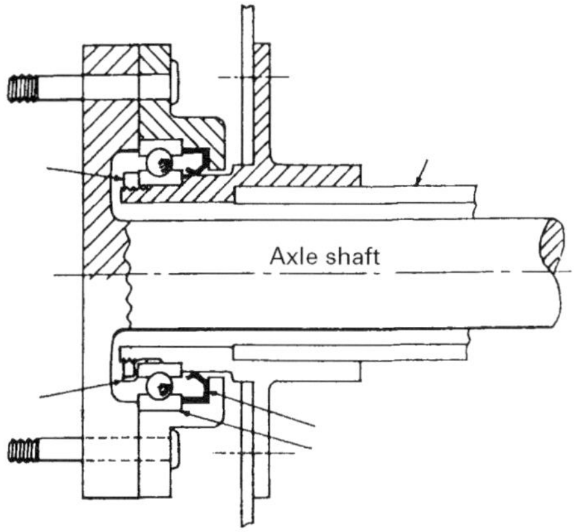

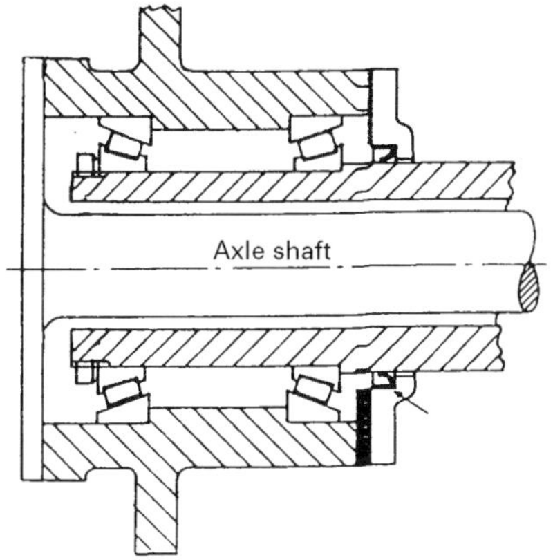

Axle shift

....

Axle shaft

....

Axle shaft

STEERED FRONT AXLE

Name the parts of the hub shown below:

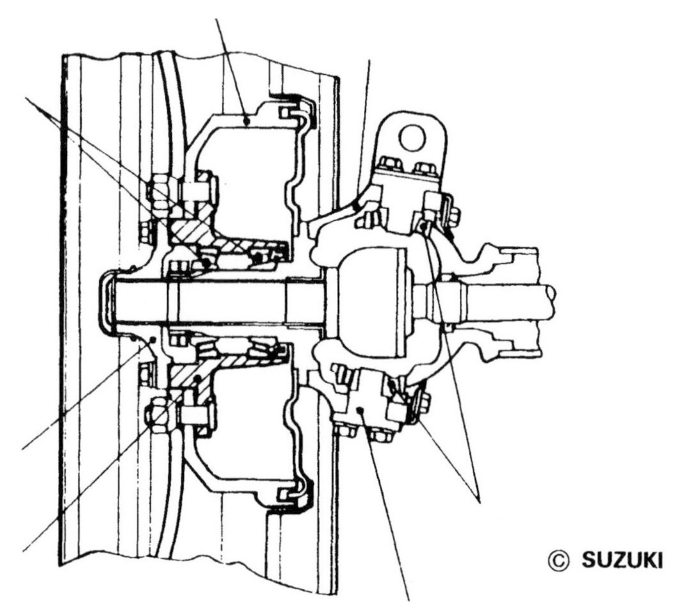

Suggest two vehicles that have a similar hub arrangement:

Make .... Model ....

Make .... Model ....

Is the hub shown for a 'live' or 'dead' axle?

....

Which class of hub is it?

....

What type of bearings are employed?

....

What is the purpose of the drive flange?

....

DEAD AXLE

What is the meaning of the term 'dead axle'?

....

....

Complete the sketch below to show a dead axle:

Name two vehicles that have dead axles:

Make .... Model ....

Make .... Model ....

Bearing arrangements in dead axle hubs are similar to those already described, however, they are not classed as semi, three-quarter or fully floating. Why is this?

....

....

....

....

DIAGNOSTICS: DRIVELINE, SHAFTS AND HUBS – SYMPTOMS, FAULTS AND CAUSES

State a likely cause for each symptom/system fault listed below. Each cause will suggest any corrective action required.

SYMPTOMS	FAULTS	PROBABLE CAUSES
Backlash on take up of drive and vibration	Worn universal joint trunnions	....
Lubricant leakage	Split rubber gaiters	....
Noise from hub; excessive hub end float; overheating of hub assembly	Worn hub bearings	....
Transmission vibration	Worn centre bearing	....
Oil leakage	Worn sliding joint oil seal	....
Knocking noise when turned on lock; vibration	Worn constant velocity joint	....
Knocking on take up of drive	Worn driveshaft splines	....
Transmission vibration and noise	Loose shaft flange bolts	....
Transmission vibration and noise	Sheared rubber (doughnut type universal joint)	....

DRIVE LINE SHAFTS AND HUBS: PROTECTION DURING USE

How can drive line systems be protected against the ingress of moisture and dirt during use and repair?

....

....

....

....

....

Describe any special tools (and their care) necessary to carry out routine maintenance and adjustments:

....

....

....

....

....

....

MAINTENANCE

Routine maintenance is essential in order to ensure reliability, maintain efficiency, prolong service life and ensure vehicle safety. List the major maintenance points:

....

....

....

....

....

....

....

....

....

....

List the general rules/precautions to be observed when carrying out routine maintenance adjustments, removal and replacement relative to the drive line, shafts and hubs:

....

....

....

....

....

....

....

....

....

....

Chapter 5

Final Drive and Differentials

Final-drive and differential location	82
Final-drive gears	83
Differential	84
Operation of the differential	84
Investigation	84
Locating the final-drive gears	85
Setting up and adjustment	86
Investigation	87
Limited-slip differential	88
Tandem (twin) drive axles	89
Inter axle (third) differential	89
Double-reduction gears	90
Maintenance and repair	91
Diagnostics	92

FINAL-DRIVE AND DIFFERENTIAL LOCATION

The final-drive and differential (normally attached to the larger of the two gears of the final drive) may be located in different positions on the vehicle, according to 'chassis' design.

Complete the sketches below to show FOUR different types and state a typical example of each layout:

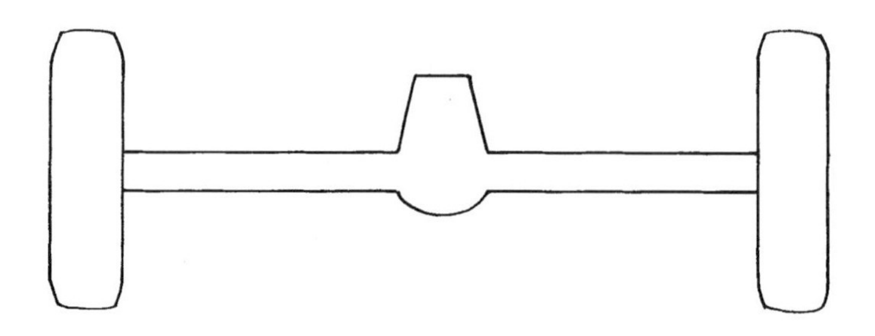

REAR MOUNTED LIVE AXLE

Make .... Model ....

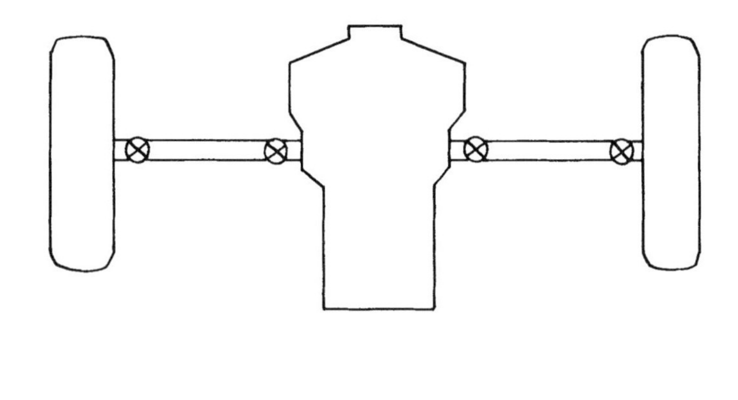

REAR MOUNTED TRANSAXLE

Make .... Model ....

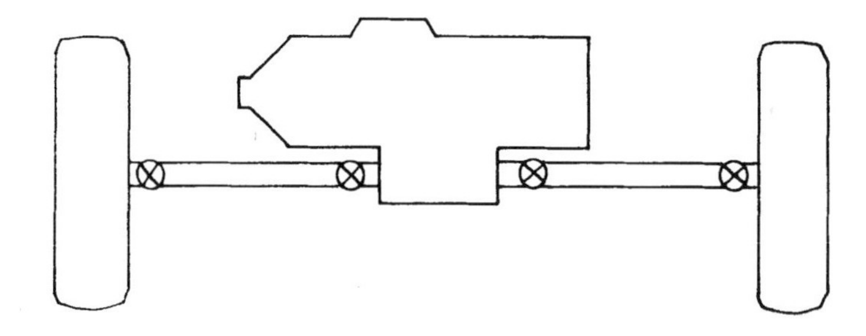

FRONT WHEEL DRIVE

Make .... Model ....

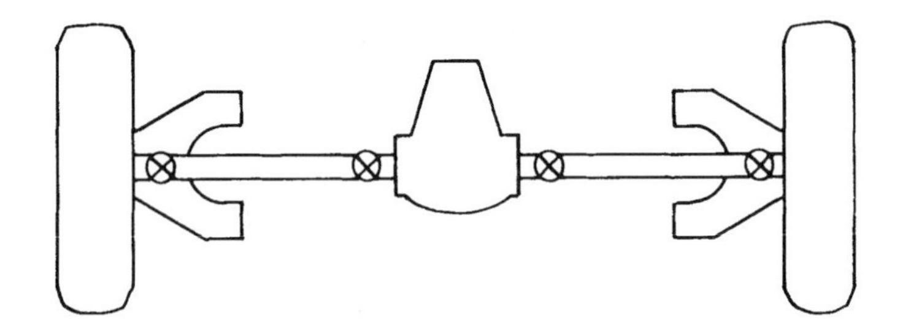

REAR DRIVE WITH IRS

Make .... Model ....

FINAL-DRIVE GEARS

The pair of gears providing the final drive are usually a bevel pinion and crown wheel or a worm and wheel. In many front-wheel-drive arrangements the final drive consists of a pair of helical gears similar to those used in most gearboxes. State the purpose of the final-drive gears:

....

....

....

....

The drawings at A, B and C opposite show three types of final-drive gears used on motor vehicles.

Name the types shown and give a brief description of each.

State the causes of axial thrust on the pinion and add arrows to the drawings to show the direction of axial thrust on drive and overrun for each type of gear:

....

....

....

....

....

Examine spiral bevel, hypoid and worm-and-wheel final-drive assemblies, count the gear-tooth numbers for each pair and state the gear ratios:

Spiral bevel ....

Hypoid ....

Worm and wheel ....

Type:

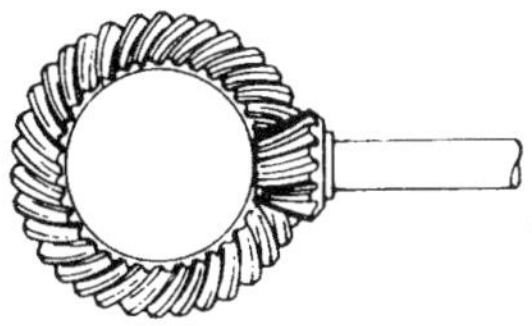

....

A

....

....

....

....

....

....

Type:

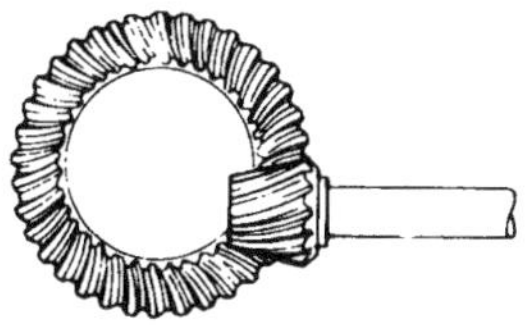

....

B

....

....

....

....

....

....

Type:

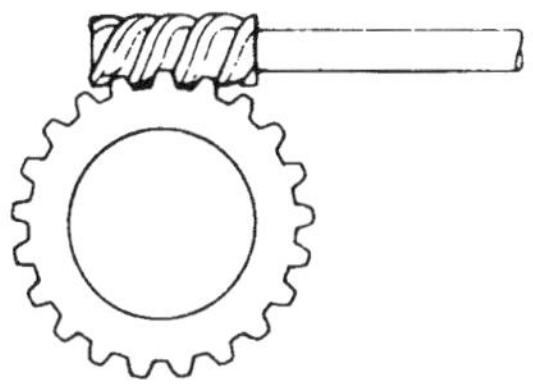

....

C

....

....

....

....

....

....

D

Make a sketch at D to show the final-drive gears for a front-wheel-drive transverse engine car and give a typical gear ratio for this arrangement.

Vehicle MAKE .... MODEL ....

Ratio ....

DIFFERENTIAL

The purpose of the differential is to transmit equal driving torques to the half shafts while allowing the shafts to rotate at different speeds when the vehicle is travelling in other than a straight line.

The drawing below shows a final-drive and differential assembly housed in an axle casing.

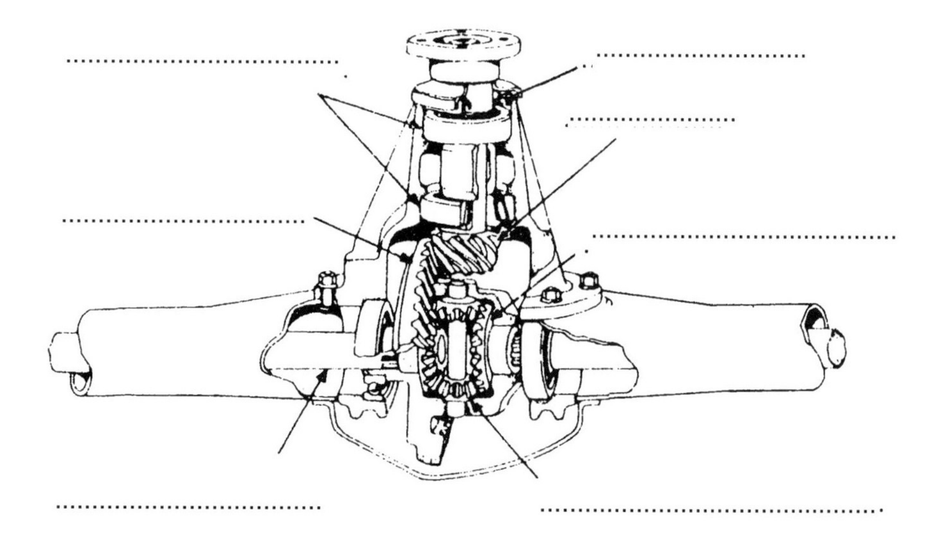

This type of axle carries the half shafts, hubs and brake assemblies and contains the lubricant. What other purpose on the vehicle does this type of axle fulfil?

....

....

....

....

....

OPERATION OF THE DIFFERENTIAL

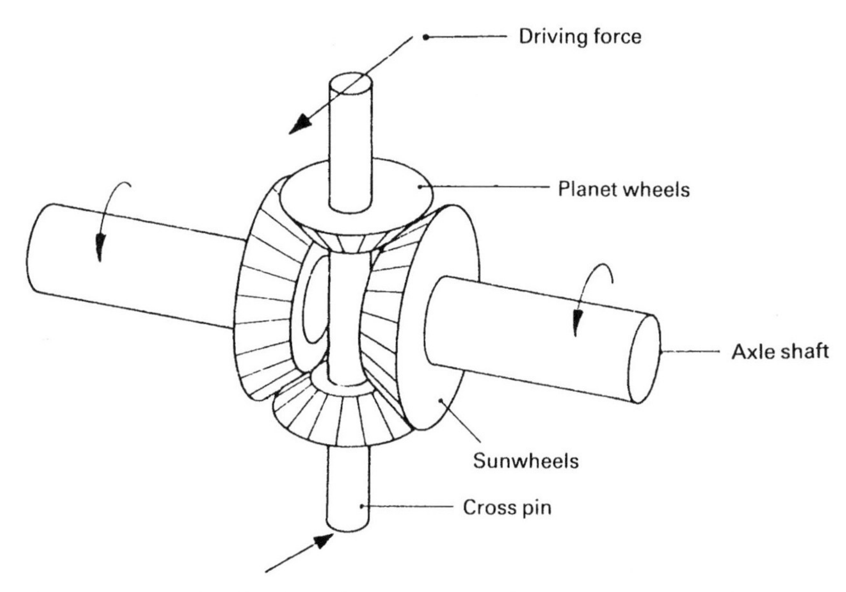

It can be seen from the drawing above that the planet wheels are pulled round by the cross pin. If the resistance to rotation at each axle shaft is equal, the sunwheels will rotate at the same speed as the cross pin, that is there is no relative rotation between planet wheels and sunwheels. Under what conditions would this occur?

....

INVESTIGATION

Observe the action of a differential by rotating the sunwheels at different speeds to each other and describe the action of the planet wheels:

....

....

....

By the action of the planet wheels rotating on the cross pin the sunwheels must rotate at different speeds (cornering), while still receiving equal driving torque from the planet wheels.

LOCATING THE FINAL-DRIVE GEARS

Precise location (this means keeping the gears in their correct position during operation) of the meshing final-drive gears is vital if they are to maintain correct contact during drive and overrun. This will ensure that, given correct lubrication, the pair of gears will operate quietly and efficiently and have maximum service life.

Describe briefly how the bearing arrangement shown below ensures accurate location of the gears, and label the drawing:

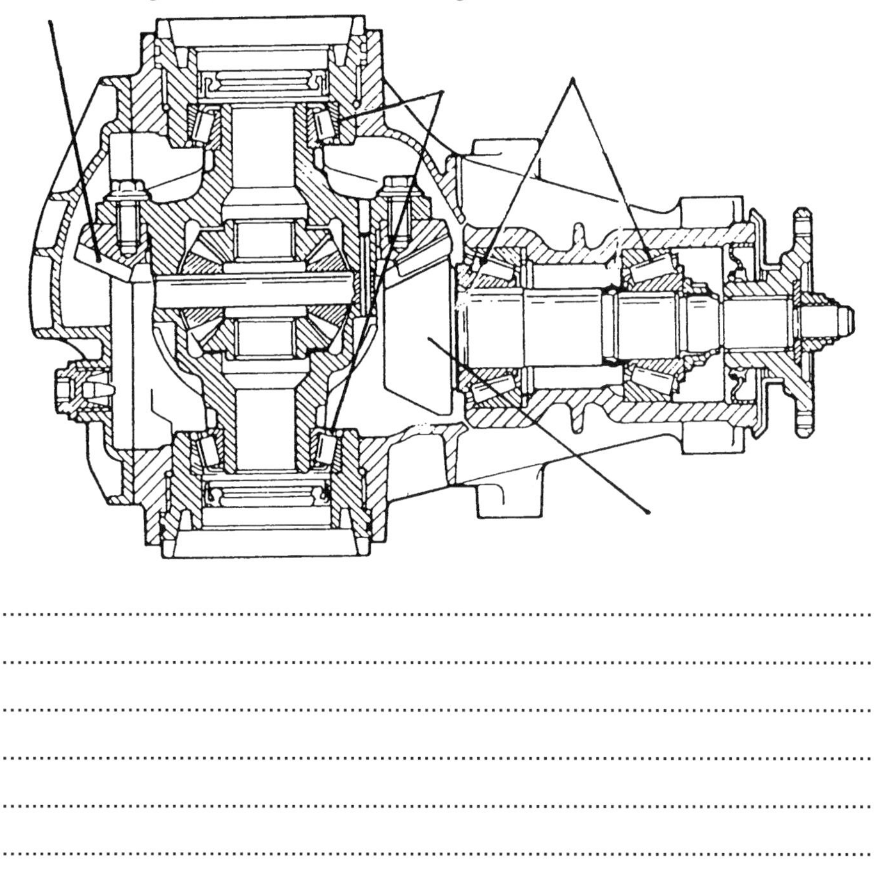

....

....

....

....

....

....

....

....

Straddle mounting of the pinion is common practice in HGV applications; make a simple sketch to illustrate this arrangement and state briefly the reason for its use.

....

....

....

When a large-diameter crown wheel is employed (HGV) a thrust block is often used to prevent any sideways deflection. Add a crown wheel thrust block to the drawing below; include the method of adjustment.

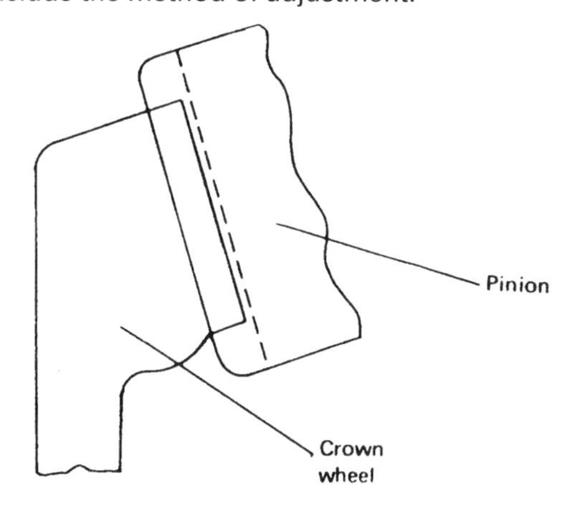

SETTING UP AND ADJUSTMENT

If the complete final-drive assembly is to operate satisfactorily, without undue noise or rapid wear, the clearances and adjustments must be as specified by the manufacturer. To meet these specifications a definite procedure must be adopted when:

(a) adjusting the bearings

(b) positioning the final-drive gears to obtain correct meshing.

Briefly outline such a procedure:

....

....

....

....

....

....

Bearing adjustment

Pinion bearings of the taper-roller type are normally moved towards each other by the adjusting arrangement until they are placed under an initial load which makes them slightly stiff to rotate.

This is known as 'PRE-LOAD'.

State the reason why such bearings are pre-loaded:

....

....

....

Name the types of bearings that should not be pre-loaded:

....

....

....

A popular method of obtaining pre-load is to use a collapsible spacer.

Show on the drawing below, where it would be used and describe how it would be used:

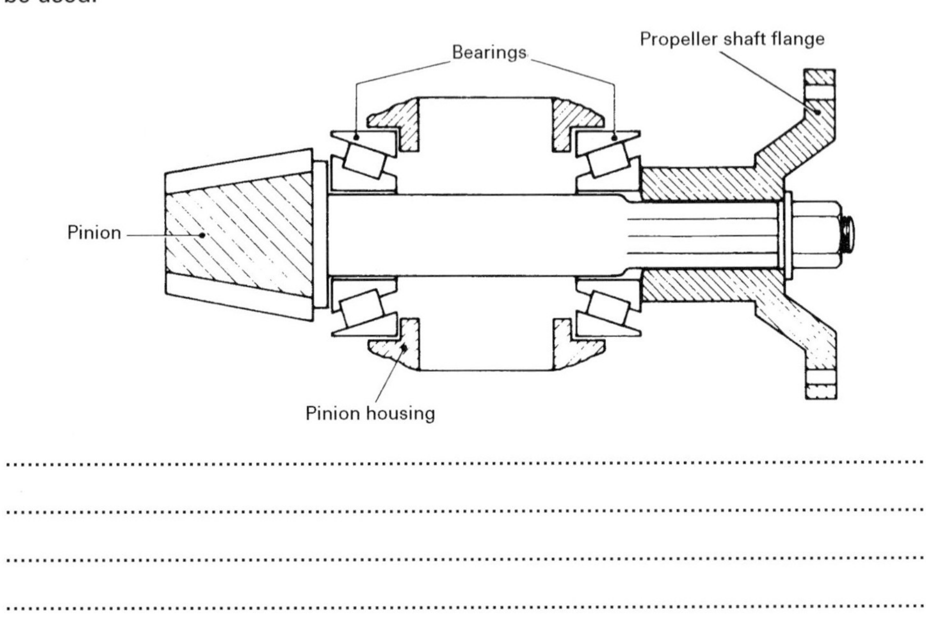

....

....

....

....

The sketch below shows a popular type of pre-loading gauge. Describe how it would be used:

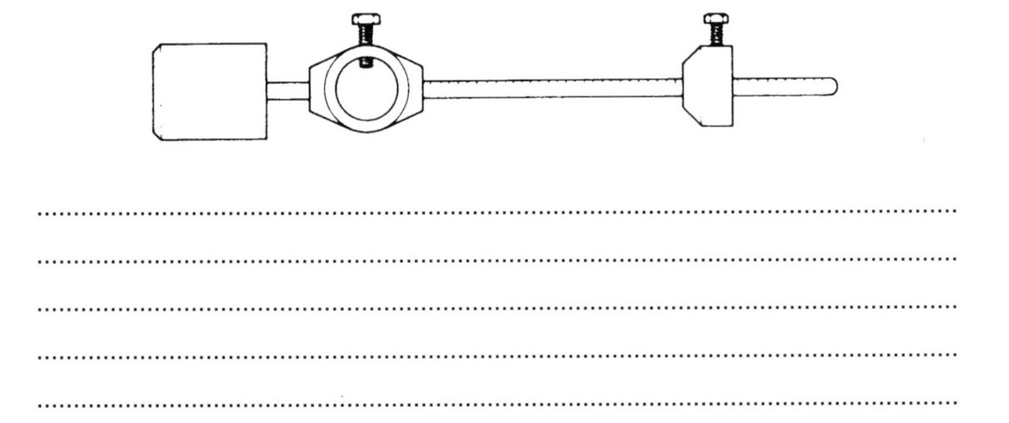

....

....

....

....

....

Tooth meshing

The correct meshing of the final-drive gears is obtained by moving the crown wheel or pinion in or out of mesh until the correct position is found.

Complete the sketch to show how the correct backlash for the final-drive gears can be checked by using a clock gauge (or dti) assuming that the pinion is held steady.

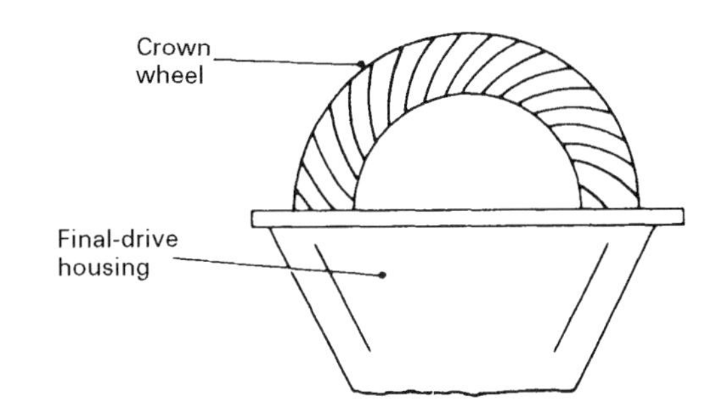

Final tooth meshing can be checked by smearing the gear teeth with marking compound. Sketch the tooth marking, for the meshings indicated, on the teeth below.

CORRECT MESHING TOO FAR IN MESH TOO FAR OUT OF MESH

State the reason why final-drive gears are supplied in mated pairs:

....

....

....

....

....

....

INVESTIGATION

1. Remove the final-drive assembly (either bevel gear or worm and wheel) from an axle casing.

   Describe the method of pre-loading the bearings.

2. Pinion bearings:

....

   Side bearings:

....

3. State the backlash setting and the pre-load setting for the assembly.

   Backlash .... Pre-load (pinion ....)

   (side bearings ....)

4. Assemble the unit and describe briefly the procedure for adjusting the final-drive gears to the manufacturer's specifications.

A. ....

....

....

B. ....

C. ....

....

D. ....

....

....

E. ....

....

LIMITED-SLIP DIFFERENTIAL

Sports and racing cars are vehicles with high power-to-weight ratios and as such can, even on good surfaces, cause a driving wheel to spin, say, during rapid acceleration or during hard cornering. Most of the higher quality or top-of the-range models incorporate an LSD.

The limited-slip device is incorporated in the differential, which automatically applies a brake to the spinning half shaft, thereby maintaining a torque in the other half shaft.

Little torque is required to drive a spinning or slipping road wheel and a normal type differential would transmit the same torque to the non-spinning wheel; thus traction is lost. Limited slip traction control and torque sensing devices all produce 'frictional torque' in the drive to a spinning wheel.

In the FRICTION CLUTCH TYPE shown at (a), when the sun gear rotates relative to the differential cage (during wheelspin), the spring loaded clutch slips generating frictional torque. Owing to the 'torque balancing' action of a differential the same torque is transmitted to the non spinning wheel.

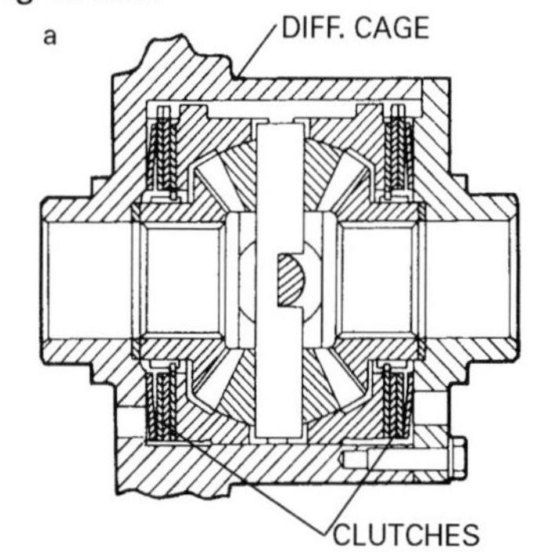

Describe briefly how to check a limited-slip differential for operation and state any safety precautions to be observed when working on a vehicle fitted with a limited-slip differential:

....

....

....

....

....

....

Frictional torque can be produced in other ways as shown in the two currently popular LSDs below.

Name the types and label the drawings.

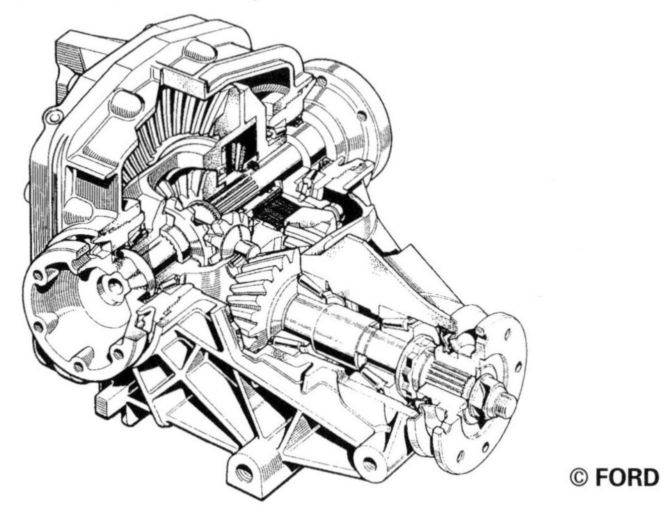

© FORD

....

© HONDA

....

TANDEM (TWIN) DRIVE AXLES

Many HGVs have two 'closely spaced' 'live' axles providing the drive, each axle having a differential to allow for wheel speed variations on that axle.

The drawing at (a) below shows a worm and wheel arrangement. Complete the drawing at (b) to show a tandem drive which uses crown wheels and pinions.

(a) Worm and Wheel Tandem Drive

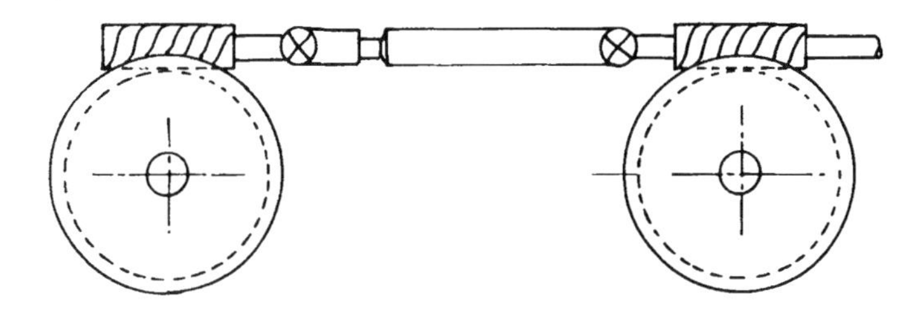

(b) Bevel Gear Tandem Drive

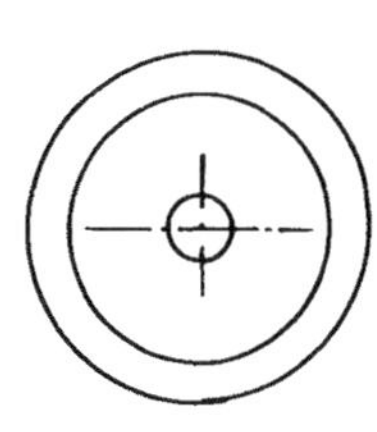

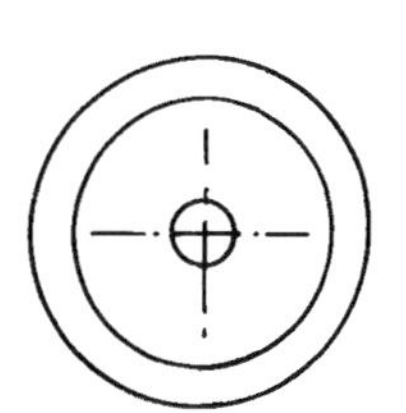

When a tandem drive is employed the forward axle differential will allow for speed variations of its inner and outer wheels, and the rear-axle differential will do likewise. It is, however, desirable to allow for a speed variation between the two final-drive crown wheels or worm wheels. Why is this?

....

....

INTER AXLE (THIRD) DIFFERENTIAL

If the drive to the two final-drive pinions or worm gears is transmitted through a third differential, speed variation between the two sets of final-drive gears can occur and the driving torque is shared equally between the two axles.

The simplified drawing below shows a worm and wheel tandem drive incorporating a THIRD differential. Describe the operation of this system.

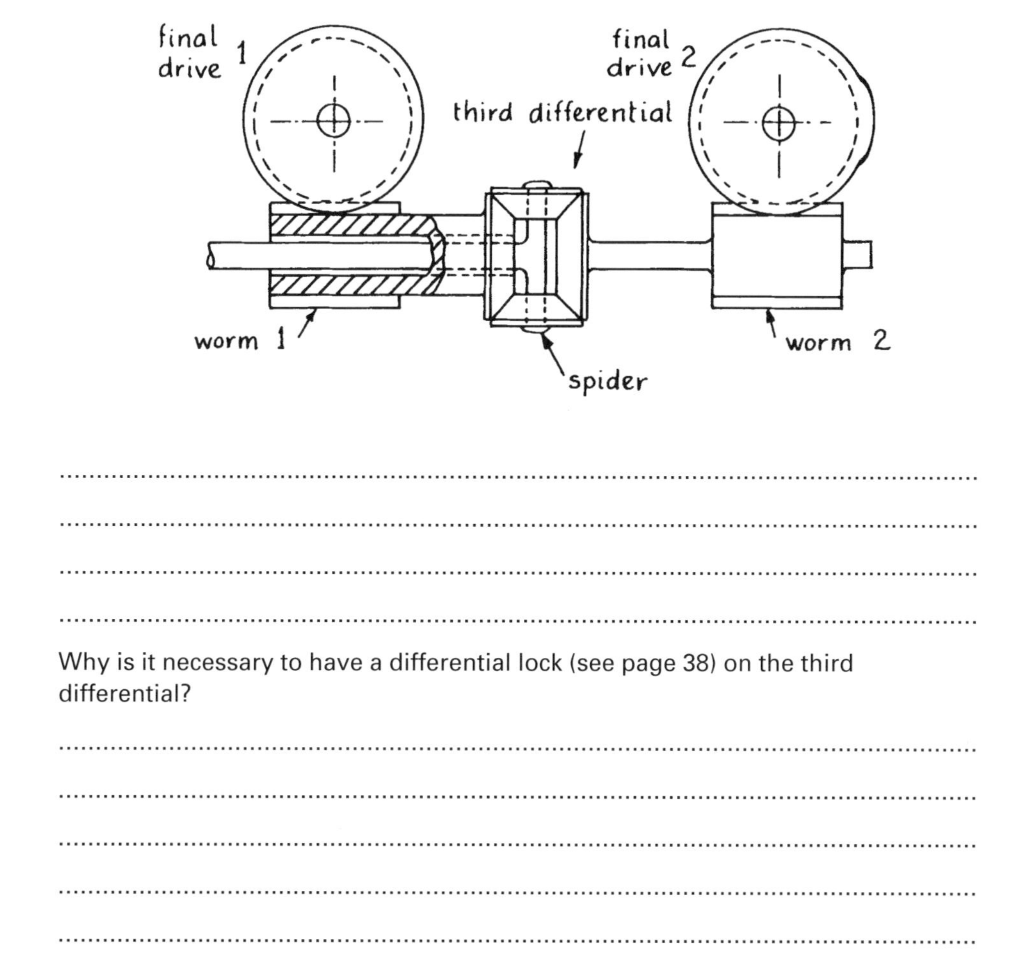

....

....

....

....

Why is it necessary to have a differential lock (see page 38) on the third differential?

....

....

....

....

....

DOUBLE-REDUCTION GEARS

A double-reduction gear used in conjunction with the final-drive gears provides, in two 'stages or steps', a large gear reduction at the driving axle.

State why double-reduction gears are used?

....

....

....

....

....

....

....

....

Complete the drawing below to show a spur-gear double-reduction arrangement:

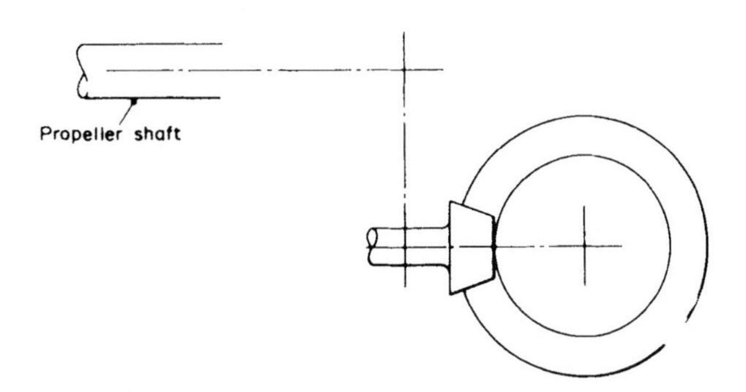

Shown opposite are two alternative double reduction gears. Label the drawings and briefly describe each arrangement giving reasons for its use.

Three alternative positions for spur-type double-reduction gears are:

1. ....

2. ....

3. ....

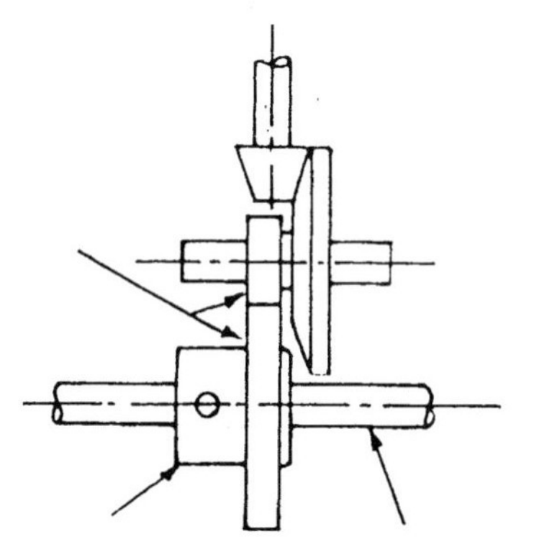

Type

....

....

....

....

....

....

....

....

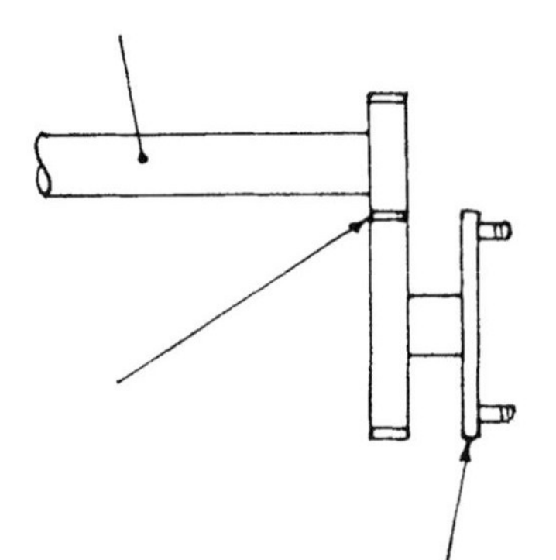

Type

....

....

....

....

....

....

....

....

Epicyclic Double-reduction Gears

The drawing below illustrates an epicyclic reduction gear in the final-drive assembly; describe its operation.

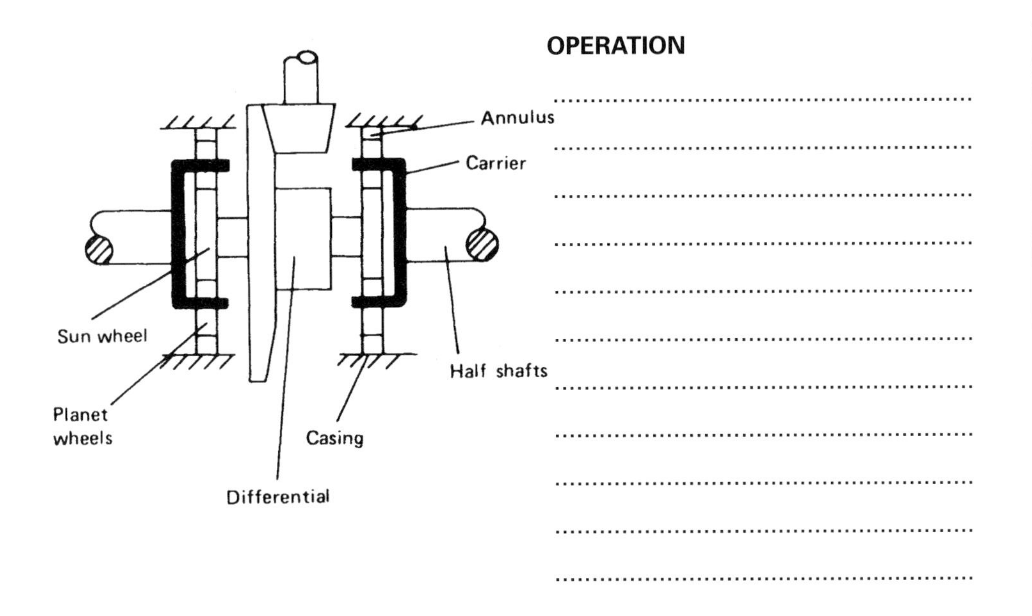

OPERATION

....

....

....

....

....

....

....

....

....

....

....

Examine an epicyclic hub-reduction gear and label the drawing.

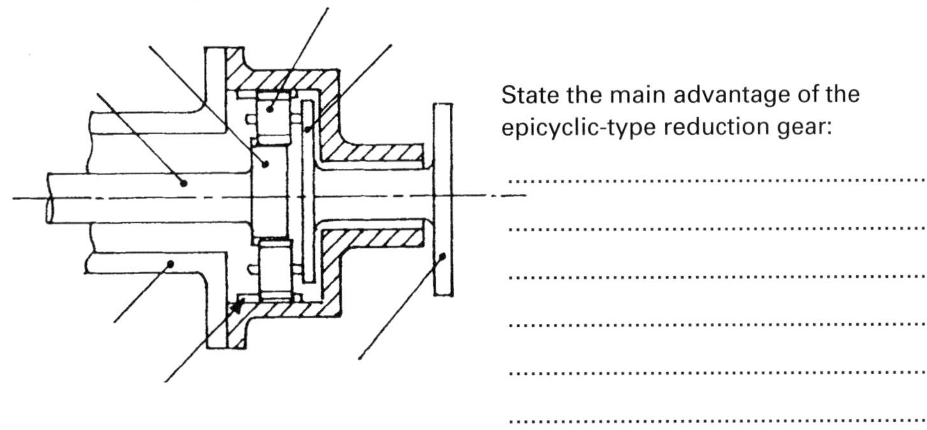

State the main advantage of the epicyclic-type reduction gear:

....

....

....

....

....

....

MAINTENANCE AND REPAIR

State general rules for efficiency and special precautions needed during maintenance and adjustments in respect of the items listed below:

ITEM	RULES/PRECAUTIONS
Vehicle lifting, supporting and checking	
Support of engine gearbox and suspension	
Cleanliness: (a) general (b) lubricant spillage (c) friction faces	

The purpose of routine maintenance is to prolong service life and ensure reliability, safety and efficiency. List below FIVE maintenance items:

....

....

....

....

....

DIAGNOSTICS: FINAL-DRIVE & DIFFERENTIAL – SYMPTOMS, FAULTS AND CAUSES

State a likely cause for each symptom/system fault listed below. Each cause will suggest any corrective action required.

SYMPTOMS	FAULTS	PROBABLE CAUSES
Oil spray on underside of floor and chassis; evidence on ground	Oil leakage	....
Transmission noise	Bearing failure	....
Noise on drive or over run; backlash on take up of drive	Excessive end float in pinion bearings	....
Transmission knock or tick; vibration	Damaged or broken final-drive gears	....
Transmission noise	Worn differential gears	....
Excessive slip; loss of traction under arduous conditions	Faulty VC or worn clutch pack	....
Knocking on take up of drive	Spline damage	....
Transmission noise; backlash	Incorrect meshing of final-drive gears	....

Chapter 6

Suspension

Suspension	94	Axle location	106
Suspension dampers	95	De-dion axle	109
Coil spring suspension	96	Rear wheel 'suspension steer'	110
Torsion bar suspension	96	Energy conversion	110
Rubber springs	97	Taper single-leaf spring suspension	111
Air suspension	98	Leaf spring HGV	111
Trailer tandem air suspension	98	Torque reaction	112
Hydro-pneumatic suspension	100	Multi-axle suspension	112
Adaptive suspension	101	Effects of torque reaction	113
Uprated suspension	102	Axle lifts	114
Uprated dampers	104	Checking suspension alignment and geometry	116
Variable rate dampers	104	Diagnostics	117
Active suspension control	105	Forces due to torque reaction: calculations	118
Checking damper operation	105		

SUSPENSION

The purpose of the suspension system on a vehicle is to:

1. *Minimise the effect of road surface irregularities on passengers, vehicle and load*
2. ....
3. ....
4. ....
5. ....
6. ....

Many different forms of suspension are employed on road vehicles – some very simple and relatively inexpensive, others highly sophisticated and expensive. Suspension systems fall into one of two main categories: Independent and Non-independent systems.

Describe with the aid of simple diagrams the essential difference between the two categories:

....

....

....

....

....

....

....

Leaf Spring Suspension

The leaf spring is widely used on light and heavy goods vehicles and is still in limited use on car rear suspension. Used in conjunction with a beam axle the leaf spring is a simple NON-INDEPENDENT suspension system. The length, width, thickness and number of leaves vary according to the load requirement. An advantage of the leaf spring is that it can provide total axle location in addition to its springing properties.

Label the multi-leaf spring suspension below:

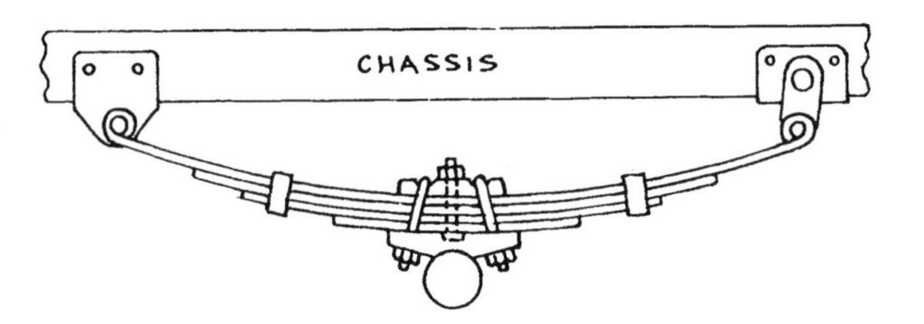

The leaf spring is positively located on the axle by a ....

and secured to the axle by .... As the spring flexes the distance between the spring eyes, that is, the length of the spring, varies.

This is accommodated by the .... or, in some cases, an open-ended spring operates in a ....

Overloading of the main leaf on rebound is prevented by ....

In relation to suspension, state the meaning of:

SPRUNG WEIGHT ....

....

....

UNSPRUNG WEIGHT ....

....

....

SUSPENSION DAMPERS

Contrary to popular belief, the function of the suspension damper is not to increase the resistance to spring deflection but to control the oscillation of the spring. The basic principle of hydraulic dampers is that of converting the energy of the deflecting spring into heat. This is achieved quite simply by pumping oil through a small hole.

The graph below shows the oscillations of an 'undamped' spring as it dissipates energy following an initial deflection.

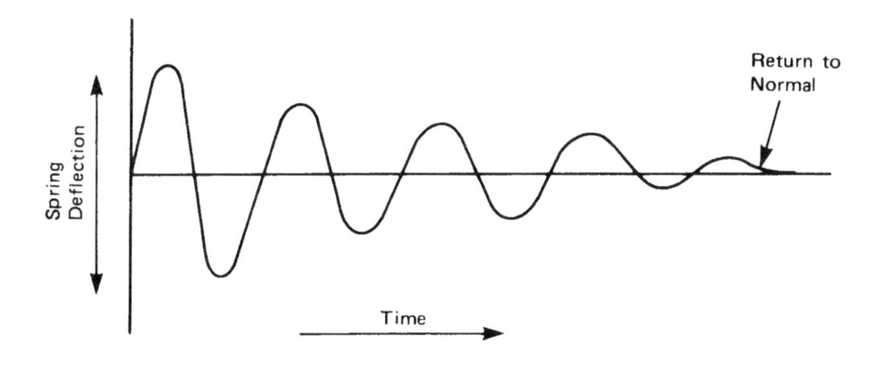

Draw a graph below to illustrate the effect of a damper on the oscillation of a spring:

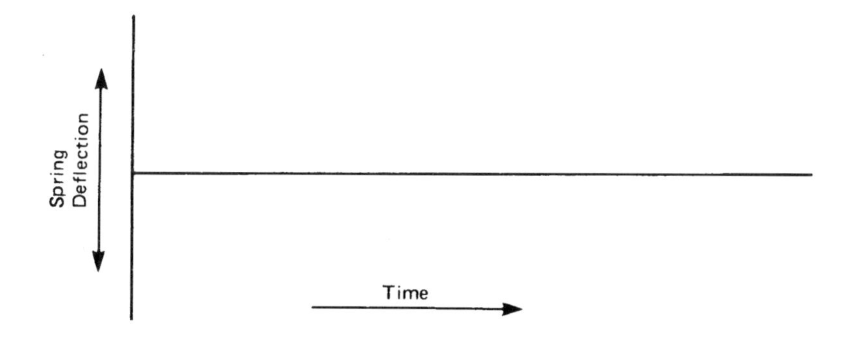

Direct-acting Hydraulic Damper

The telescopic type, as this type is known, is located directly between the frame and suspension unit or axle, thus eliminating the need for levers or links. The principle of operation is the transfer of fluid through small holes from one side of the piston to the other. Complete the sketch below by adding a sectioned view of the interior of the damper. Label the damper shown and add the bump and rebound valves to the enlarged valve assembly.

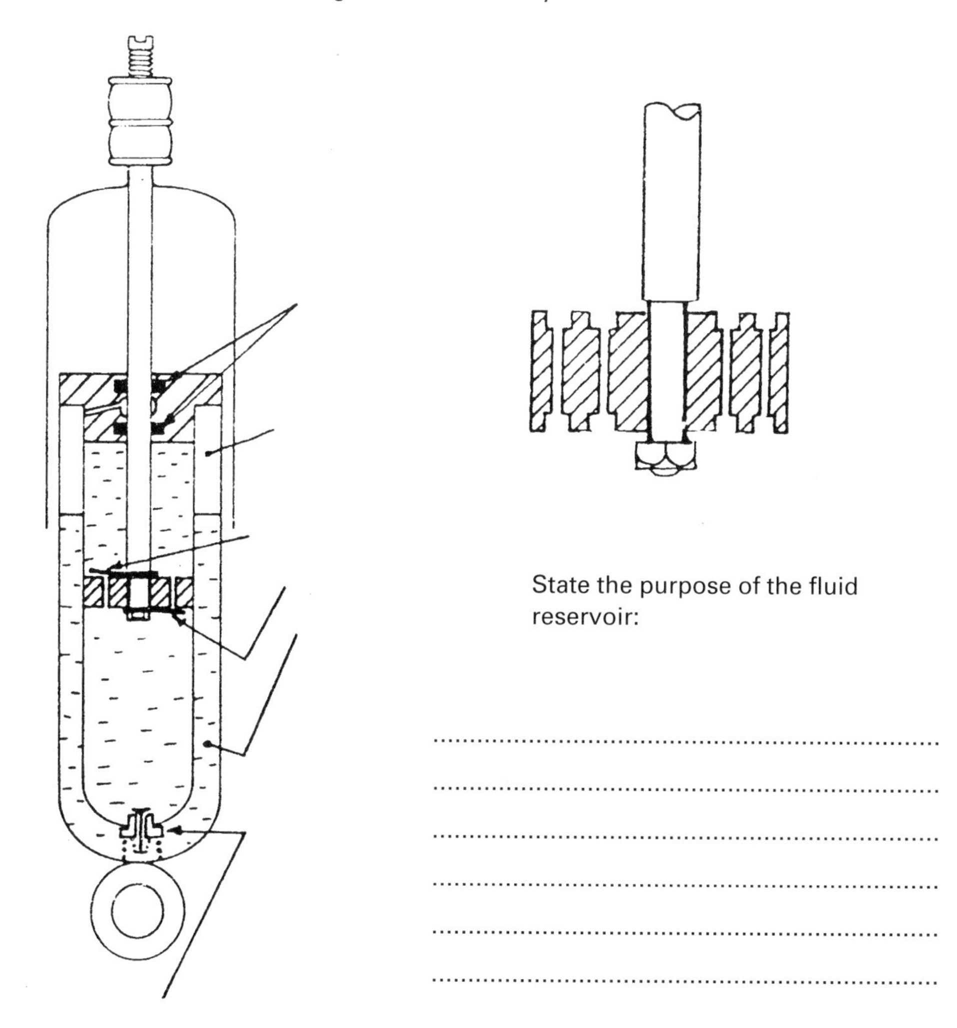

State the purpose of the fluid reservoir:

....

....

....

....

....

....

COIL SPRING SUSPENSION

The coil spring is used on many car front and rear suspension systems. When compared with the leaf spring the coil spring is:

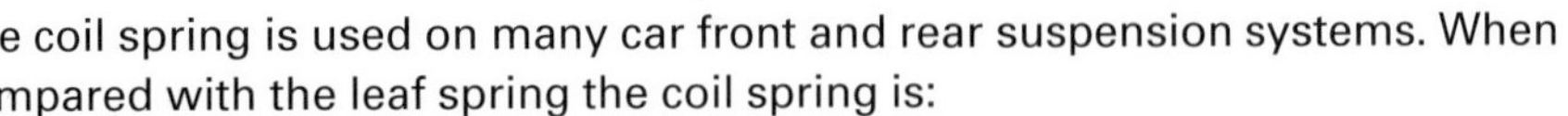

.... in weight, provides a ride which is ....

and wheel deflection can be ....

Show, on the three independent front suspension arrangements below, the position of the coil spring:

A

B

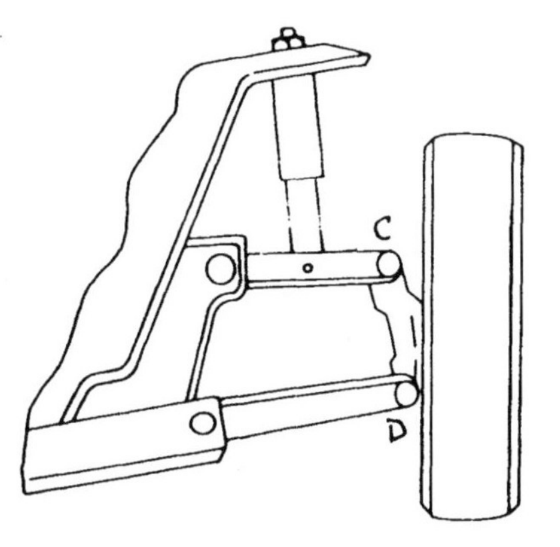

E

Some suspension ball joints are subjected to direct loading as a result of vehicle weight. It is important to appreciate this when checking a suspension system. Add arrows to the drawings to show the vehicle weight load path through the suspension and identify the loaded and unloaded ball joints.

Loaded .... Unloaded ....

TORSION BAR SUSPENSION

The torsion bar is, in effect, a coil spring which has been straightened out.

Label the drawing and describe the operation of the torsion bar suspension shown:

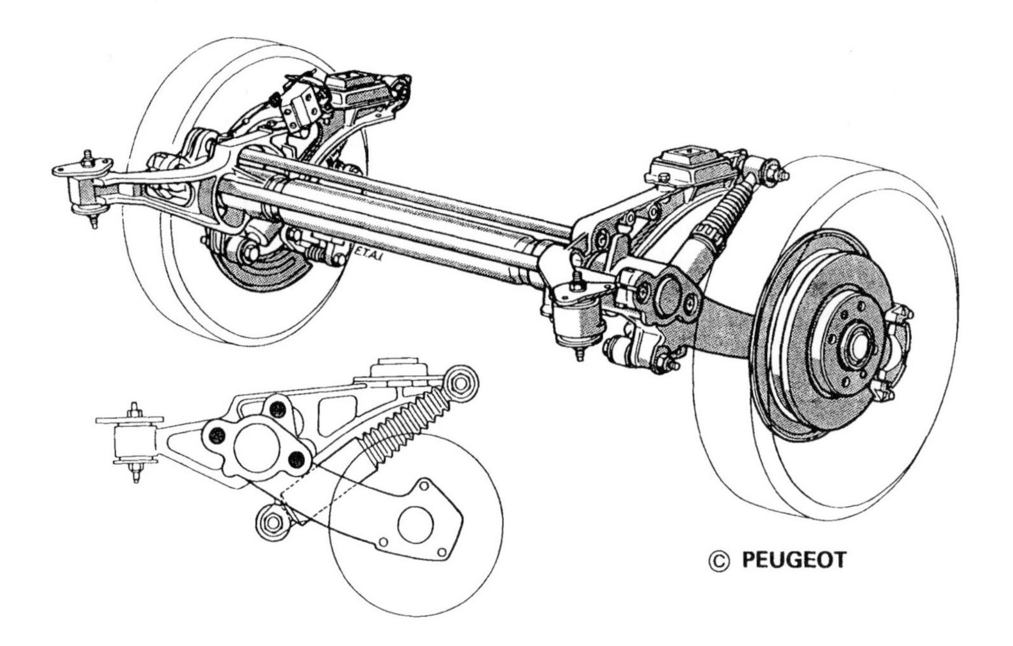

© PEUGEOT

OPERATION

....

....

....

....

....

RUBBER SPRINGS

One of the advantages of rubber springs is that for small wheel movements the ride is soft, becoming harder as wheel deflection increases. They are also small and light, and tend to give out less energy in rebound than they receive in deflection.

The deflection of the rubber spring in the suspension shown below is small when compared with the actual rise and fall of the wheel. State how this is achieved:

© ROVER

Describe the procedure for compressing the spring during dismantling of the suspension:

Heavy Goods Vehicle Application

The trailer suspension shown below employs rubber springs. The particular benefits offered by rubber springs in this application are: weight-saving and ride improvement.

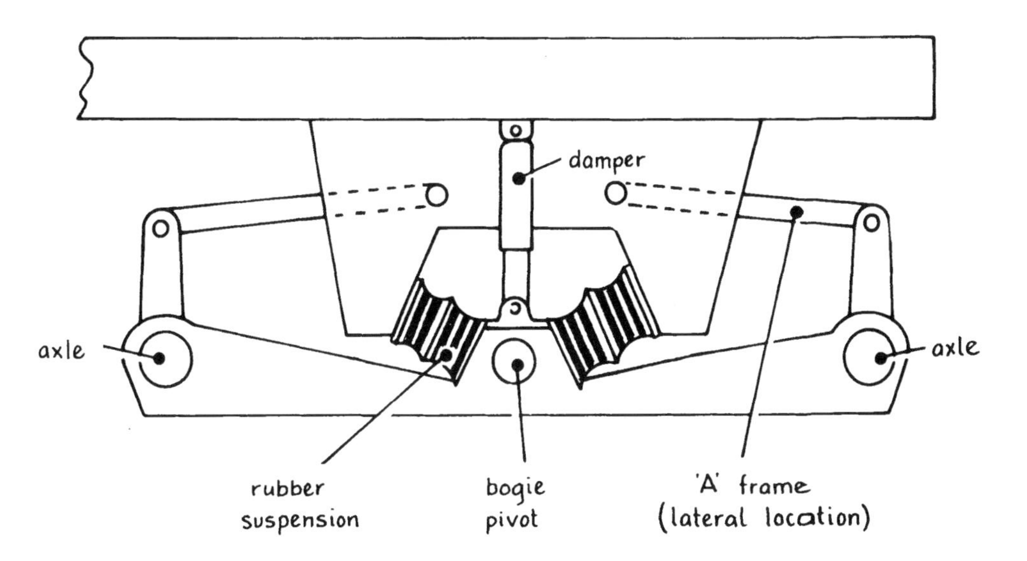

The rubber spring makes the suspension PROGRESSIVE. Explain this and describe how it is achieved:

AIR SUSPENSION

Air springs usually consist of reinforced rubber bags situated between the axle and chassis. Air suspension is a truly progressive suspension providing a soft, cushioned ride for an empty trailer, bulk liquid tanker or luxury touring coach. An important feature is the use of levelling valves in the system, which maintains a constant vehicle ride height irrespective of vehicle loading. A rolling diaphragm air spring is shown below; name the numbered parts and state the purpose of item number 4.

1. .... 2. ....

3. .... 4. .... 5. ....

6. ....

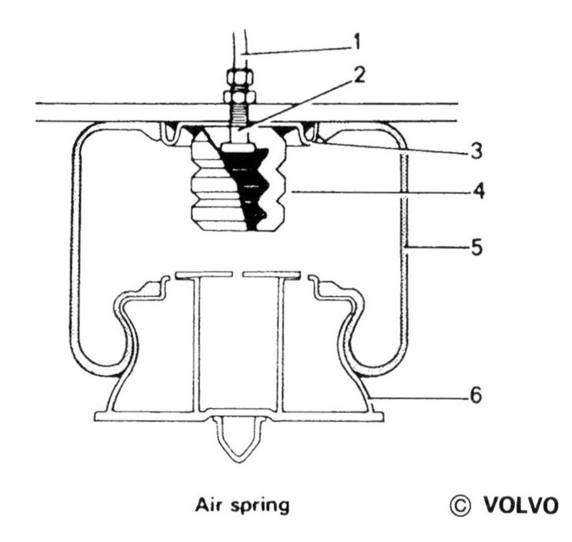

Air spring © VOLVO

It can be seen from the drawing that the air enters the piston as well as the air spring. What benefit is this to the suspension?

....

....

....

....

TRAILER TANDEM AIR SUSPENSION

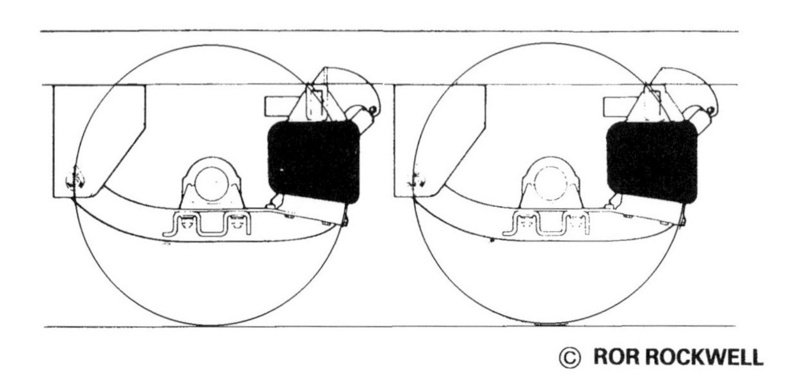

© ROR ROCKWELL

Name the arrowed components:

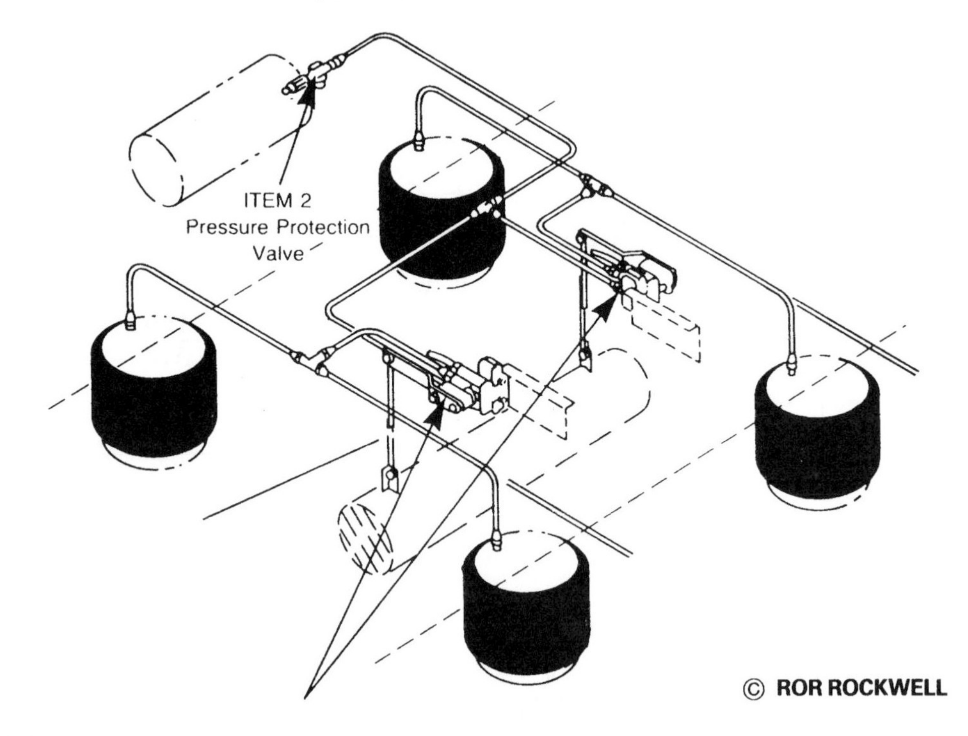

© ROR ROCKWELL

An IFS arrangement for a PSV is shown at the top opposite, and a rear suspension employing four air springs which is used on a luxury touring coach is shown below it. Describe how progressiveness and automatic levelling are achieved in an air suspension system:

Advantages of air suspension are:

(a) less trailer hop when empty;

(b) less body/chassis maintenance required;

(c) constant platform height (advantageous when using stacker trucks from a loading bay);

(d) less weight.

One drawback is the increased body sway on high, loaded vehicles.

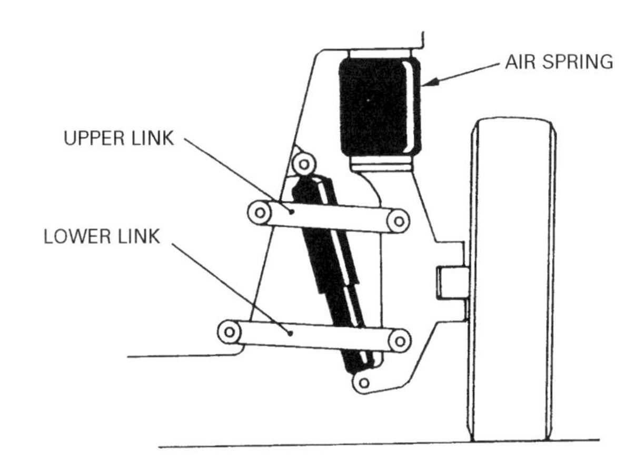

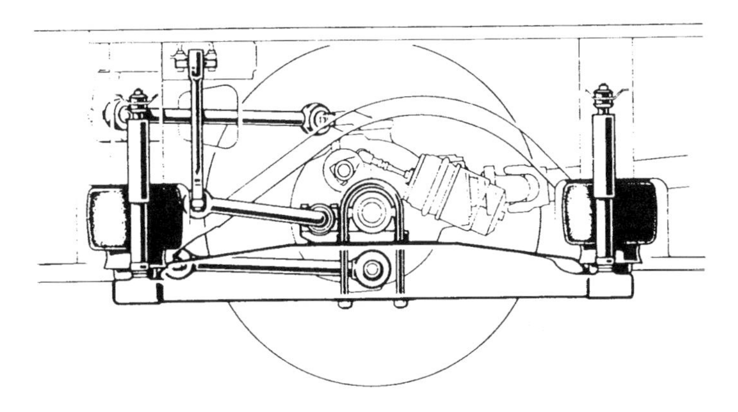

© MERCEDES-BENZ

Another feature of air suspension is manual ride height control; this allows the ride height to be increased when operating on different ground levels, for example, coaches driving on and off ferries.

HYDRO-PNEUMATIC SUSPENSION

As with air suspension, the spring is pneumatic. The main difference is that a fixed quantity of gas (nitrogen) is contained in a variable-sized chamber; liquid is used to transmit the force from the suspension link to the gas (via an intervening diaphragm).

Complete the drawing below to illustrate the principle of operation of a hydro-pneumatic suspension system:

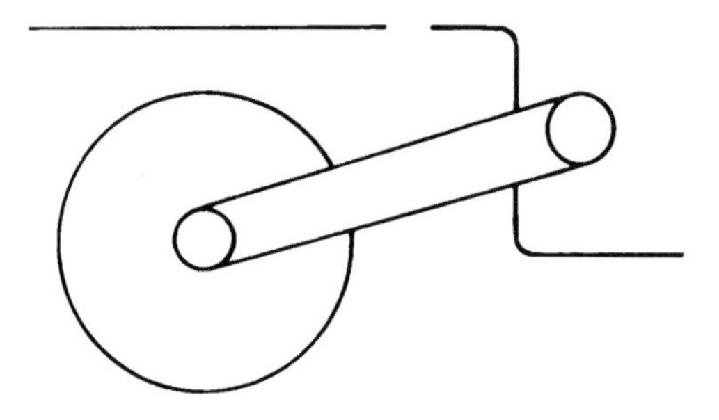

One system of hydro-pneumatic suspension incorporates a levelling valve and a hydraulic accumulator. In this system the vehicle maintains a constant ground clearance irrespective of load.

The driver can also adjust the vehicle height or ground clearance to any one of three positions.

Give examples of vehicles that use this form of suspension:

....

....

The simple layout on the right shows a suspension system with a levelling valve and an accumulator.

Add arrows to the hydraulic circuit to indicate the direction of fluid flow, label the drawing and describe how the system operates:

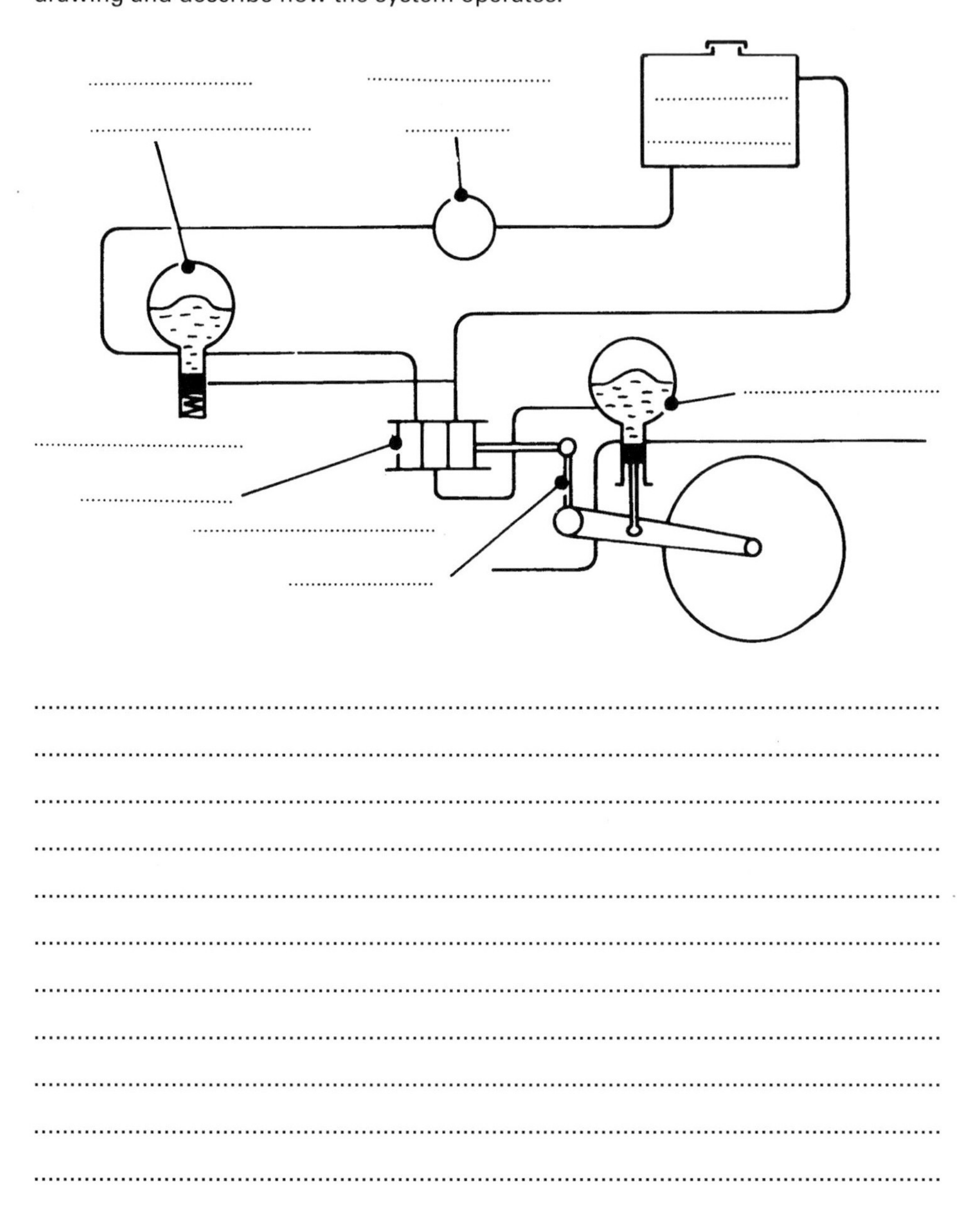

....

....

....

....

....

....

....

....

....

....

....

ADAPTIVE SUSPENSION

A development of the hydro-pneumatic suspension is shown in the diagram opposite.

Additional 'THIRD' spring units are added between the front and rear pairs. This system provides a variable spring rate and roll stiffness, that is, the suspension is ACTIVE.

What are the benefits of ACTIVE suspension?

....

....

....

....

....

....

Study the diagram, complete the key opposite and briefly describe the action of this system.

....

....

....

....

....

....

....

....

....

....

....

....

....

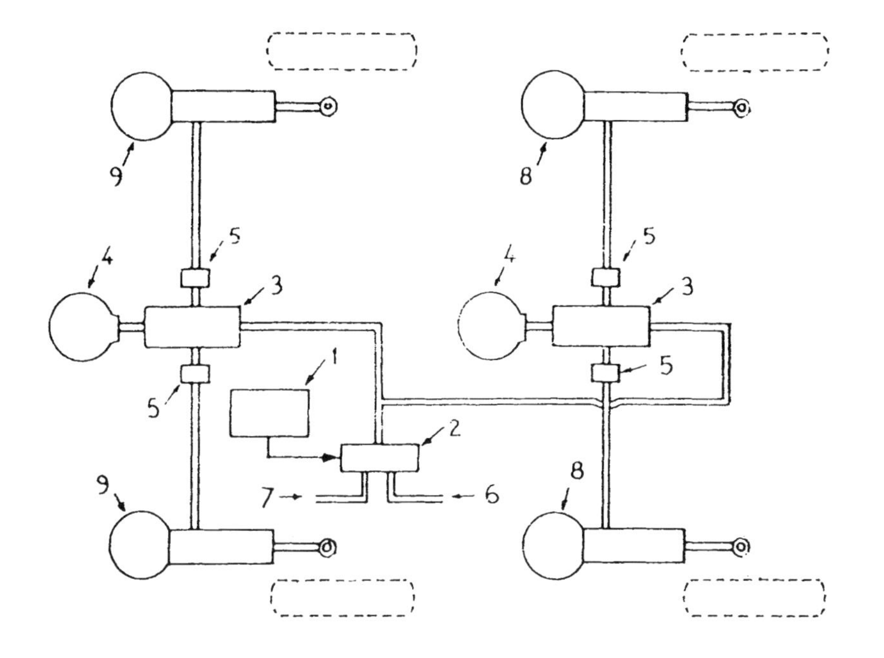

1. *ECU: senses steering wheel movement, acceleration, speed and body movement.*
2. *SOLENOID VALVE: actuates regulator valves.*
3. *REGULATOR VALVES: allow fluid to third spring units and side-to-side communication*
4. ....
5. ....
6. ....
7. ....
8. ....
9. ....

UPRATED SUSPENSION

If the suspension springing on a vehicle is uprated it offers a greater resistance to deflection, it is stiffer and will more easily cope with heavier loads.

Give reasons for using uprated springs:

1. ....

2. ....

One method of uprating or reinforcing the suspension springs is to replace the dampers with spring damper units (a) opposite, in which a variable rate coil spring increases its resistance according to the additional weight carried. The damper itself would be stiffer than the standard unit. Make a sketch at (b) to show an alternative method of uprating the springs.

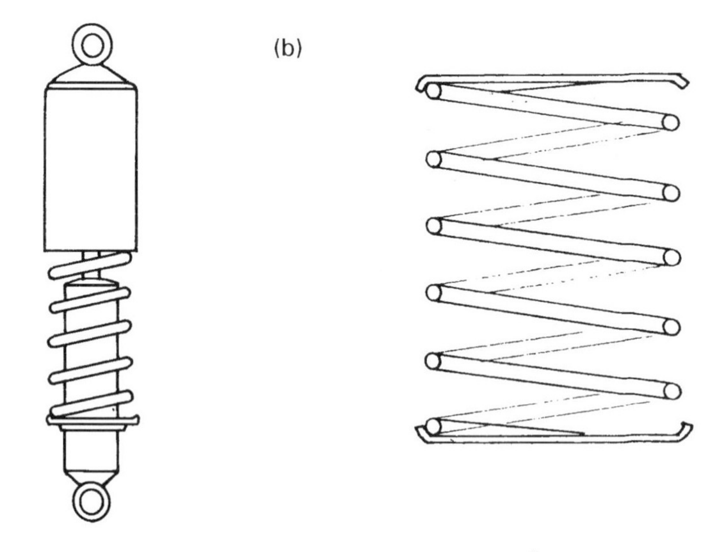

Gas pressurised dampers
The simplified drawing opposite shows a gas damper in which a floating piston separates the gas from the fluid. This is a single or mono-tube design (no fluid reservoir).

State two functions of the gas:

1. ....

2. ....

....

....

....

Note: In some gas dampers the gas is not separated from the fluid and an emulsion is formed which performs in much the same way.

State the advantages of the mono-tube design:

....

....

....

....

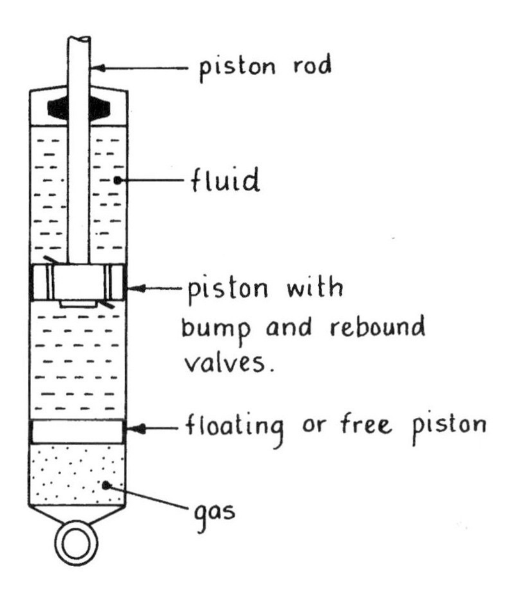

Ride leveller system

For vehicles towing trailers, caravans etc. or where a good ride is essential with a variety of loads (such as an estate car), it is desirable to uprate the rear springs.

In addition to this, certain manufacturers offer height regulating 'shock absorbers' (dampers) which will raise the loaded rear end of a vehicle back to its unloaded height.

Make a simple diagram to illustrate such a system and describe its operation:

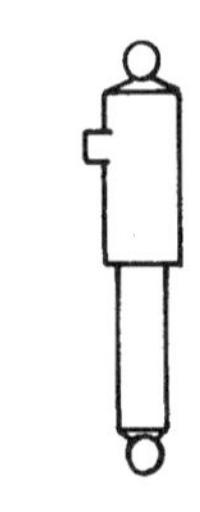

What benefits does this system offer?

Bump and rebound stops

On the drawing below show a rubber bump stop and a strap to limit rebound in a suspension system.

State the purpose of these fitments.

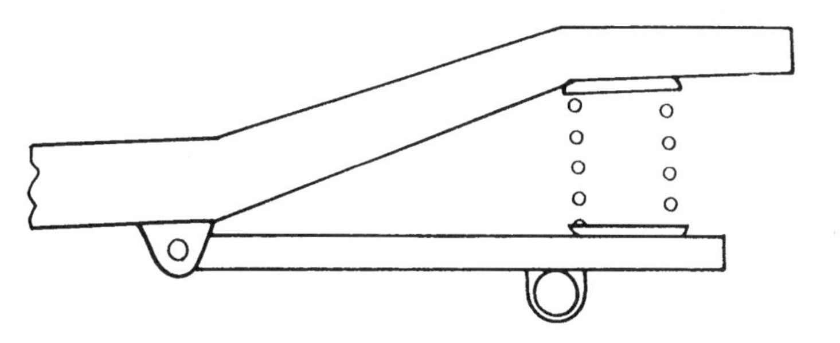

Purpose

Trim height

What is meant by 'trim height' (or standing height) and how is it measured?

UPRATED DAMPERS

How is the operation of a damper affected if it is 'UPRATED'?

....

Give TWO reasons for the use of uprated dampers:

1. ....

....

2. ....

....

On some dampers the rate can be varied by making a manual adjustment on the damper itself. With some installations a solenoid-operated adjustment can be activated by a switch on the dashboard. This type of damper is popular on vehicles which are 'driven hard' on occasions (such as rally type work). The damper rate can be adjusted to suit the usage of the vehicle or adjusted to satisfy the handling/ride characteristics desired by the driver.

Examine a damper incorporating a manual rate adjuster and, with the aid of the drawing below, describe the adjustment procedure.

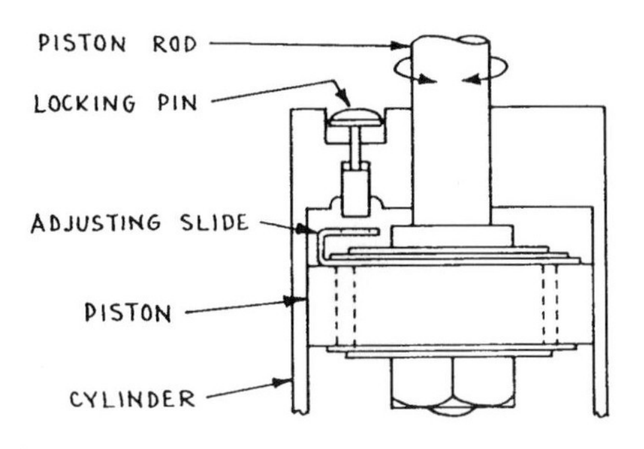

Adjustment:

....

....

....

....

VARIABLE RATE DAMPERS

The simplified damper shown is automatically uprated as the load increases. Add arrows to the drawing at (a) to indicate fluid flow during light load and complete the sketch at (b) to show high load operation.

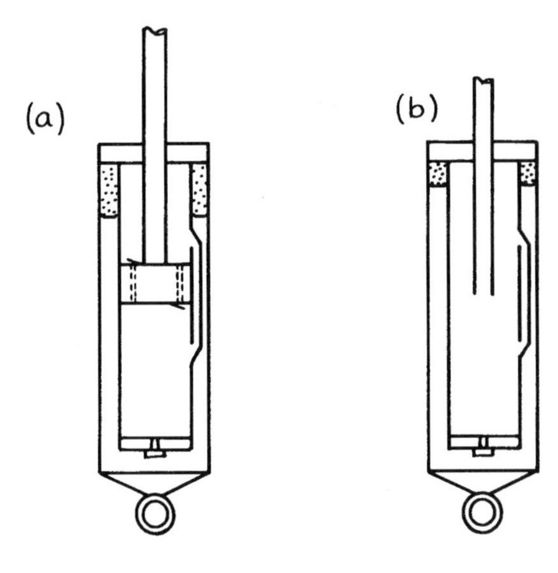

Under what other operating conditions would this type of damper offer increased damping?

....

....

....

Why does the damper rate or its resistance to movement increase under load or at extreme bump or rebound?

....

....

....

ACTIVE SUSPENSION CONTROL

In an ideal suspension system, the spring and damper rate would be automatically adjusted during vehicle operation to ensure that the ride and handling standards are optimum under all operating conditions.

In the system below, a microcomputer controls the spring rate, damper rate and ride height of the rear suspension. This system can be adapted to provide control for both front and rear suspension.

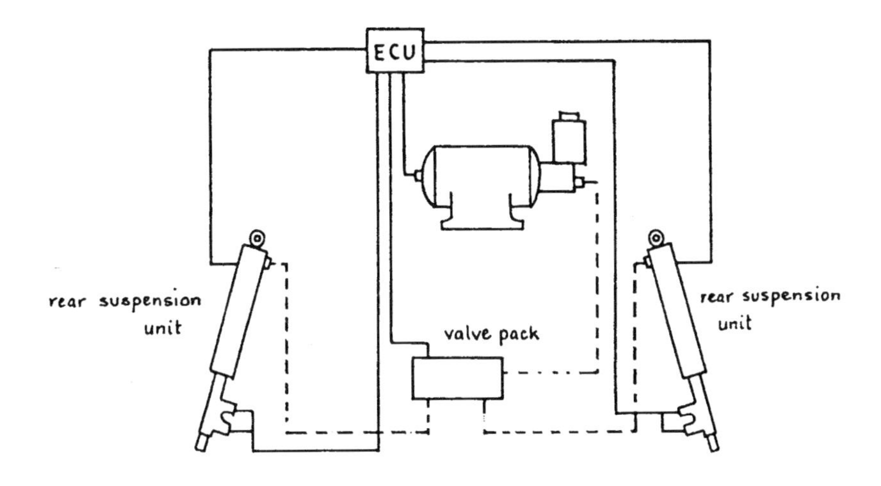

Describe the operation of the system:

....

....

....

....

....

....

....

....

....

CHECKING DAMPER OPERATION

BOUNCE TEST

This involves pushing down on the car body at each corner and noting the oscillations of the vehicle before it again becomes stationary in its normal level position. The 'rule of thumb' guidance (MUNROE) is that a vehicle taking more than one and a half oscillations has ineffective damping of the spring.

Other methods involve the use of equipment which will produce a graphical record of vehicle oscillation.

Describe the action of the damper test equipment shown below:

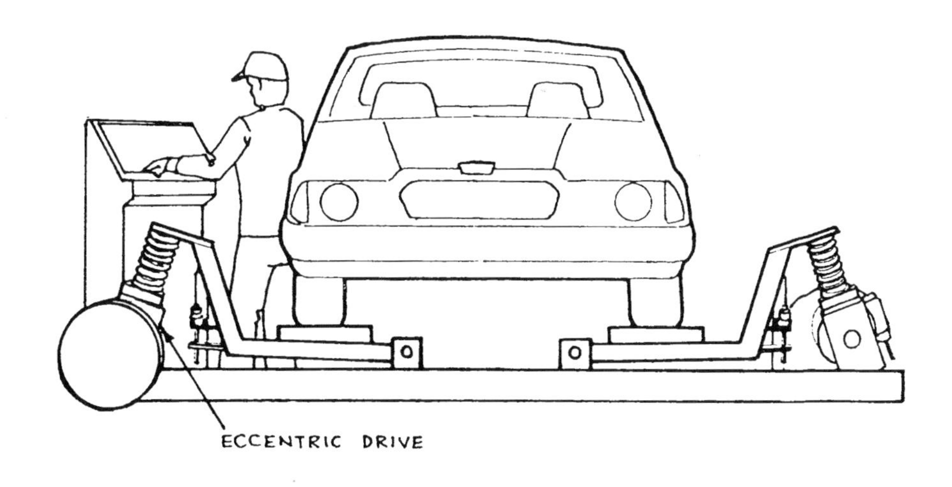

....

....

....

....

....

....

....

....

AXLE LOCATION

Coil, torsion bar, rubber and air springs serve in general merely to support and control the vertical load and do not locate the axle or wheel assembly in any way.

Additional arms or links must be provided to locate the wheels both fore and aft and laterally, and to resist the forces due to drive and brake torque reaction.

Axle location (IFS)

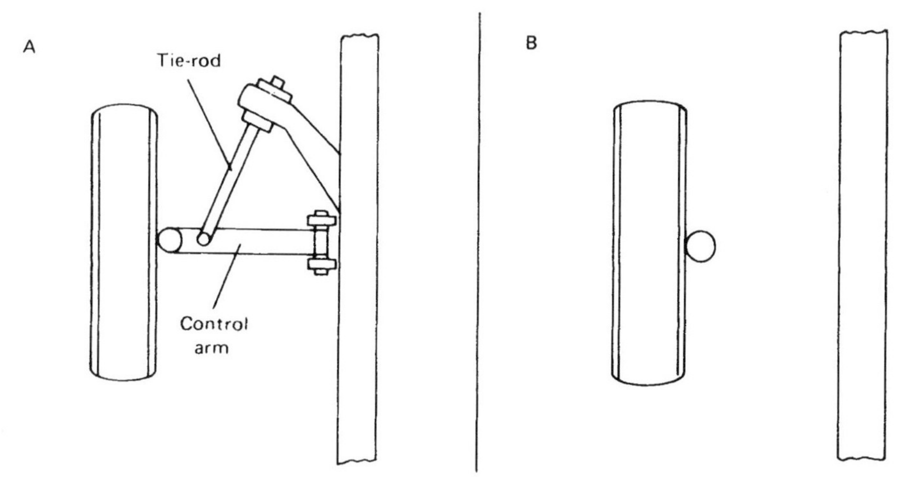

The tie-bar shown at A, above, locates the wheel assembly to prevent rearward movement of the wheel when the brakes are applied, when the vehicle accelerates or when the wheel strikes a bump.

Sketch, at B, an alternative method of providing this location.
On some suspension systems the anti-roll bar serves also as a tie-bar.

State the meaning of the term 'compliance' in relation to suspension.

....

....

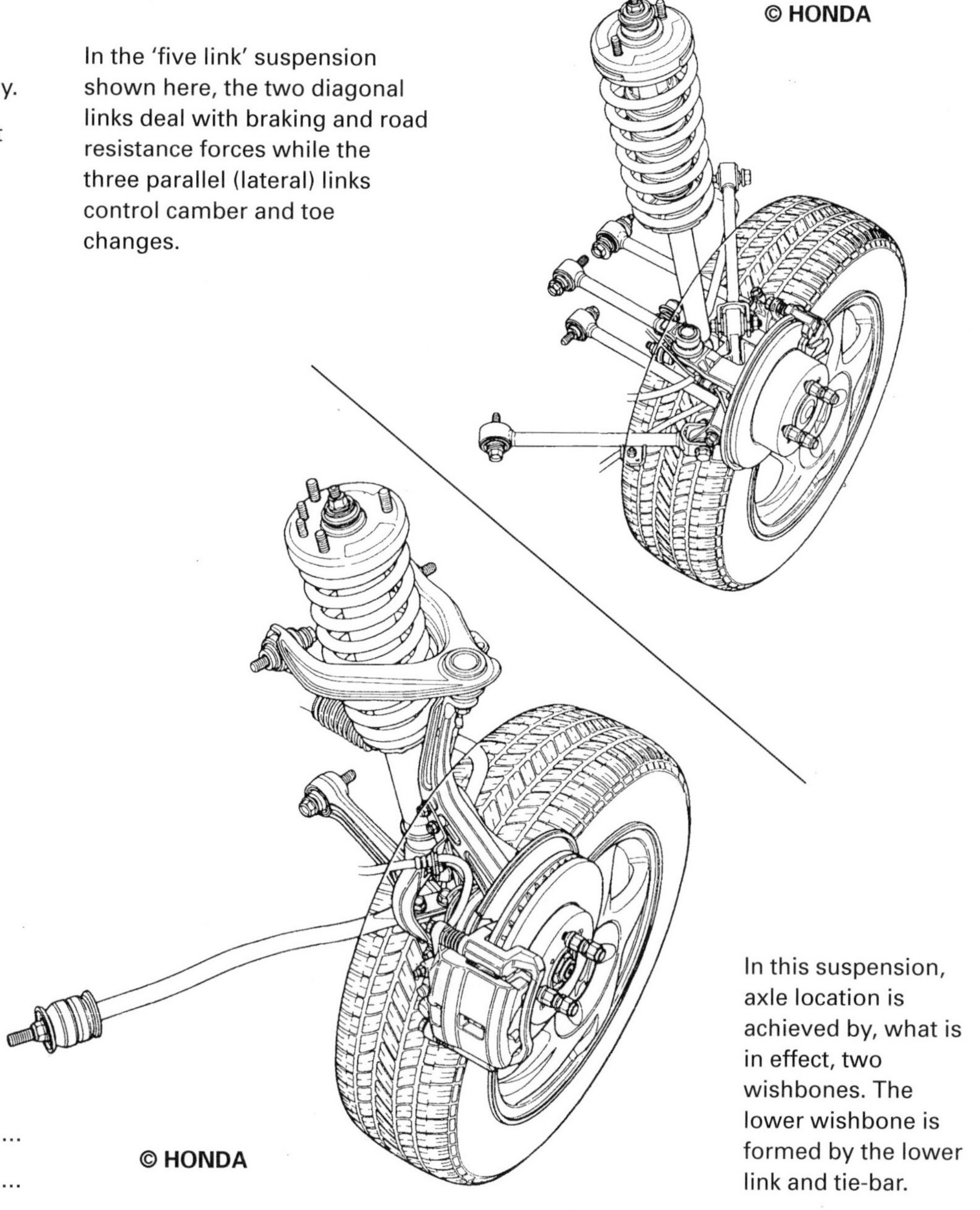

In the 'five link' suspension shown here, the two diagonal links deal with braking and road resistance forces while the three parallel (lateral) links control camber and toe changes.

In this suspension, axle location is achieved by, what is in effect, two wishbones. The lower wishbone is formed by the lower link and tie-bar.

Put appropriate descriptive headings above the suspension illustrations at (a) and (b) opposite and label the drawings.

The TORSION BEAM, as shown at (a), is now widely used. What is its purpose?

....

....

....

....

....

How is the wheel assembly located in suspension (b)?

....

....

....

....

Name component arrowed on drawing (b) and briefly describe its function and action:

....

....

....

....

The suspension/subframe assembly is attached to the chassis/body through bonded rubber mountings; indicate on the drawings where such mountings are positioned.

Why is a suspension subframe considered desirable?

....

....

....

(a) ....

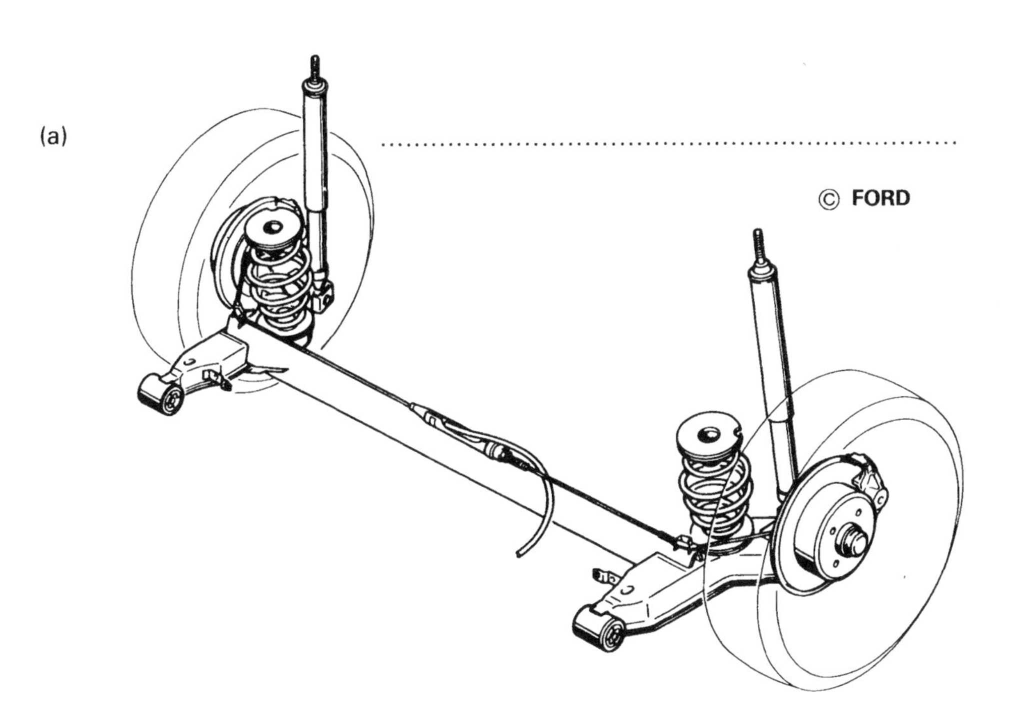

(b) ....

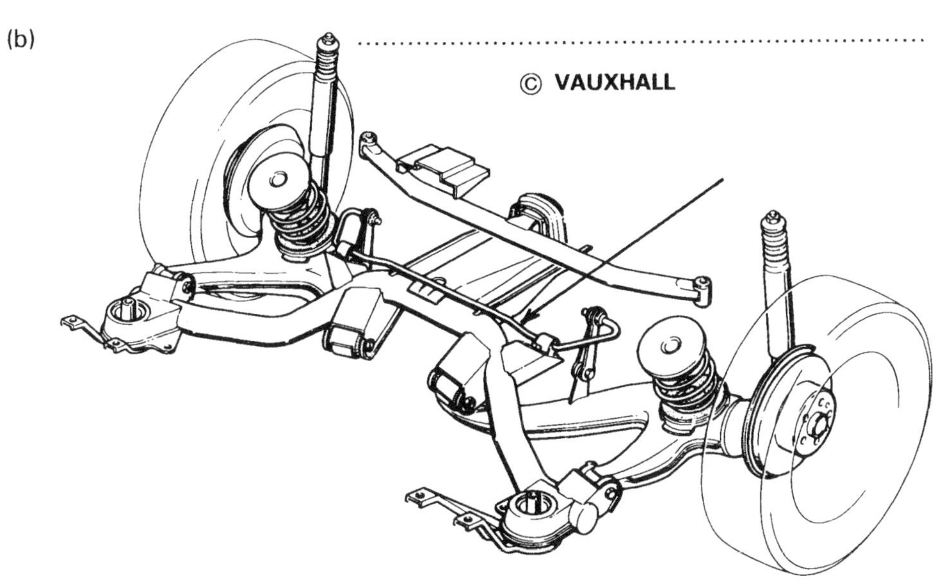

Name the popular type of suspension shown below:

....

State the purpose of component 'A':

....

....

It can be seen from the drawing that the spring and its seatings are angled relative to the strut. State the reason for this arrangement:

....

....

....

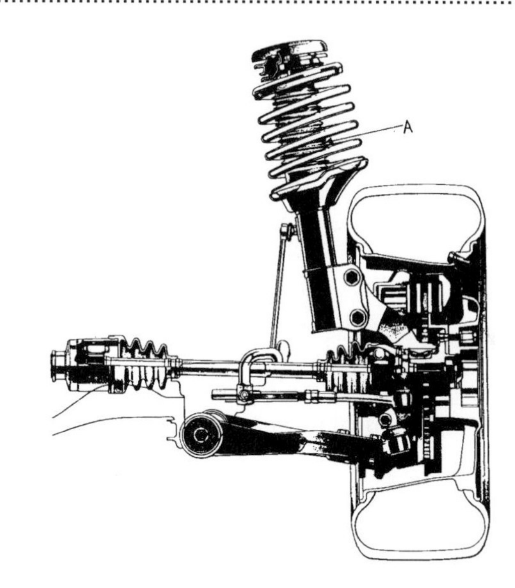

© **SUBARU**

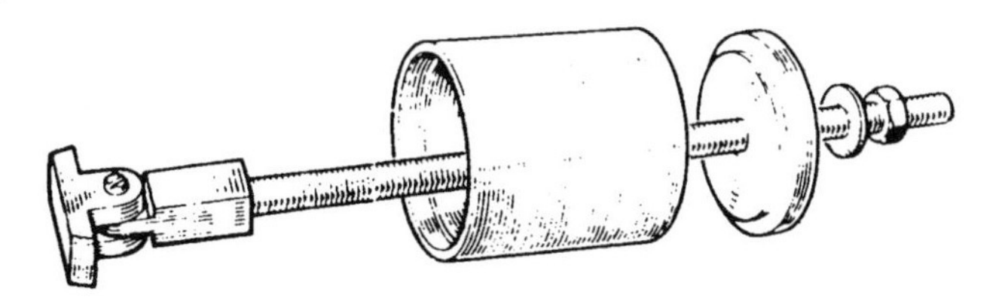

Name the rear suspension system shown below:

....

State the purpose of component 'B':

....

When replacing the shock absorber with this layout, would it be necessary to use a spring compressor, and if not, why not?

....

B

© **FORD**

One type of tool used in suspension repair is shown on the left. State the purpose of such a tool:

....

....

DE-DION AXLE

The De-Dion axle is a compromise between the live beam axle and independent suspension. The final drive and differential are mounted on the chassis and the road wheel assemblies are carried on a beam type dead axle. Coil springs are used and the axle is located by trailing links and a Panhard rod. In some systems a Watts linkage provides lateral location rather than a Panhard rod.

Complete the drawing below to show the De-Dion system:

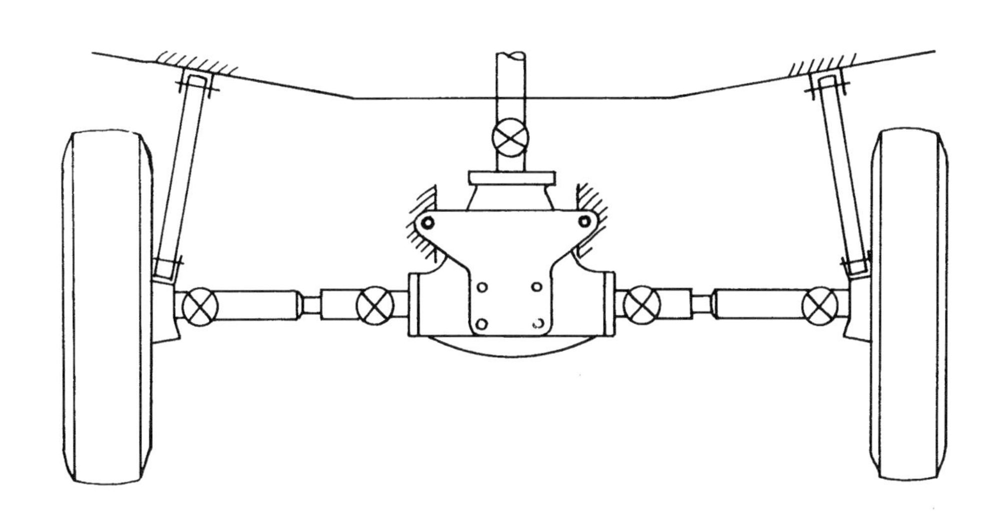

State two advantages of this layout:

1. ....

2. ....

Name a vehicle that uses the De-Dion layout:

Make .... Model ....

Influences of load variations on wheel geometry and alignment

The wheel alignment and suspension geometry can vary as the wheels rise and fall to follow the road contour, and can be affected by load. The degree to which alignment and geometry do alter depends largely on the actual suspension design and the amount of suspension movement involved.

Describe how suspension geometry is affected when the suspensions named below are involved.

Parallel arm type

Unequal length wishbone type

Macpherson type

REAR WHEEL 'SUSPENSION STEER'

The semi-trailing link IRS shown below produces a steering effect when the vehicle is cornering. Describe this:

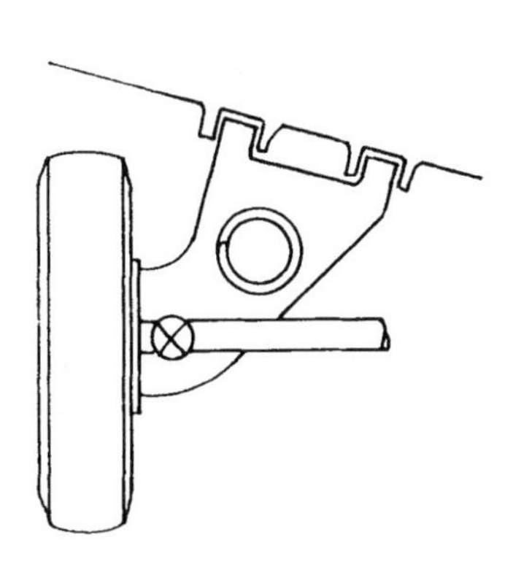

....
....
....
....
....
....
....
....
....

The radius arms in the trailer suspension shown below provide a degree of rear wheel steer when cornering. Explain this:

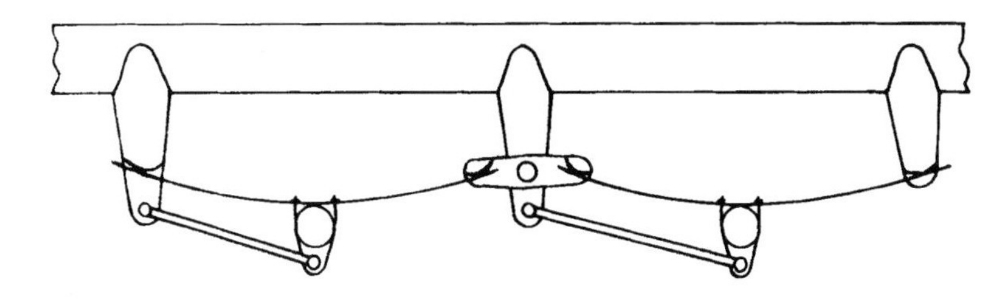

....
....
....
....
....
....

ENERGY CONVERSION

When a spring is deflected, for example

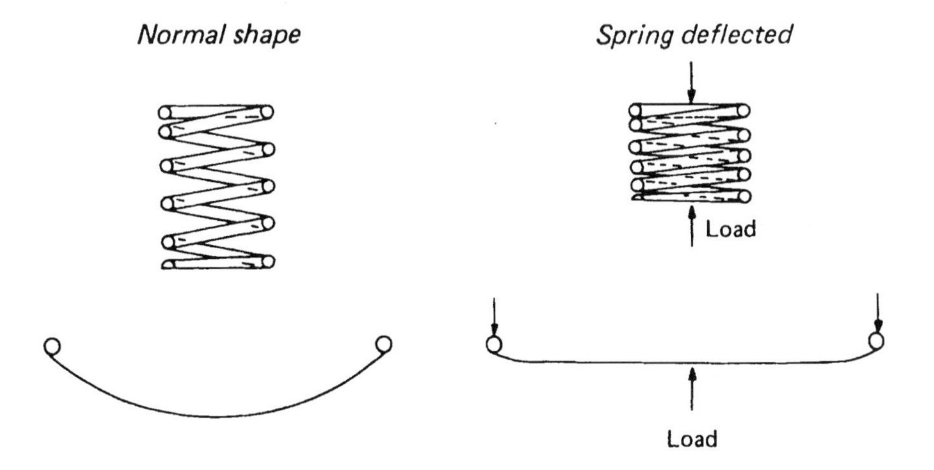

mechanical energy creates the deflection.

(a) What happens to the energy in an 'undamped' spring?

....
....
....
....
....

(b) What happens to the energy put into the spring when a hydraulic damper is used to prevent excessive spring oscillation?

....
....
....
....
....

TAPER SINGLE-LEAF SPRING SUSPENSION

The taper-leaf spring increases in thickness towards the centre and performs in much the same way as a multi-leaf spring; it is an extremely popular form of suspension on HGV trailers, where it is in open-ended form. It is also widely used on beam-type car and light commercial rear axles.

State the material from which the spring may be made:

....

Taper single-leaf spring

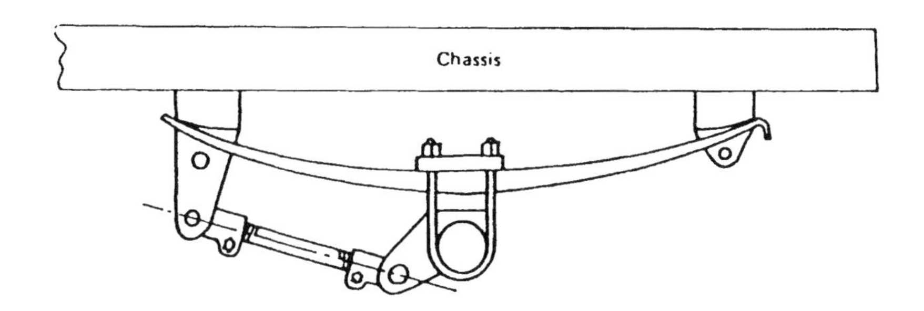

State the main advantage of the taper-leaf spring compared with the multi-leaf-type spring:

....

....

....

....

Why are multi-leaf and single-leaf springs made thicker towards the centre?

....

....

....

LEAF SPRING HGV

A modern HGV multi-TAPER leaf spring suspension is shown below. Label the drawing, state the purpose and describe the action of the lower two leaves:

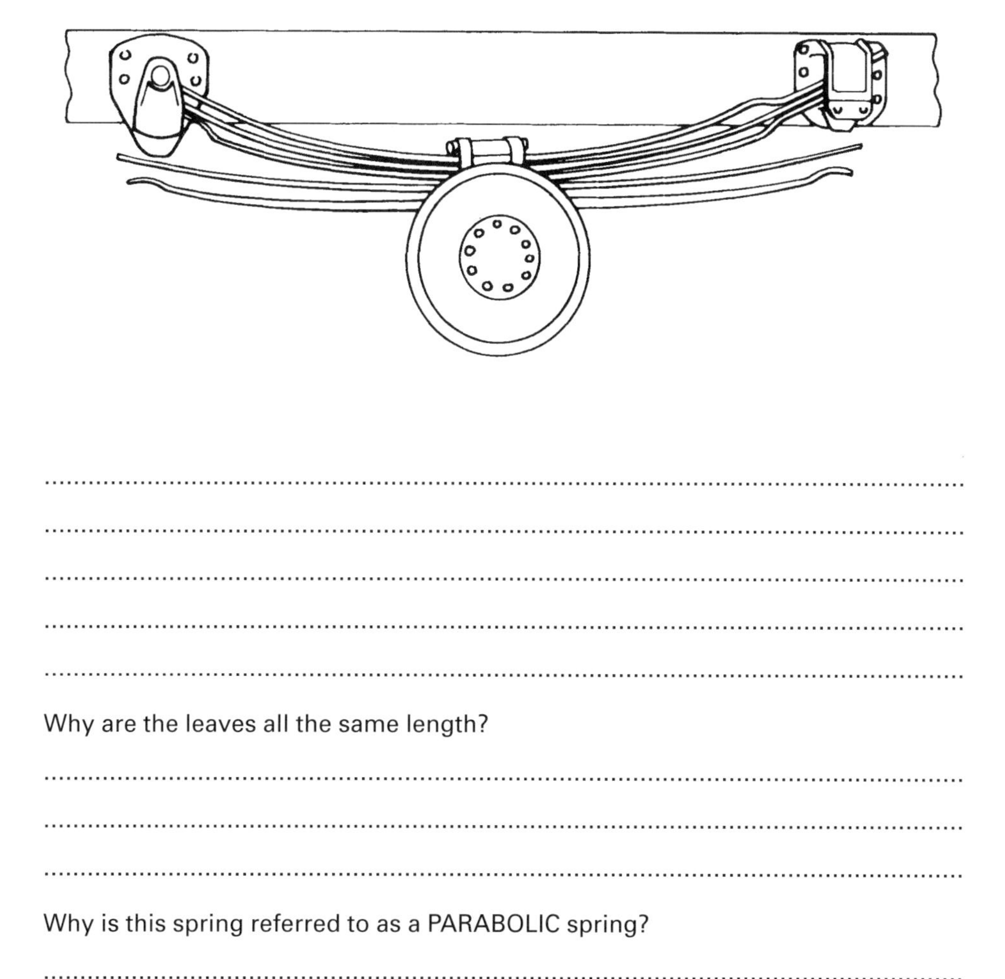

....

....

....

....

....

Why are the leaves all the same length?

....

....

....

Why is this spring referred to as a PARABOLIC spring?

....

....

....

TORQUE REACTION

In a live axle, driving and braking torques produce reactive torques, that is, torques which act equally and opposite. When an axle is located by leaf springs only, drive and brake torque reactions cause the axle casing to partially rotate until constrained by the spring.

Name the torque reaction shown below:

....

Normal load force at A acting on the chassis will .... and the force at B will ....

Normal load force at C acting on the chassis will .... and the force at D will .... as a result of .... torque reactions.

How do torque reactions affect the leaf spring?

....

....

....

....

What practical problems can be encountered as a result of drive and brake torque reactions?

....

....

....

MULTI-AXLE SUSPENSION

When two 'closely spaced' axles are employed at the rear of the HGV, the suspension design must ensure that:

(a) the payload, irrespective of position, is shared by the two axles;

(b) the forces due to road shocks are balanced out across the two axles.

To satisfy the requirements (a) and (b) above, tandem-axle suspensions are interconnected. One of the most common, and simplest, methods is to employ a 'balance beam'.

Complete the drawing below to show a balance-beam arrangement and describe briefly (use arrows on the drawing) how load balancing is achieved.

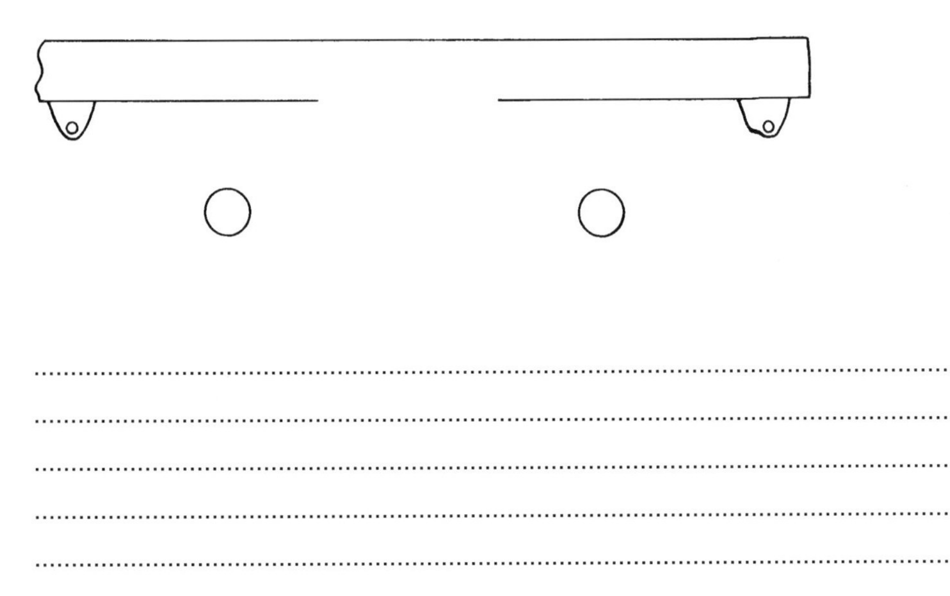

....

....

....

....

....

One disadvantage of the above system is that when a relatively short balance beam is used, wheel movement is rather limited; increasing the length of the balance beam allows a greater wheel movement at the expense of an increased axle spread, which causes more tyre scrub.

EFFECTS OF TORQUE REACTION

Drive and brake torque reaction produce load variations between the axles with certain tandem axle suspension arrangements. The effects of drive and brake torque reactions are shown (exaggerated) below. Add arrows to the drawings to indicate the direction of the forces involved and opposite describe the effects of this reaction.

.... TORQUE REACTION

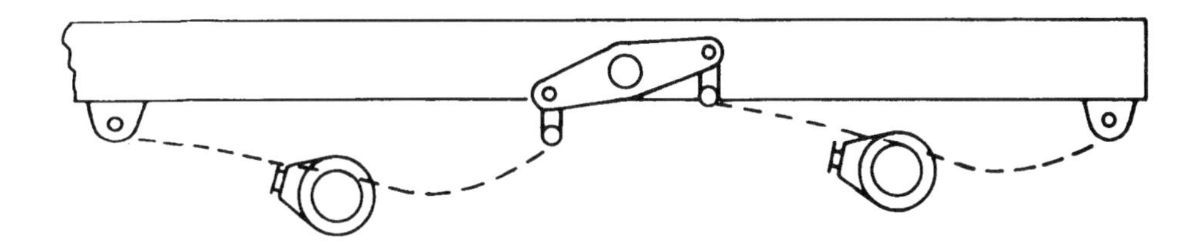

.... TORQUE REACTION

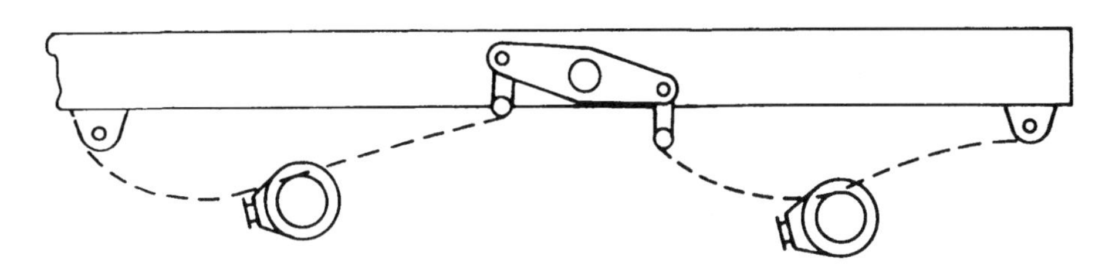

If a suspension does not counteract torque reaction and load variation occurs (as is the case with the balance-beam type) it is known as a .... suspension.

State the effects of the balance-beam action shown above.

....

....

....

....

State the meaning of the term 'non-reactive' when used to describe a multi-axle suspension:

....

....

....

....

....

The tandem-axle suspension shown below is a non-reactive type. The upper and lower torque arms locate the axles to resist drive and brake torque reaction, thereby preventing load variation between the axle and relieving the road springs of torque reaction stresses.

Complete the labelling on the drawing:

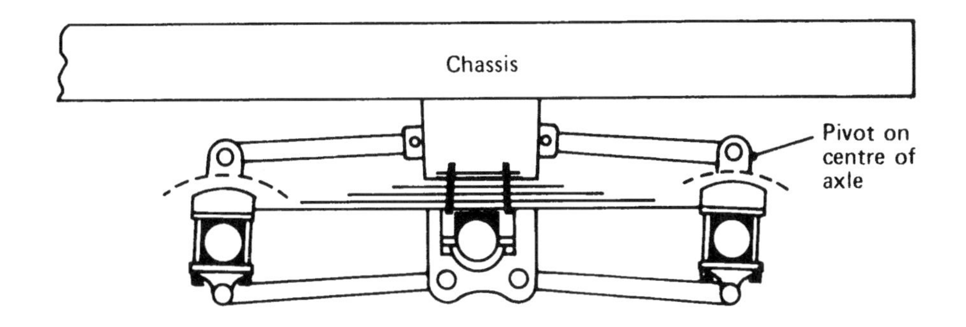

Complete the drawing below to show the bell crank lever type of non-reactive suspension and describe the action of the suspension:

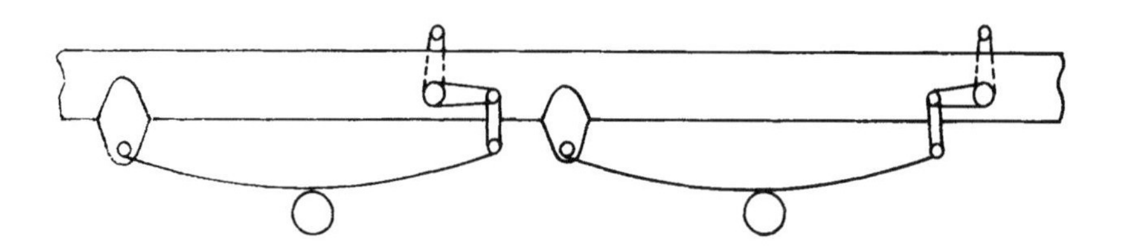

....

....

....

....

....

....

....

....

....

....

AXLE LIFTS

Axle lifts are usually employed on second steered axles or dead axles in multi-axle layouts. The purpose of an axle lift is to raise the wheels of one axle clear of the ground when the vehicle is operating unladen. The resulting benefits are:

....

....

....

Make a sketch to illustrate an axle-lift arrangement and describe briefly how it operates:

....

....

Very often the operational requirements for a particular vehicle cannot be fulfilled by the stock vehicle/chassis supplied by a vehicle manufacturer. Some companies do therefore specialise in chassis engineering in which chassis and suspensions are modified to accommodate a specific body or type of load.

Complete the sketch below to show a modified suspension for an additional axle:

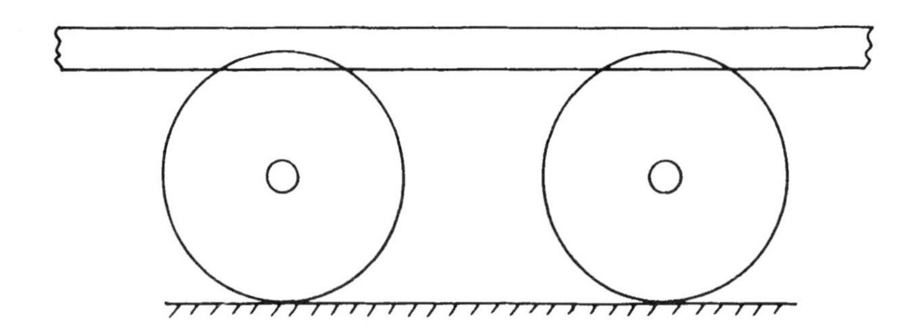

List the preventive maintenance, MOT checks/tasks associated with suspension:

What benefits are to be derived from carrying out routine preventive maintenance?

List the general rules/precautions to be observed while carrying out routine suspension system maintenance and running adjustments:

Describe the methods of protecting the system against hazards during use or repair:

CHECKING SUSPENSION ALIGNMENT AND GEOMETRY

Describe briefly the suspension alignment and geometry checks normally carried out on the rear suspension: use simple sketches on the vehicle outline to illustrate how equipment is used for this purpose:

....

....

....

....

....

....

....

....

....

....

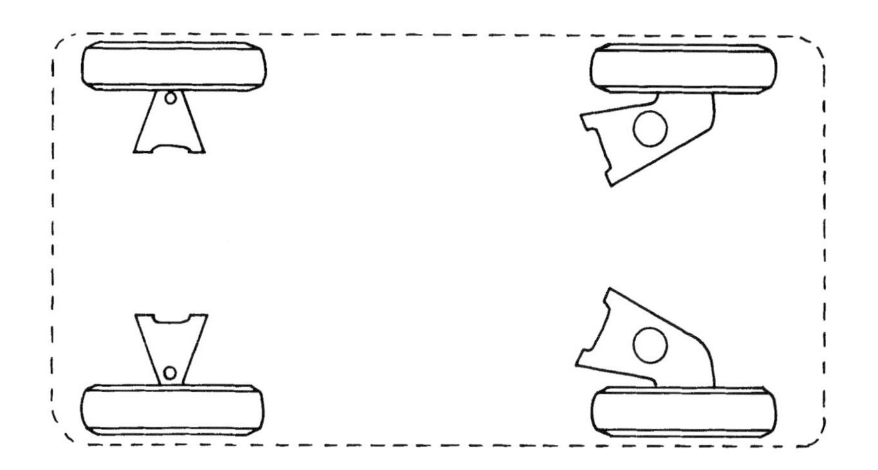

Examine a light vehicle or a HGV, and list the suspension faults found:

Vehicle Make .... Model ....

....

....

....

....

....

....

....

....

Materials used for suspension components

Complete the table below:

Component	Material	Reasons for use
Road spring		
Suspension links, radius arms, etc.		
Ball joints		
Spring eye bushes, HGVs		
Radius arm bushes		

DIAGNOSTICS: SUSPENSION SYSTEM – SYMPTOMS, FAULTS AND CAUSES

State a likely cause for each symptom/system fault listed below. Each cause will suggest any corrective action required.

SYMPTOMS	FAULTS	PROBABLE CAUSES
Pulling to one side; vibration; poor road-holding when wet	Abnormal, excessive tyre wear	....
Ride height low at one corner; axle misalignment	Broken road spring	....
Axle misalignment; crabbing	Loose 'U' bolts	....
Poor road holding	Worn dampers	....
Incorrect trim height	Corroded spring mounting	....
Steering pull; noise	Worn radius arm brushes	....
Abnormal tyre wear; heavy steering	Worn ball joints	....
Incorrect trim height	Leaking spring displacers	....
Vehicle instability under full load	Missing bump stop	....

FORCES DUE TO TORQUE REACTION: CALCULATIONS

The two forces acting on the spring eyes (where an axle is located in the centre of the spring) are said to form a 'couple', that is, they are equal and opposite forces acting on either side of a pivot. The torque exerted by a couple is found by multiplying one of the forces by the perpendicular distance between the forces, for example:

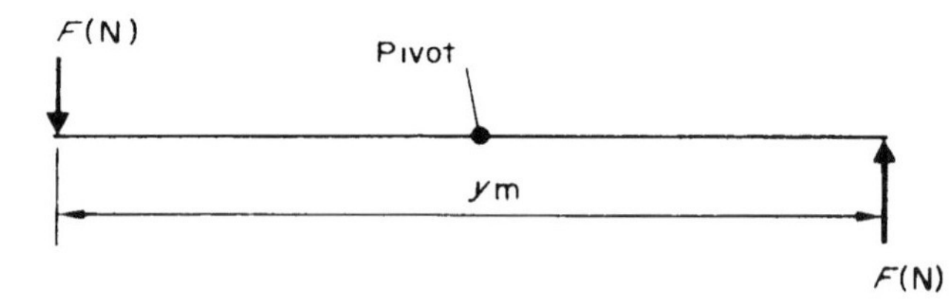

Turning moment of torque = $F \times y$ (N m)

similarly $F = \dfrac{\text{torque}}{y}$

PROBLEMS

1. A leaf spring centrally attached to a driving axle measures 1.4 m between spring eye centres. If the driving torque in the half shaft is 1000 N m, calculate the thrust on each spring eye as a result of drive torque reaction.

Upward thrust on front eye of spring $= \dfrac{\text{torque}}{\text{distance between spring eyes}}$

Upward thrust on front eye of spring $= \dfrac{1000}{1.4} = 714.28$ N *Ans.*

$\therefore$ downward thrust on rear eye of spring $= 714.28$ N *Ans.*

2. Consider the spring below, normally loaded as shown. Calculate the load supported by the front and rear spring eyes respectively when the driving torque in the half shaft is 1200 N m.

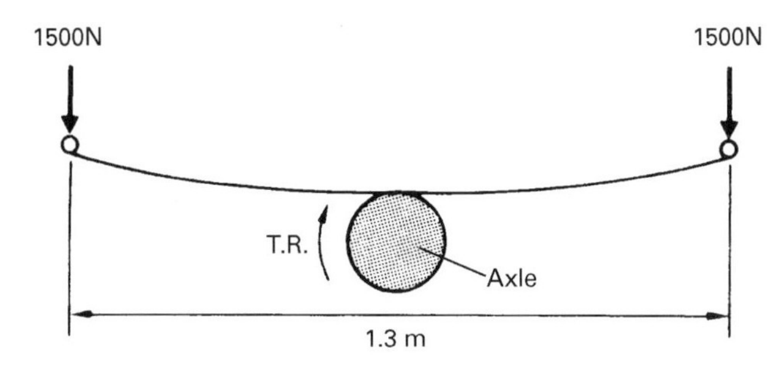

....

....

....

....

3. If the load on the front eye of a rear spring increases by 600 N and the load on the rear eye of the spring decreases by the same amount during braking, calculate the braking torque if the spring measures 1.2 m between the eyes.

....

....

....

4. During acceleration the torque in the propeller shaft of a vehicle is 1280 N m. If the rear axle ratio is 4.5 : 1, determine the change in load on each of the spring eyes, which are 1.5 m between centres, owing to torque reaction.

....

....

....

....

Chapter 7

Steering

Steering systems layout 120
Ball joints and swivel assembly 121
Front hub 122
Ackerman system 123
Steering geometry 125
Twin steer alignment 126
Castor angle 127
Camber angle and swivel pin inclination 128
Steering gear 130
Movement ratio, force ratio, efficiency 132
Calculations 133
Power-assisted steering 133
Steering system maintenance 140
PAS maintenance 141
Diagnostics 143

STEERING SYSTEMS LAYOUT

The steering system provides a means of changing or maintaining the direction of a vehicle in a controlled manner. Other main requirements of the system are:

1. *Driver effort should be minimal* ....

2. ....

....

3. ....

Beam axle, single track rod (light and HGVs)
Add the steering linkage to the layout below to illustrate a beam axle steering system:

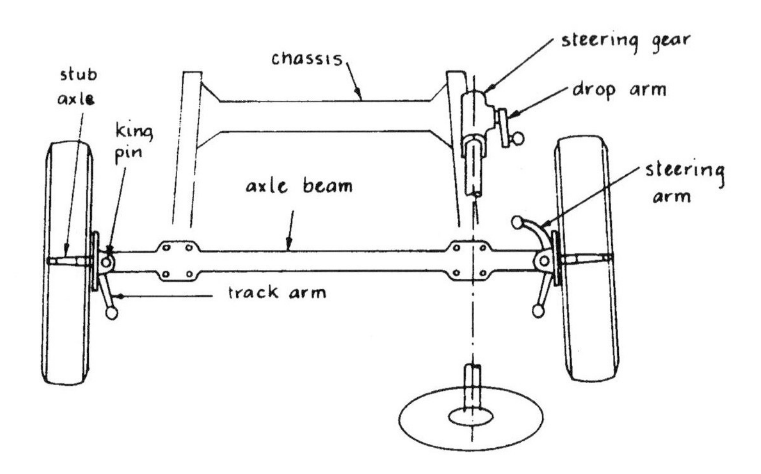

With this system, as the axle tilts owing to suspension movement, the single track rod moves correspondingly and the steering is unaffected. It is unacceptable however to employ a single track rod linkage with independent front suspension (IFS). Why is this?

....

....

....

....

Divided track rod car steering system
Virtually all cars and many light goods vehicles have independent front suspension (IFS). The steering linkage must therefore be designed to accommodate the up and down movement of either steered wheel without affecting the other steered wheel. In most systems two short track rods which pivot in a similar arc to the suspension links are connected to the stub axles.

One system in use can be described as: a steering gearbox with 'idler' and three-track-rod layout.

Complete the drawing of such a system shown below by adding the linkage and labelling the parts:

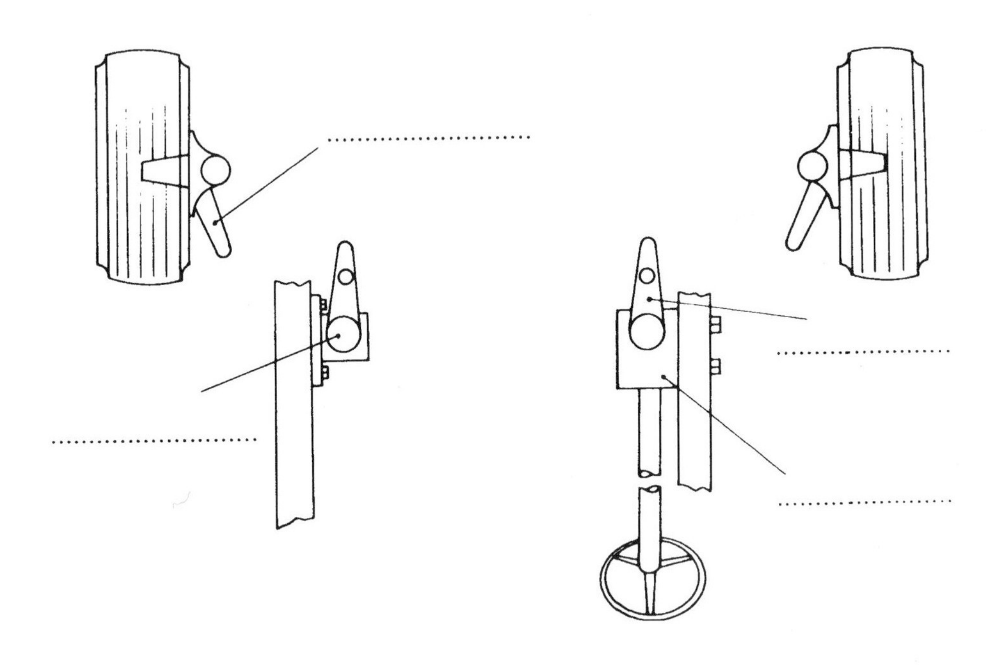

What are two main disadvantages of this layout?

....

BALL JOINTS AND SWIVEL ASSEMBLY

Name the steering system shown below and give reasons for using this in preference to a three piece track rod arrangement:

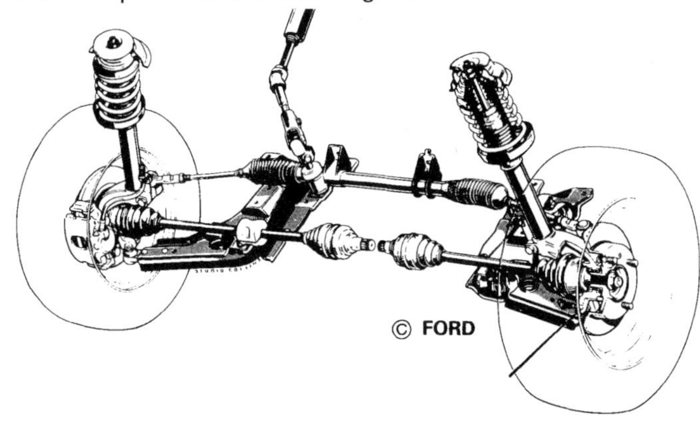

....

....

....

Linkage ball joints

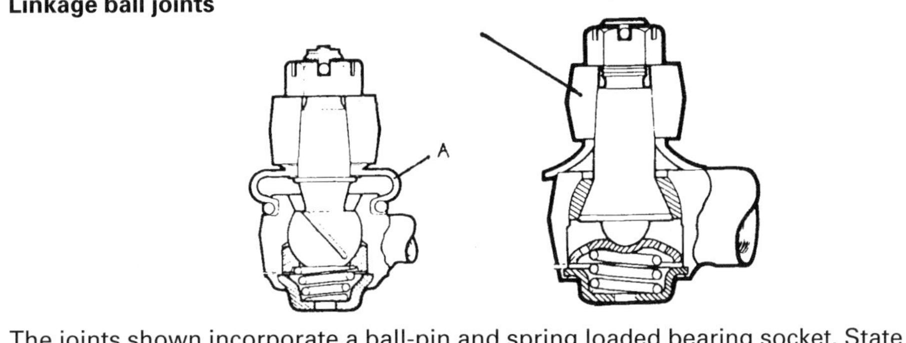

The joints shown incorporate a ball-pin and spring loaded bearing socket. State the purpose of the spring and label the drawings:

....

....

State the purpose of part A ....

....

Steering swivel assembly (stub axle)

Examine a beam axle steering system and complete the drawing below by adding the king pin and thrust bearing:

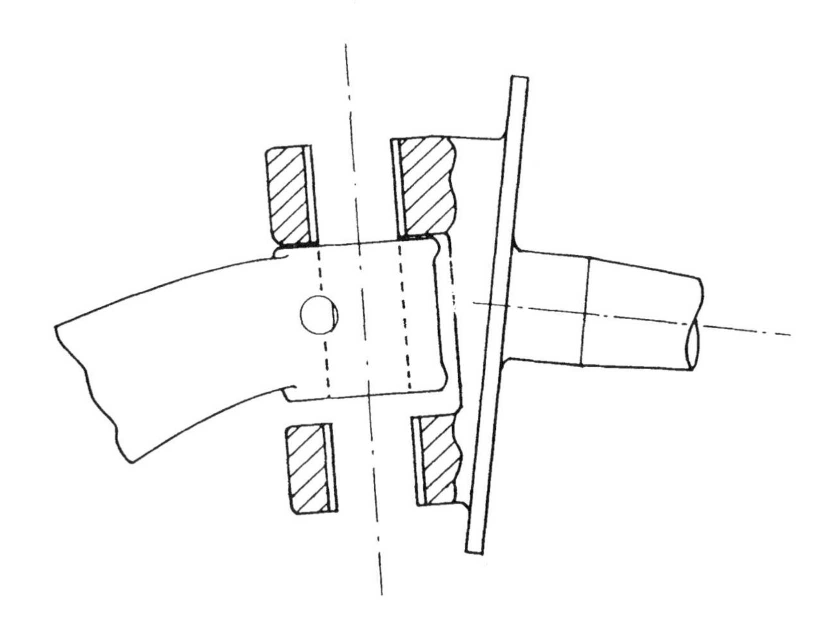

State the purpose of the king pin thrust bearing:

....

....

How is the king pin located in the axle beam?

....

....

How are the king pin bushes lubricated?

....

....

Describe the steering action in the suspension systems shown below. Indicate the steering pivots on the drawings.

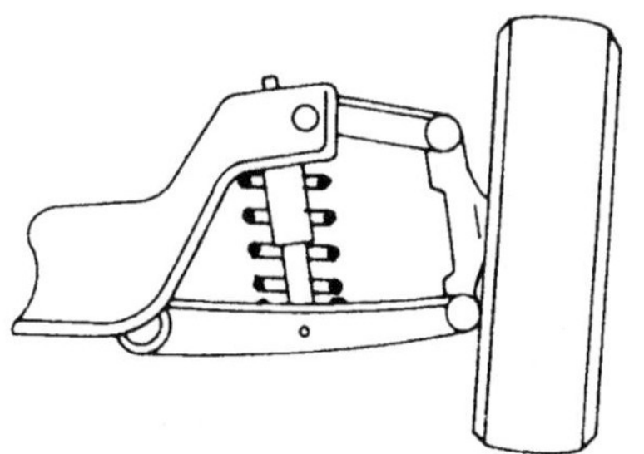

Double wishbone

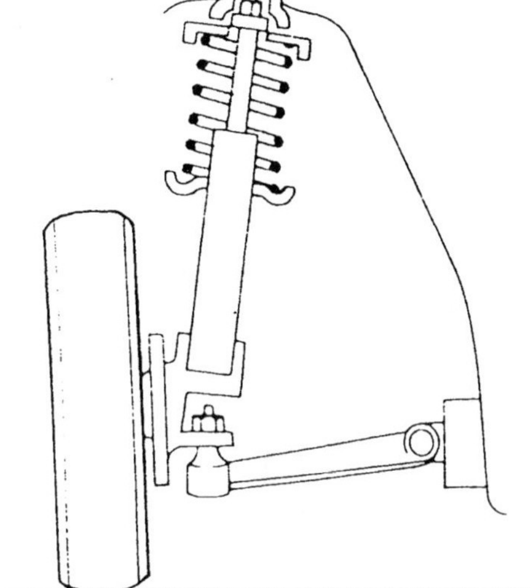

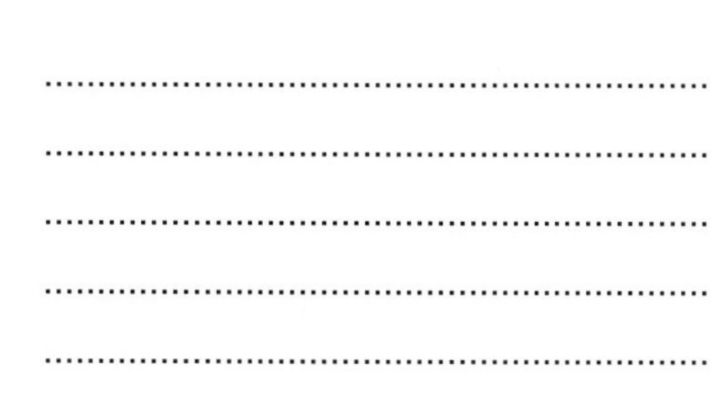

Macpherson strut

FRONT HUB

INVESTIGATION

Examine a front hub and complete the drawing below to include bearings and grease seal. Label all parts.

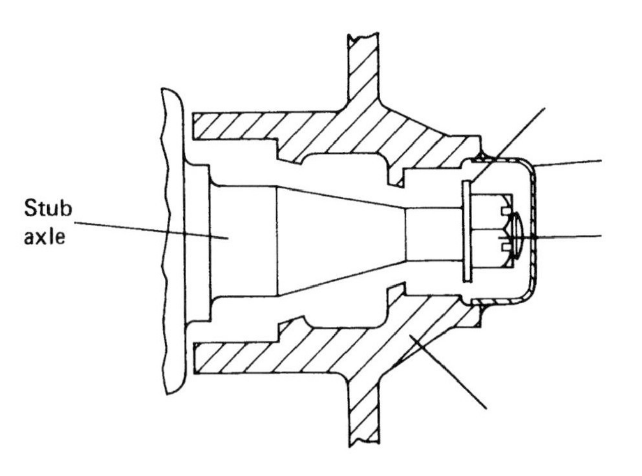

What type of bearings are used?

Name two other types of hub bearing:

1. 2.

A popular type of grease seal used in front hubs is the spring loaded rubber seal. Name two other types of grease seal:

1. 2.

State three reasons why grease can leak past a serviceable seal:

1.

2.

3.

True Rolling

When a vehicle travels on a curved path during cornering, true rolling is obtained only when the wheels roll on arcs which have a 'common centre' or common axis. Show the common centre on the drawing below:

State why true rolling of the wheel is necessary:

....

....

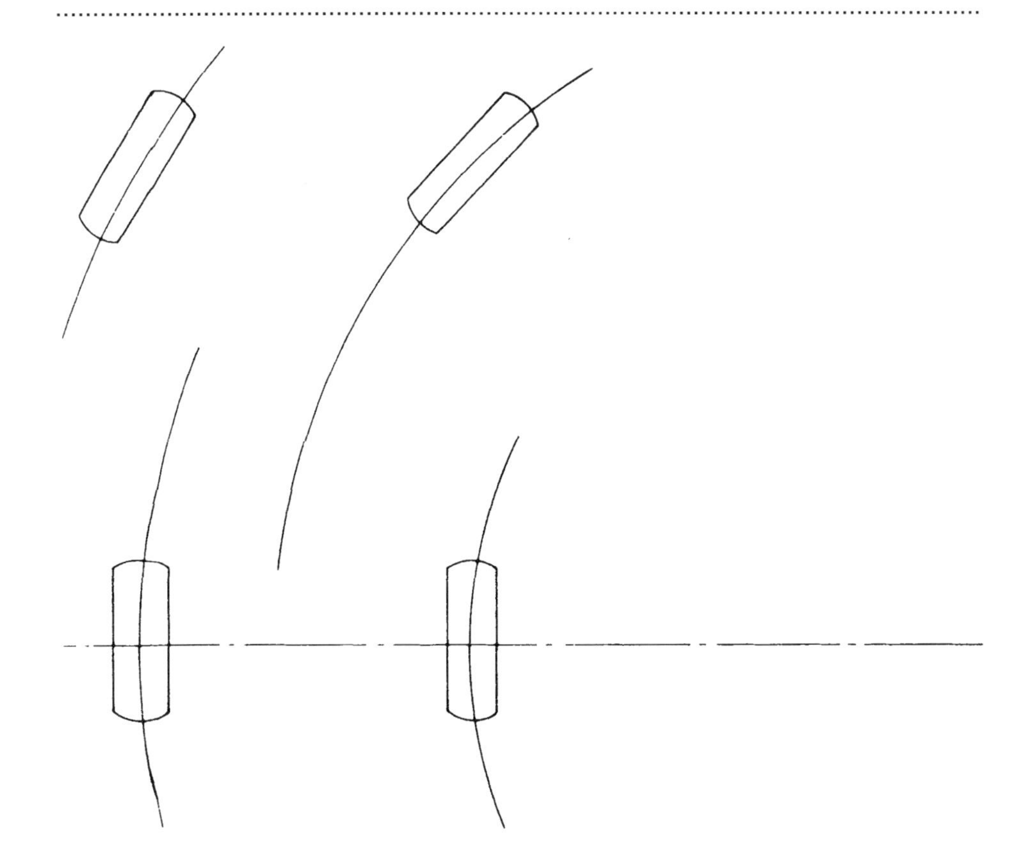

To achieve true rolling while cornering, the wheels are steered through different angles; that is, they are not parallel; the inner wheel on a turn is turned through a .... angle than the outer wheel.

ACKERMAN SYSTEM

The difference in steering lock angles to give true rolling of the wheels while cornering can be achieved by making the track rod shorter than the distance between the king-pin centres; that is, the steering arms are usually inclined inwards. If the track rod is in front of the axle, as on many cars, then it is longer than the distance between the king-pin centres and the steering arms must be inclined outwards.

Add a track rod and steering arms to the drawing below and extend lines through the steering arms to show the point of intersection on the vehicle centre line if the linkage is to satisfy conditions for true rolling:

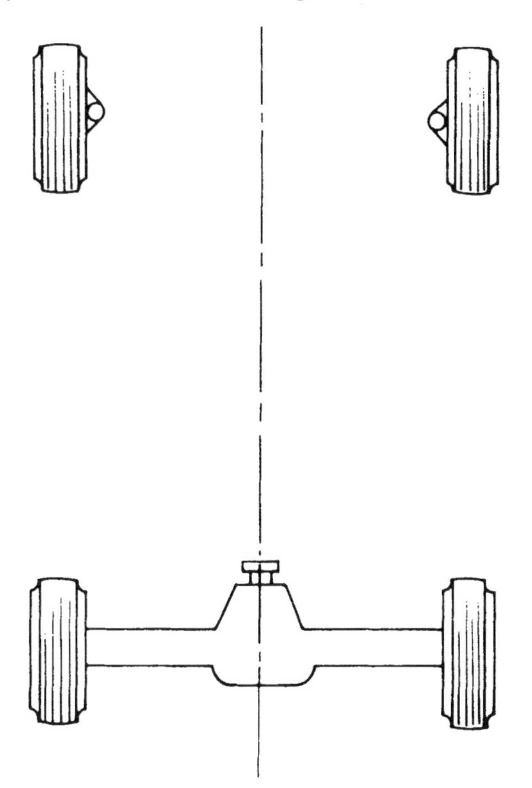

This type of steering linkage is known as:

....

The diagram below shows the position taken up by the track rod when the wheels are on lock. Indicate on the drawing below a typical angle for the inner wheel for the angle of lock shown on the outer (left-hand) wheel.

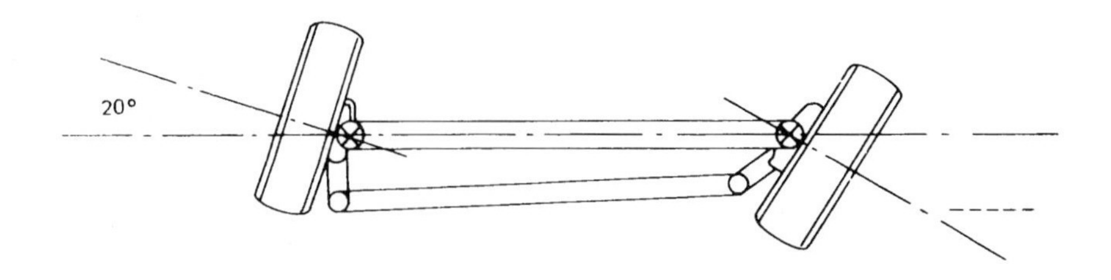

The difference in lock angles shown above is known as:

....

Ackerman as applied to heavy goods vehicles

Using twin steering axles spreads the load and reduces axle loading.

To obtain true rolling of the steered wheels on a twin-steering axle arrangement, the first axle wheels are steered through a greater angle of lock than the second axle wheels. If a tandem-axle arrangement is used at the rear, tyre scrub is inevitable owing to the fact that the rear wheels can never roll about a common centre.

To obtain as near true rolling as possible, the 'common centre' lies on a line extended from a point midway between the rear axles.

State how the twin-steer linkage shown at A opposite produces true rolling and sketch an alternative linkage at B:

....

....

....

....

Add dotted lines to the drawing below to show the steering geometry to give (as nearly as possible) true rolling for the eight-wheel vehicle layout shown:

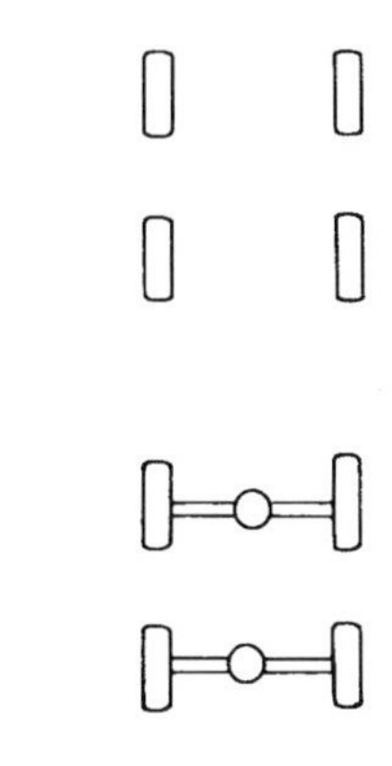

State how the rear-axle spread (that is, their distance apart) affects tyre scrub:

....

....

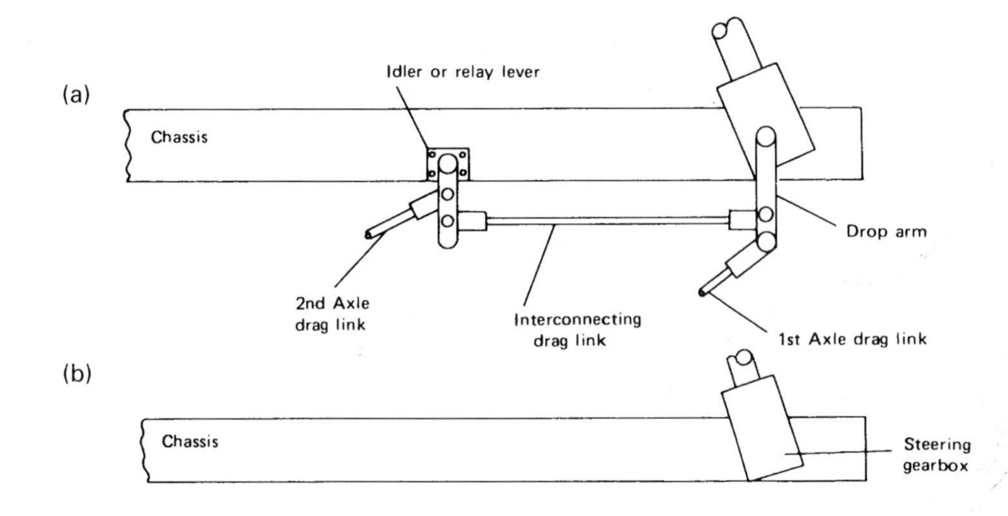

STEERING GEOMETRY

The road wheels of a vehicle should always rotate with a true rolling motion (no tyre scrub). To ensure true rolling the steered wheels must be correctly aligned. This usually entails adjusting the effective length of the track rod to give a static 'toe-in' or 'toe-out' of the steered wheels.

The wheels shown below are ...., that is, dimension A is slightly less than dimension B.

When dimension B is slightly less than dimension A the wheels would be said to have ....

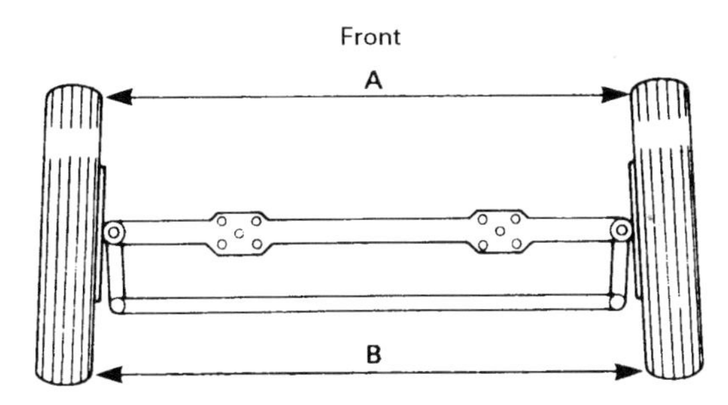

State the purpose of toe-in or toe-out:

....

....

....

....

....

State typical wheel alignment settings:

Vehicle make	Model	Toe-in	Toe-out

State the main effects of incorrect wheel alignment:

....

....

Describe briefly the method of using the equipment shown below and (on the next page) how to adjust alignment.

MIRROR
FEET
ADJUSTABLE TRAMMELS
ROAD WHEEL
SCALE
REFLECTOR PLATE
VIEWING TUBE

Checking wheel alignment

....

....

....

....

....

....

....

Adjusting wheel alignment

TWIN STEER ALIGNMENT

One type of optical alignment gauge is shown below in position to check the rear axles for parallelism. This is part of the wheel alignment check.

Each front axle is checked individually for alignment as in single axle steering.

Describe the further checks necessary with this steering arrangement:

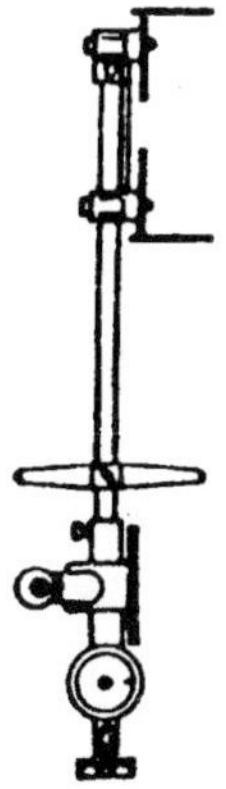

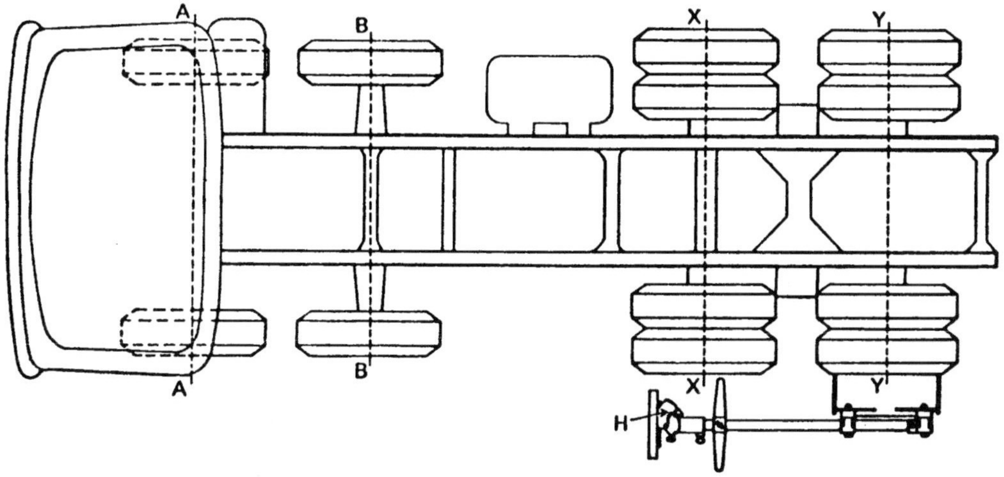

Toe out on turn

The inner steered wheel turns through a greater angle than the outer wheel when cornering (Ackerman) to give an amount of 'toe-out on turn'.

Describe a method of checking and adjusting toe-out on turn, and state a typical setting for this:

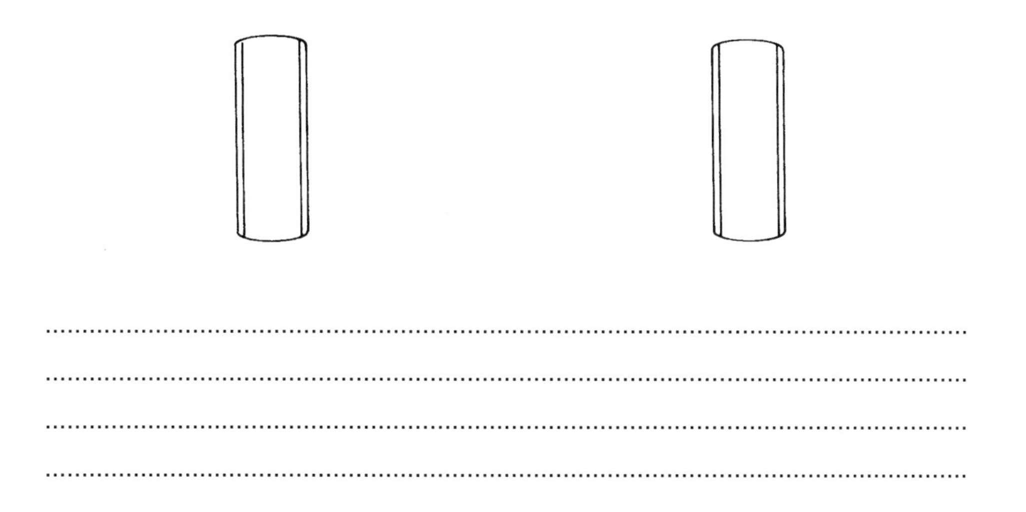

Lock stops
Adjustable lock stops in the steering mechanism limit maximum lock to avoid tyres fouling chassis and prevent extreme travel of the steering gear. Examine a car or HGV and sketch the lock stop arrangement:

CASTOR ANGLE

One of the desirable features of a steering system is the ability of the road wheels to self-centre after turning a corner. This self-centring effect can be achieved by designing the wheel and steering swivel assembly so that the wheel centre trails behind the swivel axis.

Make a simple sketch on the right below to show how the castor is applied to the vehicle; indicate the castor angle.

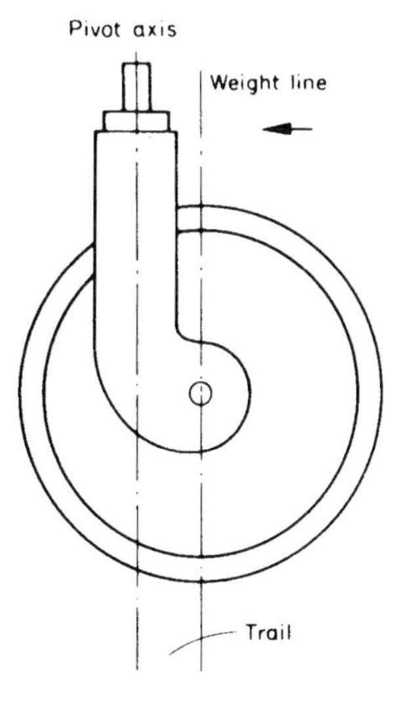

Principle of castor action

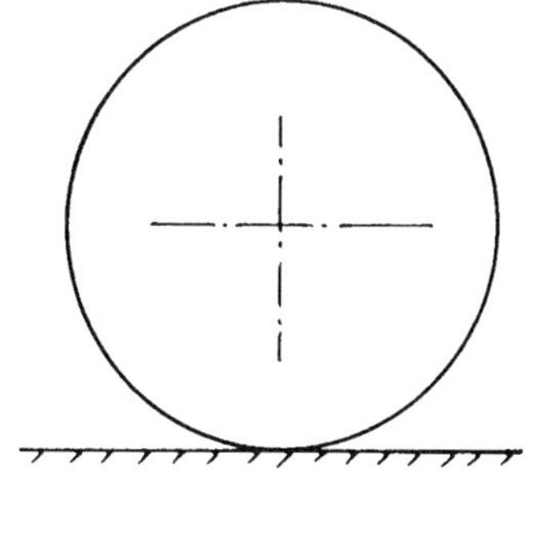
Castor applied to vehicle

Complete the table below by adding typical castor angles:

Car	Light commercial vehicle	Heavy commercial vehicle

CAMBER ANGLE AND SWIVEL PIN INCLINATION

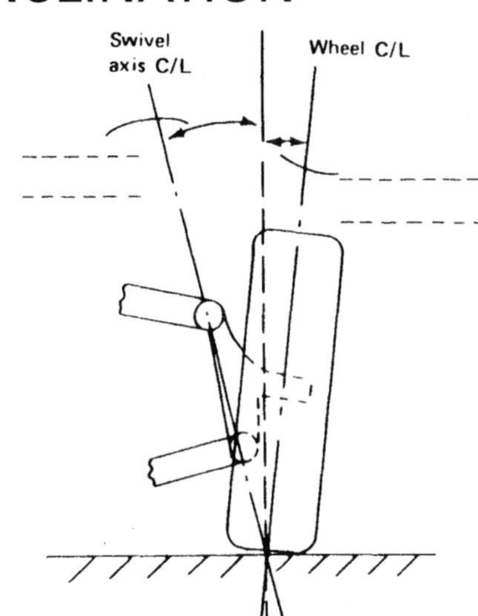

Name the angles shown here and give typical values for each.

Camber angle is the amount that

the .... is tilted out of the

....

KPI is the amount that the

.... is inclined from the

....

State the purpose of the two angles shown above:

Camber

....

....

....

....

....

KPI

....

....

....

....

....

Centre-point steering

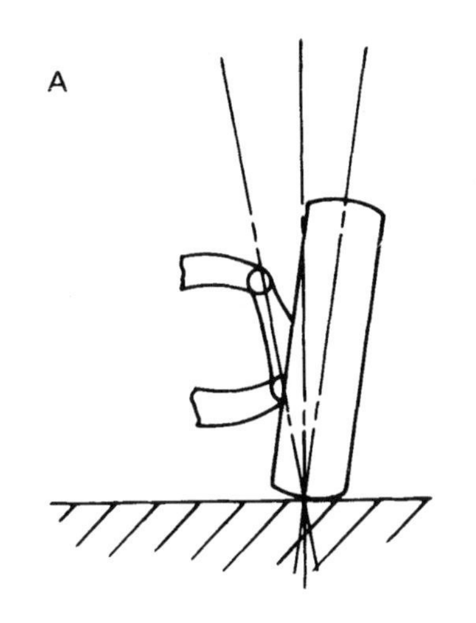

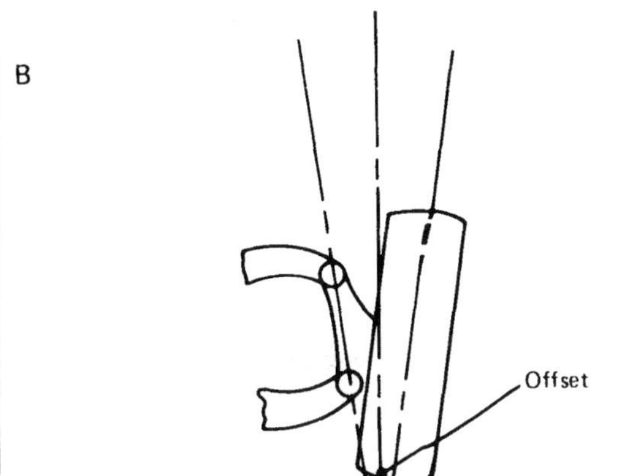

The drawing at A above illustrates true centre-point steering (the wheel centre line and swivel axis centre line converge at ground level). Many vehicles, however, adopt the arrangement shown at B; this shows the wheel and swivel axis centre lines converging below ground level to give a 'positive offset'; this give a better steering action.

What steering feature is illustrated below?

....

Give reasons for using this arrangement:

....

....

....

....

....

....

....

....

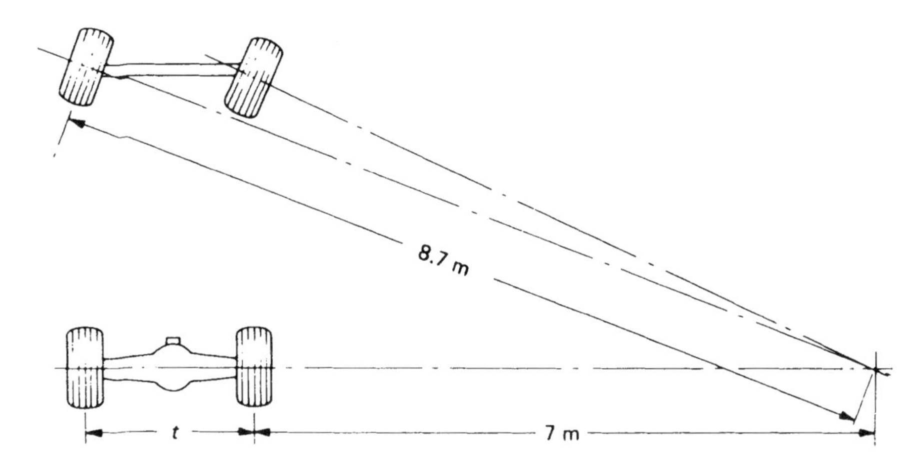

The diagram above illustrates the conditions necessary for true rolling when a vehicle is cornering. Determine the track measurement of the rear wheels (t) for the vehicle shown if the wheelbase is 2.5 m, and the steered angles for both steered wheels.

When checking the toe-out on turn on a vehicle the following readings were recorded:

Outer wheel angle	8°	15°	25°	30°
Inner wheel angle	8.5°	16°	28.5°	35°

Plot a graph using the results and from the graph determine the steered angle of the inner wheel for an outer wheel lock of 20°.

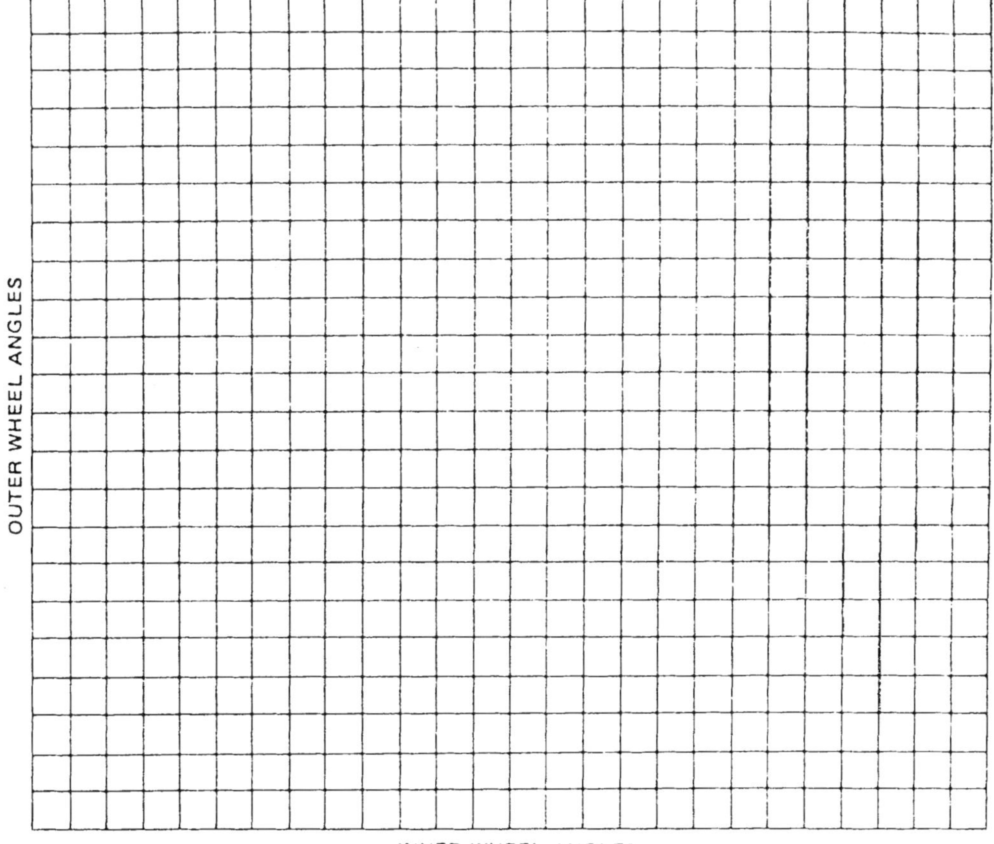

STEERING GEAR

The steering gearbox is incorporated into the steering mechanism for two main reasons:

(a) *To change the rotary motion of the steering wheel into the 'to and fro' movement of the drag link or rack.*

(b) ....

State the desirable characteristics of steering gearboxes:

....

A number of different types of steering gears are in use; complete the list below to name six types:

1. *Cam and peg.*
2. ....
3. ....
4. ....
5. ....
6. ....

Irrespective of the type of steering gearbox employed, rotation of the worm or screw by the steering wheel results in a limited rotation of the rocker or cross shaft, causing the drop arm to swing through an arc and move the drag link to and fro.

Name the types of steering gear shown below:

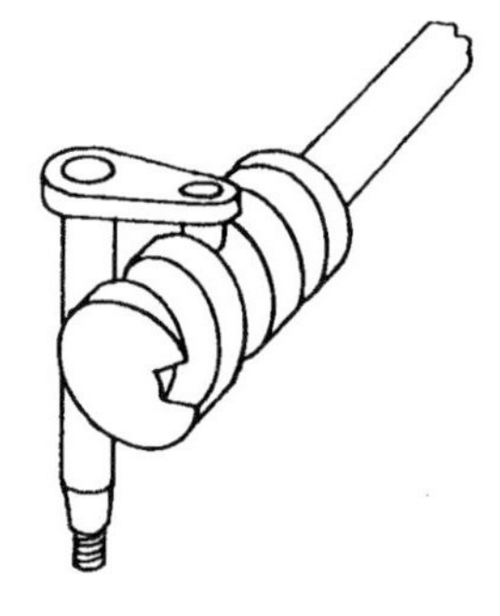

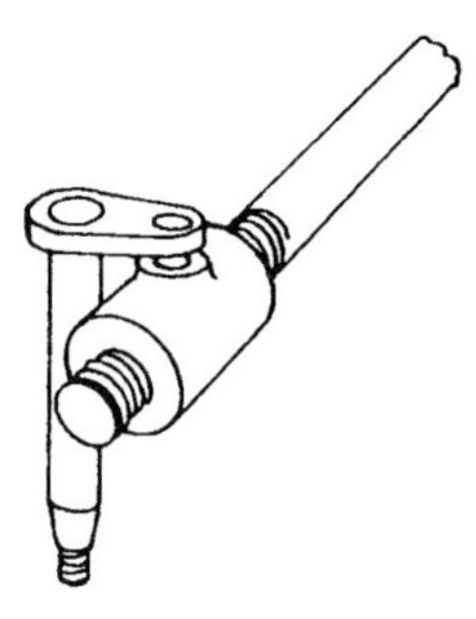

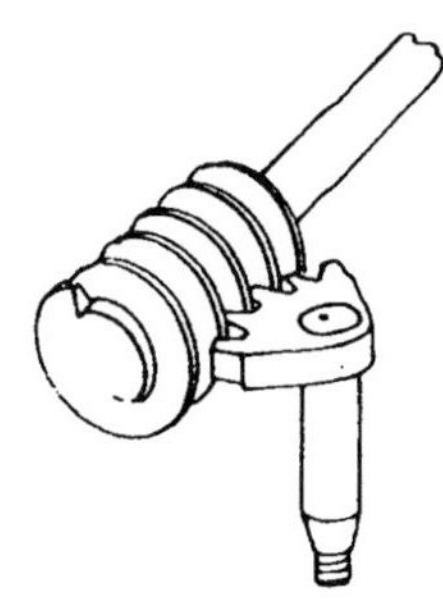

....

The steering gears shown above have now been superseded by more efficient types. In what way have steering gear mechanisms become more efficient?

....

Examine any type of steering gearbox, observe the rotation of the steering wheel and rocker shaft and state the gear ratio.

Type of steering gearbox: ....

Gear ratio: ....

Name the steering gear shown below and label the drawing:

....

State why the worm is 'hourglass' shape:

....

....

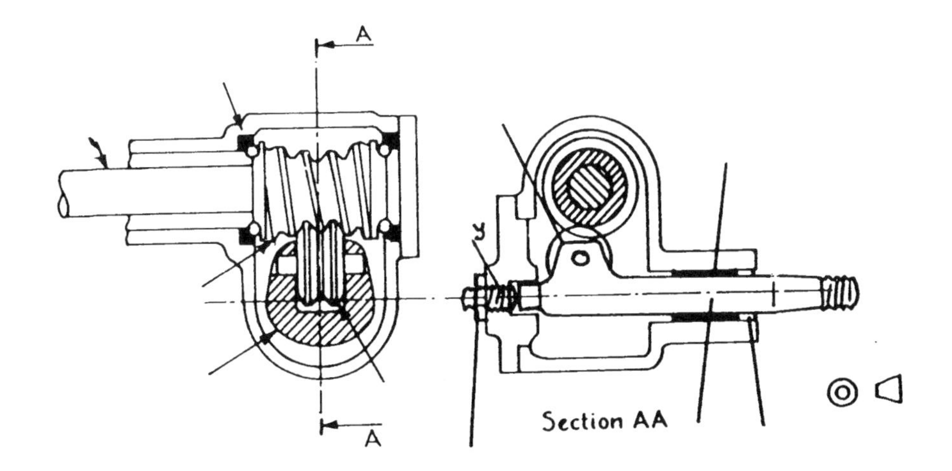

The bearings used are either combined radial and axial thrust ball, or taper roller to withstand both radial and axial forces as the worm rotates.

How is the running clearance of the bearings adjusted?

....

....

State the effects of incorrect adjustments of screw 'y':

too far in ....

....

too far out ....

....

Recirculating-ball Steering Box

The recirculating-ball steering box is a development of the worm and nut type. A row of ball bearings operates in grooves between the nut and worm to form in effect a screw thread; in this way, rolling contact between the nut and worm replaces sliding contact.

Complete the simple drawing below to include the ball bearings, and name the parts:

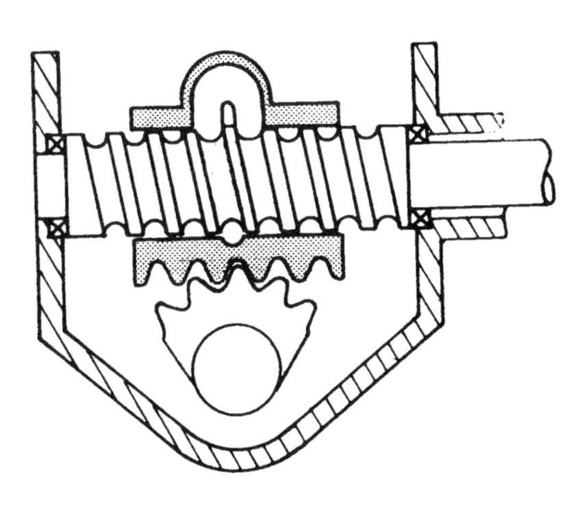

State the purpose of the return passage:

....

....

....

....

....

Name two vehicles, a car and a goods vehicle, which use this type of gear:

....

....

Rack and Pinion

The steering gearbox operating a drop arm and draglink type linkage is seldom used on modern cars. Almost all now employ rack and pinion steering gear.

Name and state the purpose of the numbered components on the rack and pinion assembly shown below:

1
2
3
4

....

....

....

....

....

....

....

....

....

....

....

....

One slight drawback of a rack and pinion steering gear is its reversibility. Explain this:

....

....

MOVEMENT RATIO, FORCE RATIO, EFFICIENCY

The efficiency of a steering gearbox is largely dependent on the type of gear employed. Two factors to be considered when calculating the efficiency are:

TORQUE RATIO and GEAR RATIO

Torque ratio =

Gear ratio =

% efficiency = $\times \frac{100}{1}$

When considering force ratio and movement ratio two other factors must be taken into account:

1. ....
2. ....

Force ratio =

Movement ratio =

% efficiency = $\times \frac{100}{1}$

State the effect of efficiency on torque ratio:

....

....

....

CALCULATIONS

A worm and sector type steering gearbox requires a torque applied by the driver of 12 N m to produce a rocker shaft torque of 135 N m. Calculate the efficiency of the steering gear if the gear ratio is 15:1.

Given the information below for a recirculating-ball type steering mechanism, calculate:

(a) Movement ratio.

(b) Force ratio.

(c) The force in the drag link when a force of 70 N is applied at the steering-wheel rim.

Steering gearbox ratio = 20:1

Efficiency = 90%

Drop-arm length = 0.2m

Steering-wheel dia. = 0.5m

POWER-ASSISTED STEERING

To keep the steering wheel load to an acceptable level without increasing the number of turns required from lock to lock, a system of power assistance can be incorporated in the steering mechanism. This will reduce the effort required by the driver and also, owing to the smaller gear ratio required, keep the number of turns required from lock to lock to a reasonable figure.

Three hydraulic systems in use are:

1. *External ram type.* ....
2. ....
3. ....

The power assistance on most vehicles is hydraulic, the system consisting basically of the following main parts:

....

....

....

....

The most popular hydraulic system used for this purpose is one which operates on the .... principle.

....

....

....

....

....

....

....

....

State the functional requirements of PAS:

....

....

....

....

What is the main difference between a Power-Assisted and a Power Steering system?

....

....

....

External Type PAS

In the external type PAS the power ram and control valve assembly are incorporated into the steering linkage.

Add the power ram, control valve and piping to the HGV system below and describe briefly the operation of the system.

HGV layout (external)

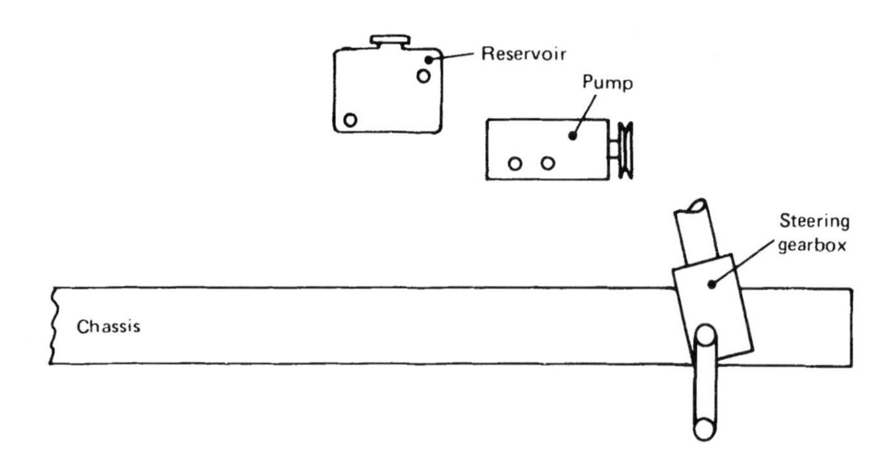

OPERATION (external type)

....

....

....

....

....

....

....

Control

Add a spool valve to the simplified drawing below and describe the operation of the system:

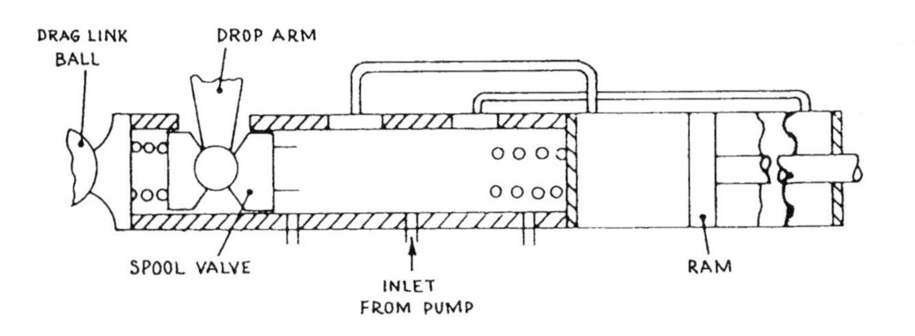

OPERATION

....

....

....

....

....

Semi-integral PAS

The drawing below shows the layout for a twin-steer HGV. In this system the control valves are in the steering gearbox and the power ram operates directly on to the linkage. Complete the drawing to show this arrangement by adding the power cylinder and pipes:

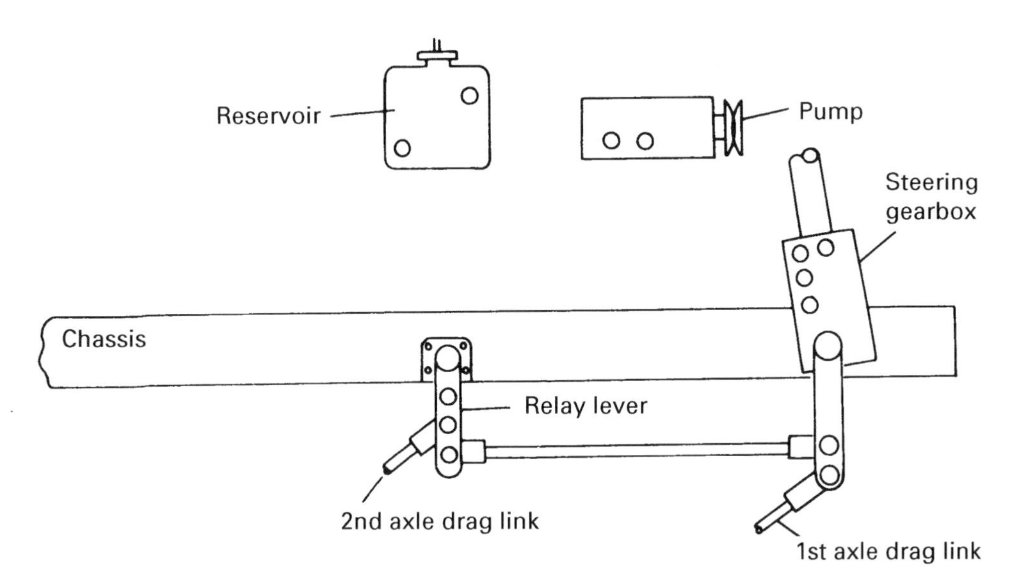

State the average pull required at the steering wheel rim to overcome the control valve spring pre-load on:

(a) a car ....

(b) an HGV ....

Slight angular movement of the power cylinder takes place during operation. How is this accommodated?

....

....

Integral PAS

With the integral system, the servo piston or ram and control valve assembly are located in the steering gearbox or rack.

A power-assisted recirculating ball-type steering box is shown below.

Complete the labelling on the simplified arrangement shown below and describe briefly how power assistance is achieved:

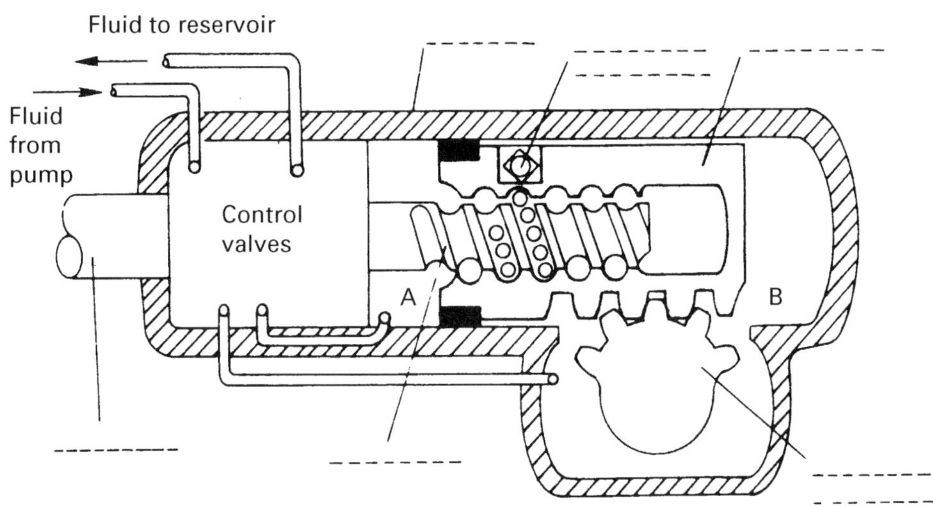

OPERATION

....

....

....

....

....

....

....

Rack and Pinion PAS

Name the main parts of the system layout shown below:

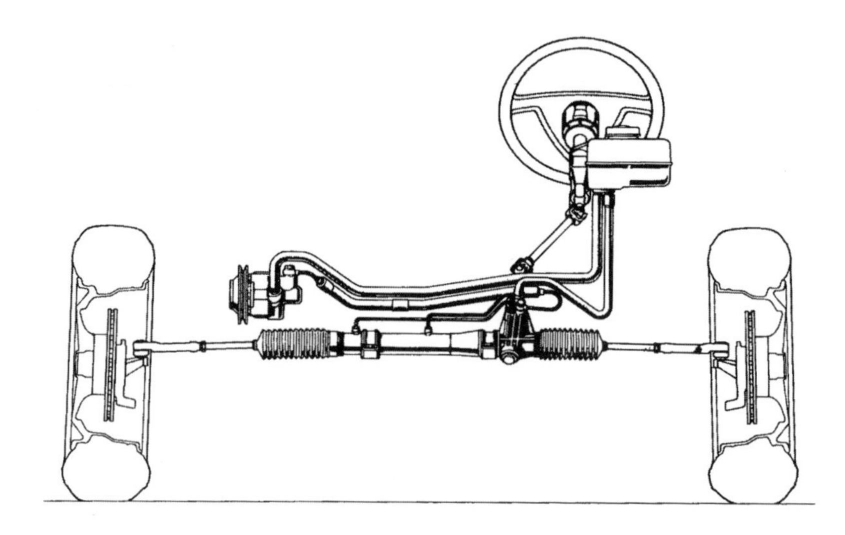

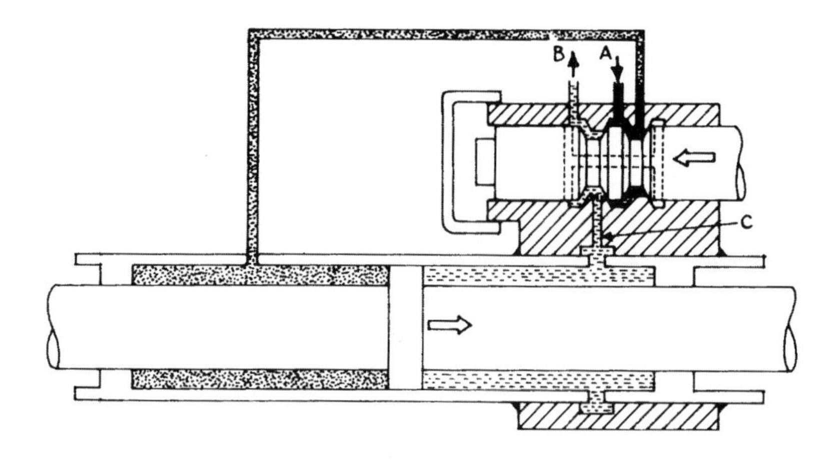

With the spool valve held to the left as shown above, incoming fluid A, from the hydraulic pump can enter the left-hand pressure chamber to move the rack to the right.

With reference to the drawing, describe the operation of the power-assisted steering rack when it operates opposite lock:

....

....

....

....

....

....

In addition to the control valve spring or torsion bar providing sensitivity or driver feel, it is desirable to increase the required driver effort in proportion to the power assistance demanded; this gives a degree of natural feel to the system.

How is this normally achieved?

....

....

....

....

....

....

....

PAS systems are described as 'speed sensitive'. What does this mean and why is it employed?

....

....

....

....

....

Control of power assistance

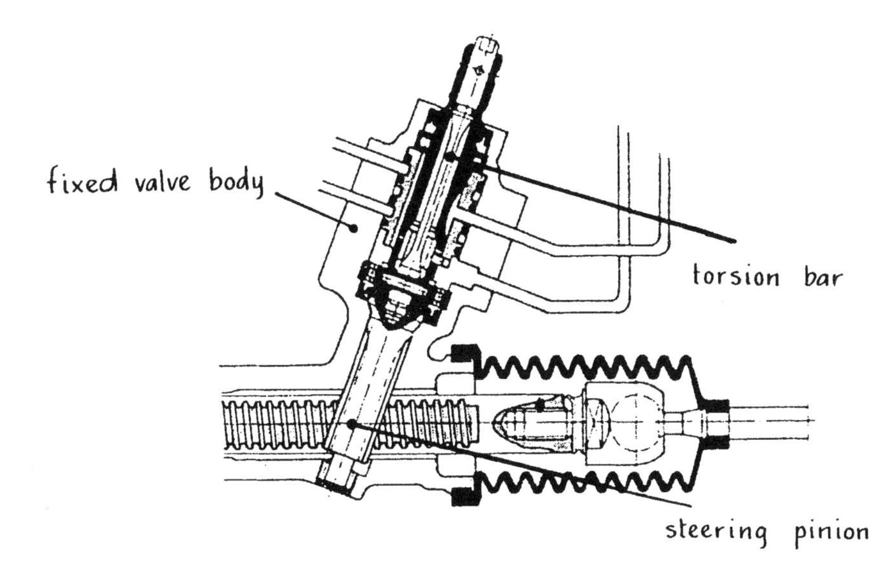

On initial rotation of the steering column, the TORSION BAR twists to allow relative movement between the inner and outer fluted sleeves. This relative movement of the sleeves changes the alignment of the ports formed by the flutes. The fluted sleeves do therefore serve as a directional control valve for the fluid. Stops limit the relative angular movement of the sleeves and the whole assembly rotates as the steering is operated.

Describe how the steering would operate in the event of a failure in power assistance:

....

....

....

....

....

....

Valve operation

Add arrows to the drawings at (b) and (c) to indicate fluid flow through the valve assembly.

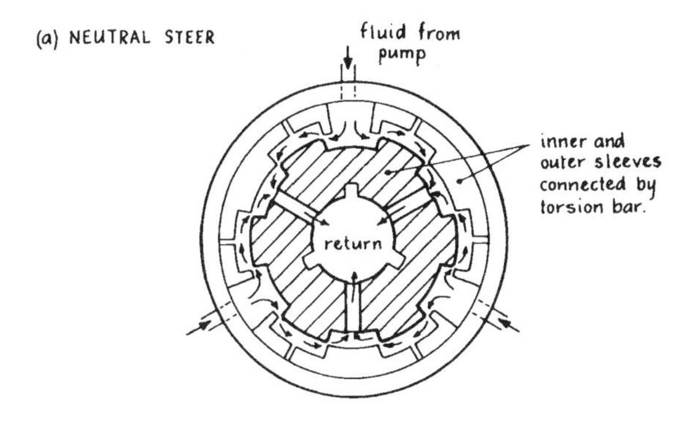

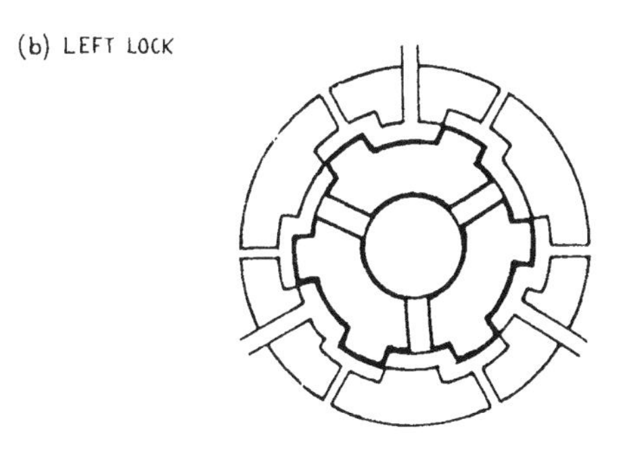

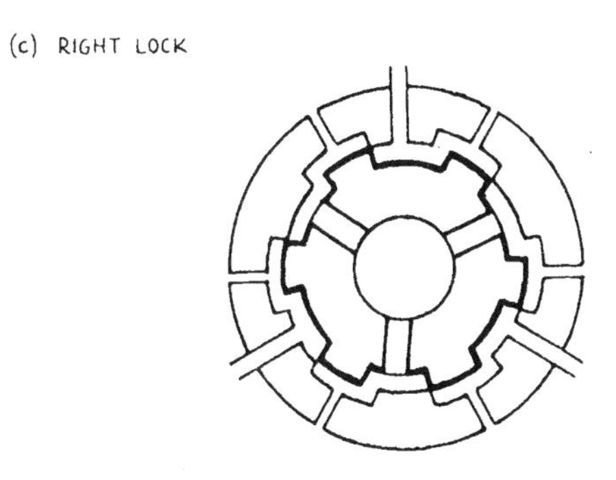

Hydraulic pump and flow relief valves

The type of hydraulic pump shown below is used in many PAS systems. Name the pump, indicate on the drawing the fluid inlet and outlet points, and describe its operation. Label the drawing.

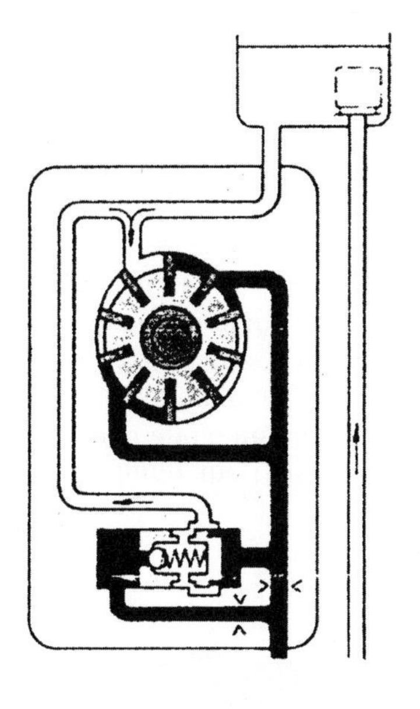

....

....

....

....

....

....

State the purpose and describe the operation of:

(a) the pressure relief valve and

(b) the flow control valve.

Pressure relief valve

Purpose:

....

....

....

....

....

Operation:

....

....

....

....

....

Flow control valve

Purpose:

....

....

....

....

....

Operation:

....

....

....

....

....

Electro-hydraulic PAS

Some PAS systems utilise an electric motor to drive the hydraulic pump rather than a 'V' belt from the crankshaft pulley. One benefit of this arrangement is that it provides a relatively simple and compact installation.

State two other advantages of an electric pump in a PAS system.

....

....

....

....

SPEED SENSITIVITY can be controlled electronically. In the system shown below a microprocessor evaluates speed signals from the electronic speedometer and computes the degree of hydraulic power assistance required. The power assistance varies in proportion to road speed. This provides minimum driver effort at the steering wheel when parking and at high speeds the steering action is almost as direct as a mechanical steering gear, providing precise and responsive steering.

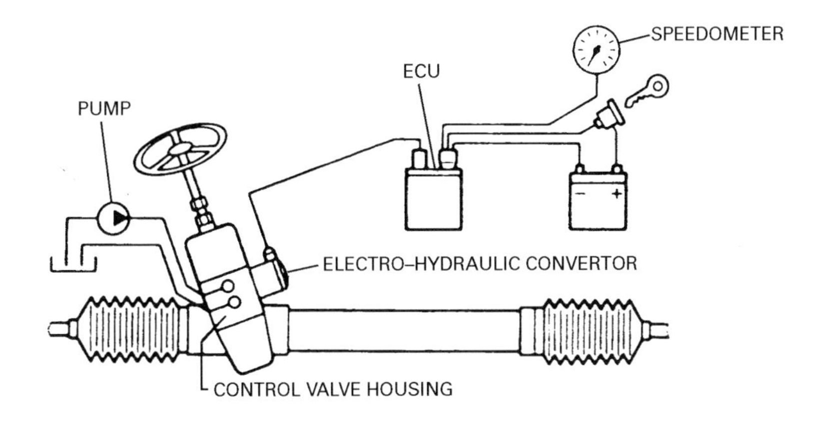

The level of hydraulic reaction transmitted to the hydraulic control valve is varied by the electro-hydraulic convertor, this directly affects the actuating force required at the steering wheel.

Electrical PAS

In electrical power assisted steering an electric motor is used to assist in the actuation of the steering system, rather than a hydraulic power cylinder. One electrical PAS arrangement is shown below.

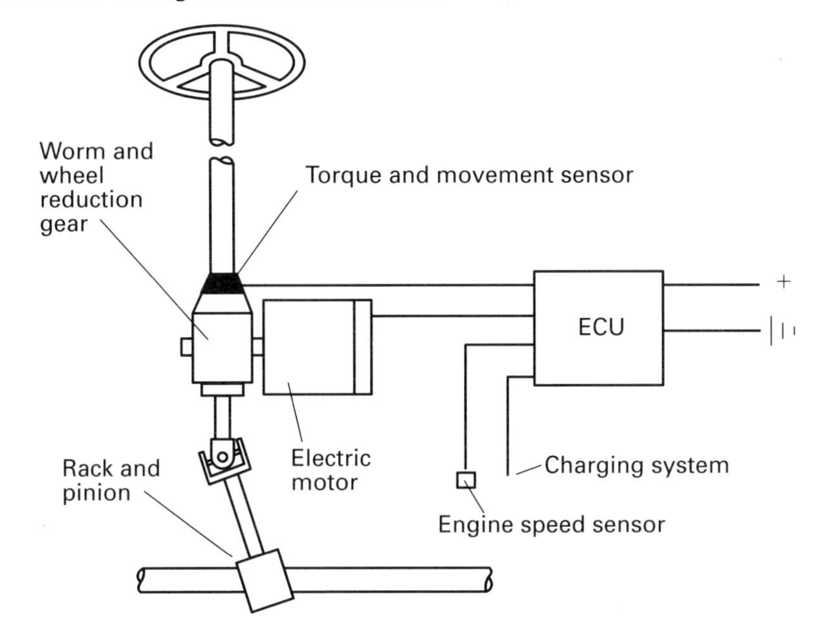

Study the drawing and describe briefly how the system operates.

....

....

....

....

....

....

....

....

....

STEERING SYSTEM MAINTENANCE

State the benefits of routine preventive maintenance:

How is the steering system protected against the ingress of moisture and dirt during repair and use?

List the general rules for efficiency and any special precautions to be observed when carrying out maintenance and repair:

List typical preventive maintenance tasks and Department of Transport (MOT) checks associated with the steering system:

PAS check:

Note: during maintenance, 'check' also means adjust if necessary.

PAS MAINTENANCE

Name a type of oil suitable for use in a PAS system:

....

....

Name two types of hydraulic pump used in PAS systems:

....

State the effects of air in a hydraulic PAS system and describe the procedure for bleeding air out of the system:

....

....

....

....

....

....

....

State the effects of a sticking pressure relief valve in a PAS pump assembly:

....

....

....

....

....

....

....

....

....

The illustration below shows a test being carried out on a PAS system. Label the items indicated and describe:

(1) a pressure balance test and (2) a pump pressure test.

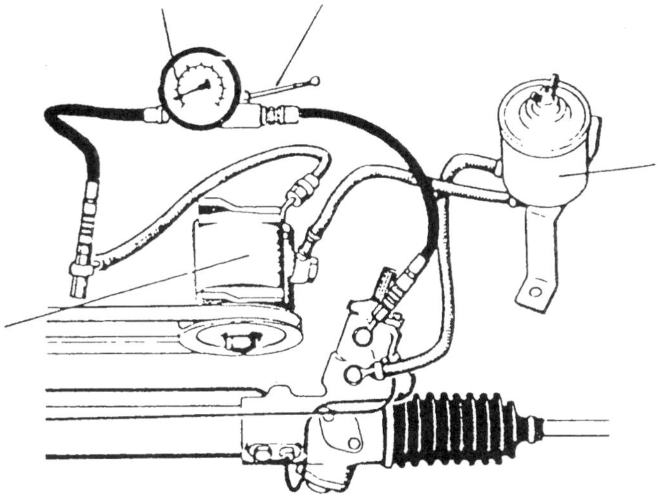

....

....

....

....

....

....

....

....

....

....

How much free play is allowable at the steering wheel rim?

Describe briefly how to check steering linkage ball joints for wear:

Show on the sketch below how king pins (or ball joints), king pin thrust bearings and wheel bearings are checked for wear:

State the possible causes of axial movement in the steering column:

Give examples of the use of the following items of equipment in steering system maintenance.

TURNTABLES:

BELT TENSION GAUGE:

CRACK DETECTOR:

DIAL TEST INDICATOR:

Name and state the purpose of the tool shown below:

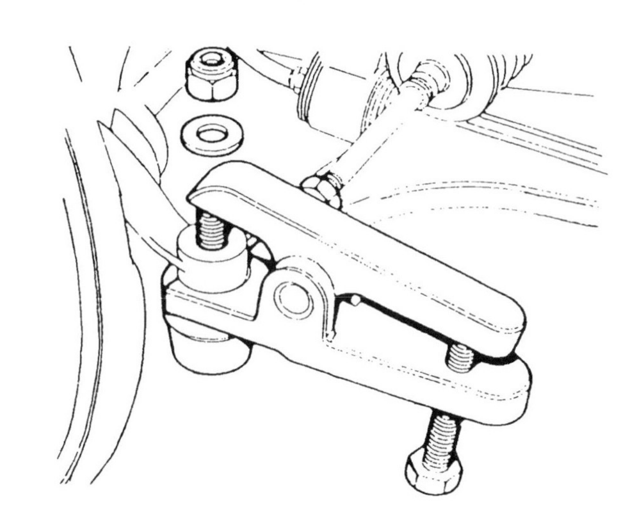

DIAGNOSTICS: STEERING SYSTEM – SYMPTOMS, FAULTS AND CAUSES

State a likely cause for each symptom/system fault listed below. Each cause will suggest any corrective action required.

SYMPTOMS	FAULTS	PROBABLE CAUSES
Heavy steering; poor self-centring	Partially seized king pins	....
Excessive free play (backlash); misalignment of wheels; abnormal tyre wear	Worn track rod end	....
Pulling to one side; abnormal tyre wear	Loose tie bar	....
Misalignment of steering wheel	Incorrectly fitted drop arm	....
Loss of power assistance	PAS fluid loss	....
Noise as steering is operated	Faulty PAS pump	....
Intermittent power assistance and squeal	Slack PAS belt	....
Excessive play or backlash at steering wheel; wandering	Excessive rocker shaft end float	....
Heavy steering; wandering; knocking noise over bumps	Worn or collapsed king pin thrust bearing	....

Chapter 8

Tyres and Wheels

Tyre construction	145
Tubed and tubeless tyres	146
Tyre tread design	146
Load index	149
Tyre speed symbol markings	149
Tyre profiles	150
Legal requirements	151
Tyre combinations	152
Tyre faults	153
Materials used in tyre and wheel construction	154
Road wheels	155
Wheel attachment and location	157
Safety precautions	159
Wheel-balancing	160
Tyre pressure	162
Gas law calculations	162

TYRE CONSTRUCTION

The modern tyre has evolved over the past 100 years from the simple pneumatic cycle tyre to the sophisticated tread and cord structures of today.

State the purpose and functional requirements of tyres:

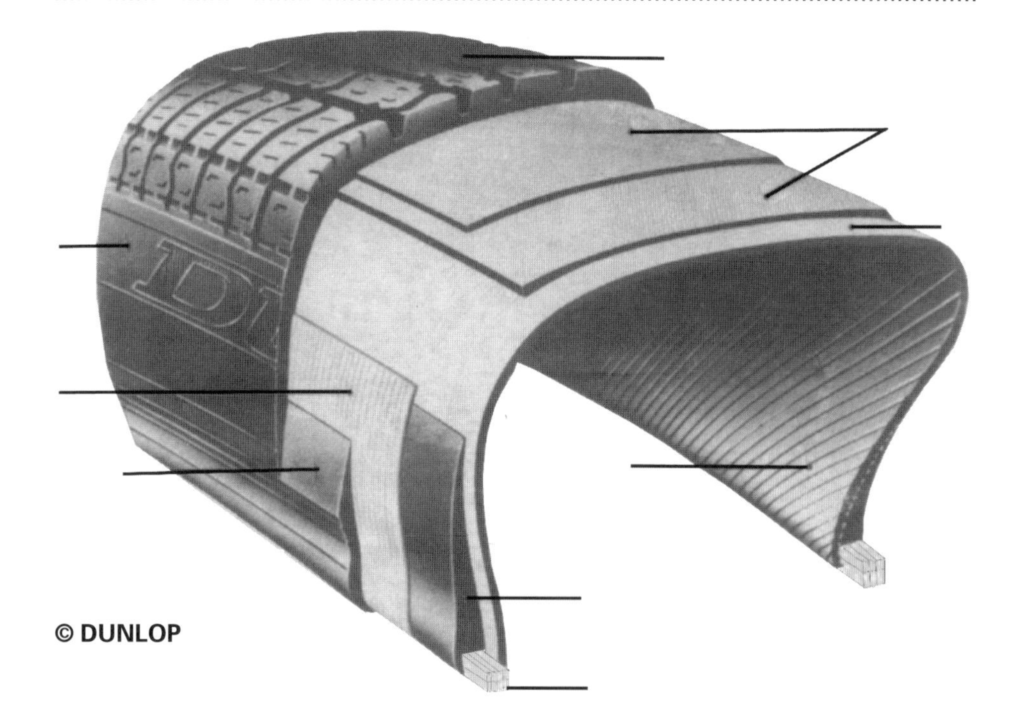

© DUNLOP

The structure of a modern car tyre is shown above; name the type of tyre construction and label the drawing.

State the purpose of the breaker strips:

Increased tyre life is one advantage of a radial-ply tyre when compared with a cross-ply. State the main reasons for this reduction in tyre-wear rate:

The relatively rigid casing of a cross-ply tyre creates a reduction in tread contact area when the tyre is subjected to a side force during cornering. This is illustrated below. Make a simple sketch to show the action of a radial-ply tyre when cornering:

Cross-ply | **Radial**

State why cross-ply tyres are not as suitable as radial-ply tyres for sustained high-speed operations:

TUBED AND TUBELESS TYRES

The drawing below illustrates tubed and tubeless design for truck applications.

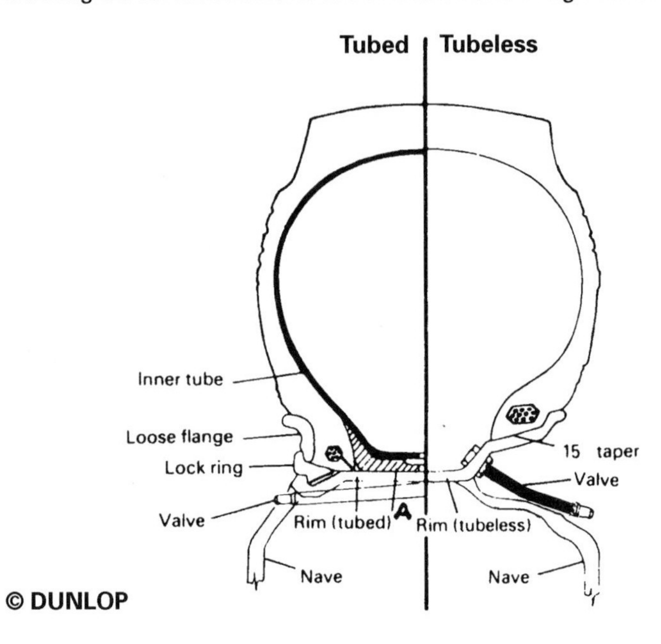

State the particular advantages of tubeless construction when employed on trucks:

State the purpose of part 'A' on the section through an HGV tyre and rim assembly shown above.

TYRE TREAD DESIGN

The design of a tyre, particularly the tread pattern, is dictated by its application i.e., the type of vehicle it is fitted to and how the vehicle will be used. The variation in application is obviously extreme e.g. family saloon, high powered sports car, F1 racing car, HGV, 4x4 off the road, etc. Features of tread design are:

Tread compound
This affects: grip, wear rate, flexibility, rolling resistance, etc.

Circumferential grooves
These affect steerability and water dispersal.

Cross ribs (lateral notches)
For traction and mud evacuation.

Blocks
Provide high mileage performance and good grip.

Sipes or blades
What are these and how do they contribute to the performance of a tyre?

A tyre tread pattern can be made up of a combination of the above features or it may have one as a dominant feature depending on the application.

State the operational suitability for the car and HGV tread designs shown:

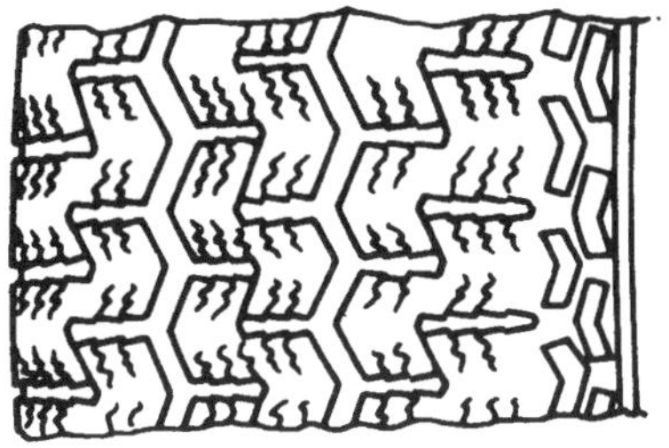

....

....

....

....

....

....

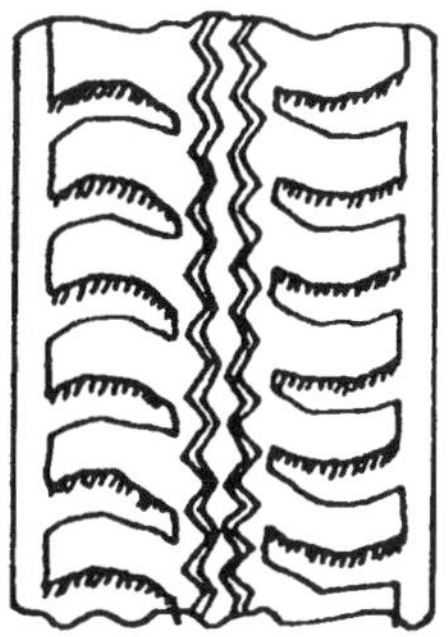

....

....

....

....

....

....

Name and state the purpose of the tread design shown at (a):

....

....

....

....

....

....

....

....

(a)

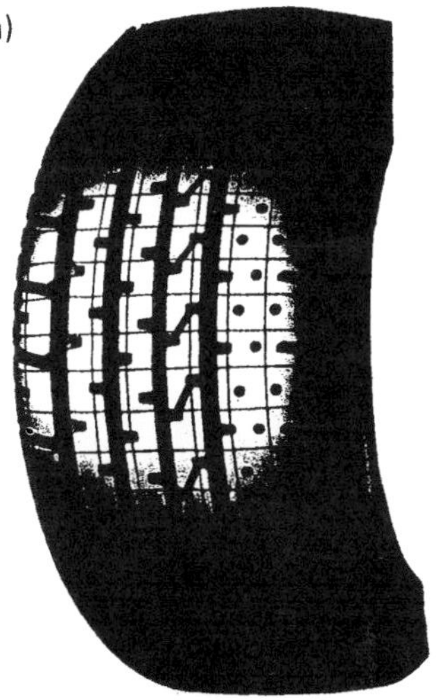

© YOKOHAMA HPT

Each tyre manufacturer uses a coding system relating to tread design and compounds.

Shown opposite is part of the system used by the Michelin Tyre Company. Letters indicate particular tyre applications and are included, along with other information, in the sidewall markings.

Label the arrowed part of the tread on the adjacent sketch and state its purpose:

....

....

....

....

....

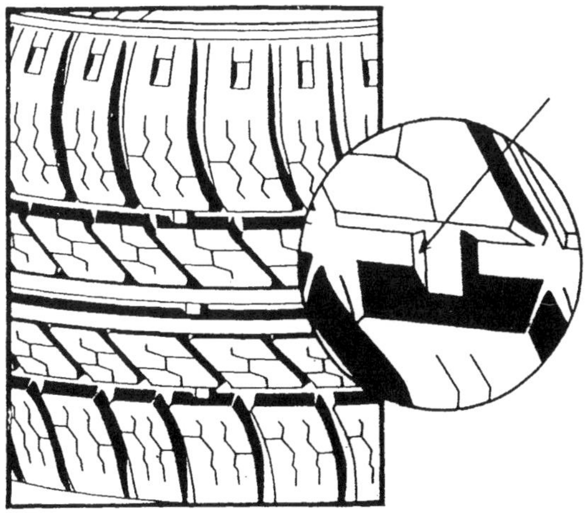

TYRE TREAD DESIGN

These are examples of tyre tread designs and their applications. (GOODYEAR TYRES)

Low noise levels and long life for normal road use.

Good traction in wet conditions; ideal winter tyre.

High performance sports tyre; offers resistance to aquaplaning.

4x4 off the road tyre; good traction in mud and snow; selfcleaning tread.

Truck steered axle tyre, long distance work.

Truck drive axle tyre, long distance work.

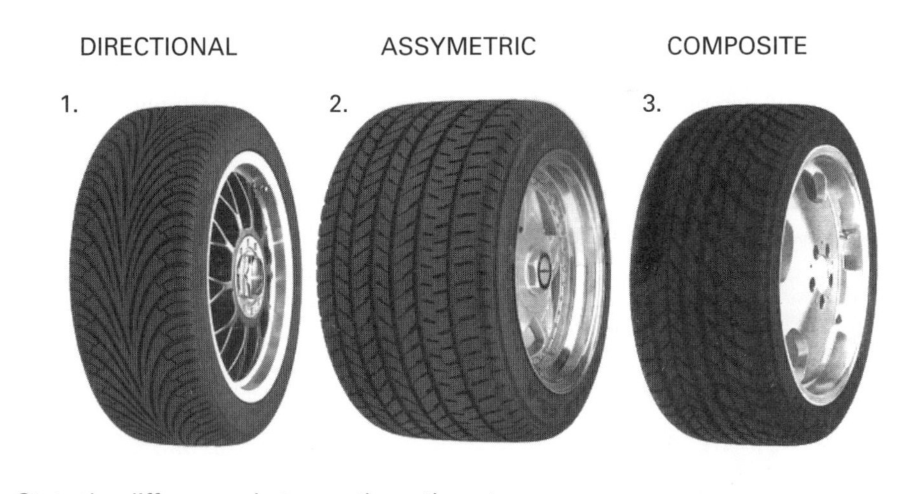

State the differences between these three types.

....

....

....

....

LOAD INDEX

A 'load index' table is shown below. State its purpose.

LT	kg	LT	kg	LT	kg	LT	kg	LT	kg	LT	kg
0	45	20	80	40	140	60	250	80	450	100	800
1	46.2	21	82.5	41	145	61	257	81	462	101	825
2	47.5	22	85	42	150	62	265	82	475	102	850
3	48.7	23	87.5	43	155	63	272	83	487	103	875
4	50	24	90	44	160	64	280	84	500	104	900
5	51.5	25	92.5	45	165	65	290	85	515	105	925
6	53	26	95	46	170	66	300	86	530	106	950
7	54.5	27	97.5	47	175	67	307	87	545	107	975
8	56	28	100	48	180	68	315	88	560	108	1000
9	58	29	103	49	185	69	325	89	580	109	1030
10	60	30	106	50	190	70	335	90	600	110	1060
11	61.5	31	109	51	195	71	345	91	615	111	1090
12	63	32	112	52	200	72	355	92	630	112	1120
13	65	33	115	53	206	73	365	93	650	113	1150
14	67	34	118	54	212	74	375	94	670	114	1180
15	69	35	121	55	218	75	387	95	690	115	1215
16	71	36	125	56	224	76	400	96	710	116	1250
17	73	37	128	57	230	77	412	97	730	117	1285
18	75	38	132	58	236	78	425	98	750	118	1320
19	77.5	39	136	59	243	79	437	99	775	119	1360

The table at top right shows 'speed ratings' for tyres. These are maximum safe speeds of operation.

State the relationship between LOAD INDEX and SPEED RATING:

TYRE SPEED SYMBOL MARKINGS

These are some of the letters which indicate the speed category for a vehicle.

SYMBOL	km/h	mph
Q	160	100
R	170	105
S	180	113
T	190	118
H	210	130
V	240	149
VR	210+	130+
Y	300	186
ZR	240+	150+

How do tyre pressures relate to vehicle LOADING and HANDLING?

How do the tyre pressures vary from, say, the pressure before the start of a long, fast journey to the pressure immediately after, and what is the reason for this variation?

If sustained high-speed driving is to be undertaken, what adjustment should be made to the tyre pressures?

TYRE PROFILES

Before about the early 1950s, tyres were designed to a basic circular cross-section – the tyre width being approximately equal to the radial height.

Development in racing clearly demonstrated that a tyre with a low profile gives many advantages.

This wider, lower type of tyre is said to have either a low height/width, low profile or low aspect ratio.

The height to width is quoted as a percentage, such as 70%; for a tyre with a 70% aspect ratio, therefore, the height dimension is 70% of the width.

The aspect ratio is normally indicated on the sidewall marking, if, however, it is not indicated, it usually means that the tyre is of the now normal aspect ratio which is about 82%.

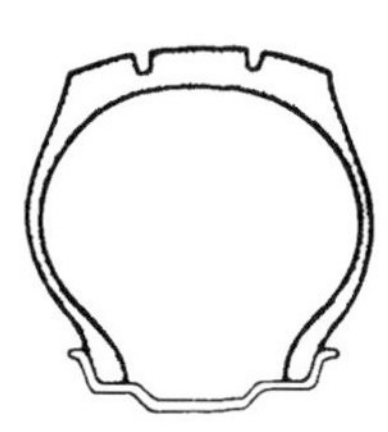

Tyre having 100% profile ratio

Draw a tyre having a 70% profile ratio

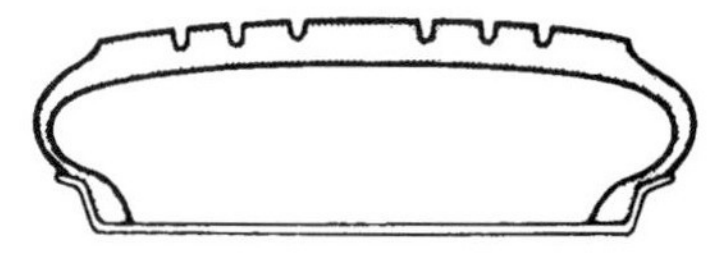

Racing-car tyre 35% profile ratio

State the advantages of low-profile tyres when compared with 100% profile tyres:

....

....

....

....

....

A tyre marking of 165/70 R 13 79T means:

165 = ...

70 = ...

R = ...

13 = ...

79 = ...

T = ...

Examine two cars and two HGVs and complete the table below:

Vehicle make/model	Tyre marking	Specified pressure	Tubed/ tubeless	Radial/ cross-ply

On certain tyres, load-carrying capacity is shown as a PR number, such as car: PR 4. Truck: PR 14.

PR means: ....

The manufacturer's specification for original equipment is the best guide when choosing tyres for a vehicle. Deviations from this, however, may be when a vehicle is used for particular applications, such as rallying, cross-country operation, travelling in different climatic conditions abroad, etc.

LEGAL REQUIREMENTS

The tyres fitted to a vehicle must conform to the current legal requirements as laid down in the construction and use regulations.

Tyres selected for a vehicle must be the correct size and type in relation to:

1. *Type of wheels used*
2. ...
3. ...
4. ...

Information relating to: rim fitment, maximum loads and speeds, inflation pressures and minimum dual spacing, can be obtained from tyre manufacturers' tables.

State the legal main requirements with respect to wear, condition and usage:

...

RECUT TYRES

Tread recutting or regrooving is a process carried out using an electrically heated blade to restore a tread pattern as it approaches illegality, thereby increasing the tyre life.

There are two different recutting processes:

1. *The original tread depth is increased by cutting into the base rubber between the lowest point of the original grooves and the casing plies.*
2. ...

To what depth is recutting normally carried out?

...

On which vehicles can re-cut tyres be legally used?

...

TYRE COMBINATIONS

Various legal and illegal combinations of radial- and cross-ply tyres are shown on this page. Write legal or illegal, as appropriate, under each combinaton:

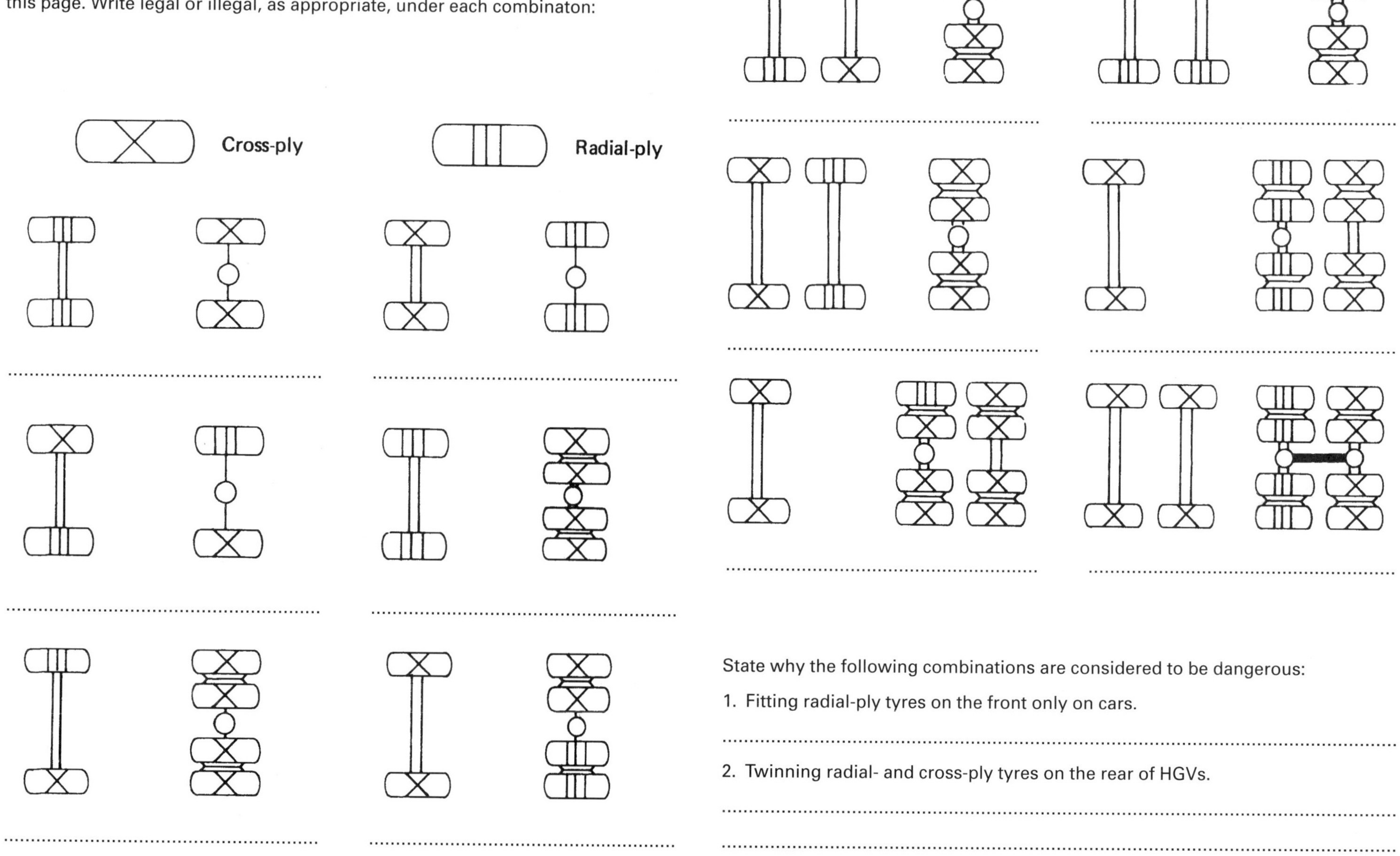

State why the following combinations are considered to be dangerous:

1. Fitting radial-ply tyres on the front only on cars.

....

2. Twinning radial- and cross-ply tyres on the rear of HGVs.

....

....

TYRE FAULTS

Some common tyre faults are illustrated on this page; identify the faults in each example and state the possible causes:

Fault ....

....

Cause ....

....

....

....

....

....

Fault ....

....

Cause ....

....

....

....

....

....

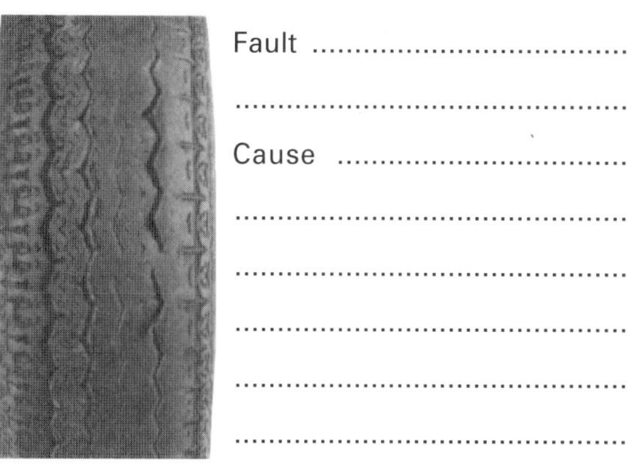

Fault ....

....

Cause ....

....

....

....

....

....

Fault ....

....

Cause ....

....

....

....

....

....

Fault ....

....

Cause ....

....

....

....

....

....

Fault ....

....

Cause ....

....

....

....

....

....

© DUNLOP

MATERIALS USED IN TYRE AND WHEEL CONSTRUCTION

Carcass (casing)

The carcass is the framework of the tyre. It must be strong enough to hold high pressure air, yet flexible enough to absorb load changes and impact. It consists of PLIES (layers) of tyre CORDS (sheets of parallel strands of strong material) bonded together with rubber.

CORD materials include: steel and polyester.

Name two other cord materials.

....

What material is used to reinforce the tyre bead?

....

The ply material has its main threads all running in one direction. This means that those threads will very easily separate if the material is pulled sideways. How is this overcome (a) on cross-ply tyres and (b) on radial-ply tyres?

(a) Cross-ply

....

....

....

....

(b) Radial-ply

....

....

....

....

....

The compound used for the tread is a mix of natural and synthetic rubber in which additives are included to improve its properties of grip, wear and response. The most essential single additive is carbon black which helps to give the tread resistance to wear.

Name any other important tread rubber additives and state their function:

....

....

....

....

What features must tyre treads embody in order to reduce the following?

Squeal...

....

....

....

Pattern noise ....

....

....

....

The material used in the construction of most car and HGV wheels is ....

................., which is strong enough to resist the torsional forces and bending loads imposed on the wheel. Name one other material used in the construction of vehicle wheels and give reasons for its use:

....

....

....

....

....

ROAD WHEELS

State the main purposes and functional requirements of road wheels:

....

....

....

....

....

....

....

....

....

....

....

....

....

....

....

Wheel Rim Design

There are many different configurations of wheel rim design for cars, vans, HGVs and buses. However, three design features are particularly important, these are the parts of the rim which:

1. allow fitting and removal of the tyre;
2. tightens the tyre bead on to the rim during inflation to provide an air tight seal and to secure the tyre on to the rim;
3. prevents the tyre bead becoming dislodged in the event of a sudden deflation. Indicate on the car rim profile shown at the top opposite these three rim features, i.e. 1, 2 and 3.

Car wheel rim

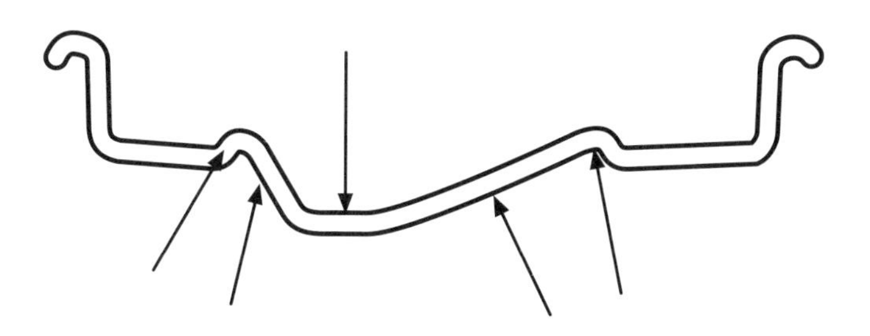

Most cars, light and heavy goods vehicles use tubeless tyres on 'one piece' wheel rims. Below is a selection of car wheel rim profiles.

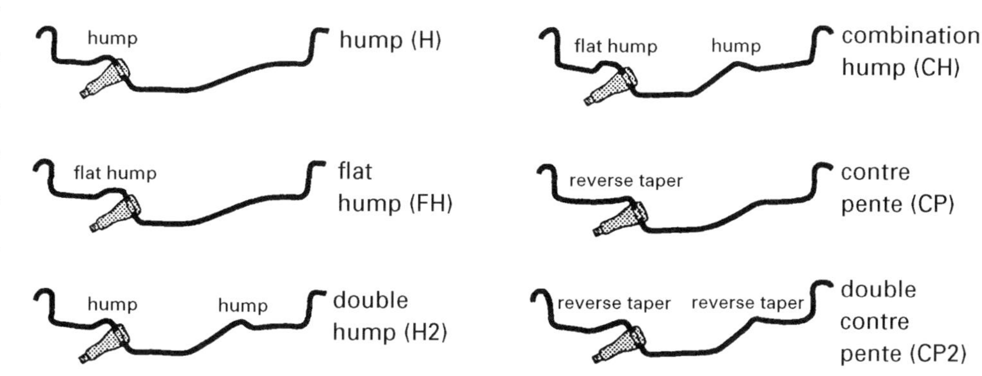

Additional 'runflat' stability is offered by the special rim shown below. The 'toe' of the tyre bead locks into the groove. TD Rim (Dunlop/Michelin).

ROAD WHEELS

Wheel size is indicated by measurements of rim width and diameter shown here. A wheel rim size may be $4\frac{1}{2}$ J x 13. What does the letter J refer to? Show this on the drawing.

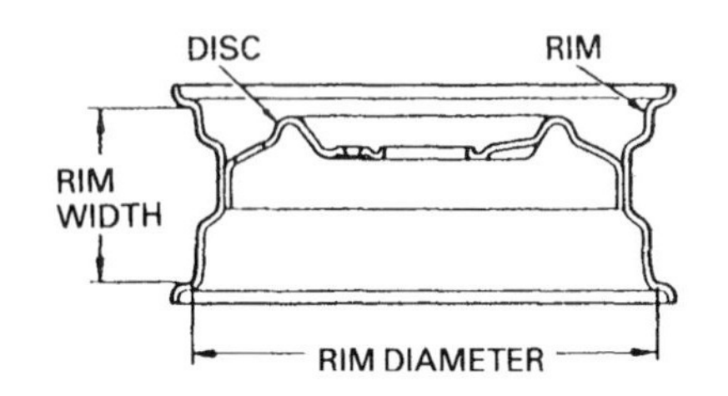

Apart from being heavier and stronger than car wheels, commercial vehicle wheels are designed to allow the fitting and removal of tyres with more rigid sidewalls and beads; many wheels are two-, three- or four-piece structures.

Name the type of rims shown below:

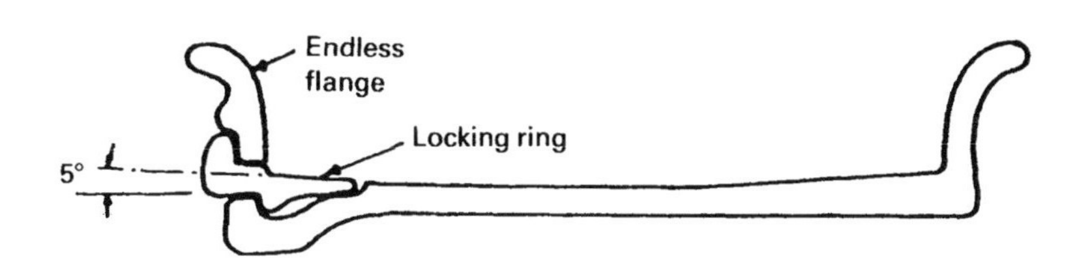

The DROP-CENTRE one-piece rim is used on commercial vehicles with tubeless tyres. Sketch this type of rim profile.

Divided-type rim

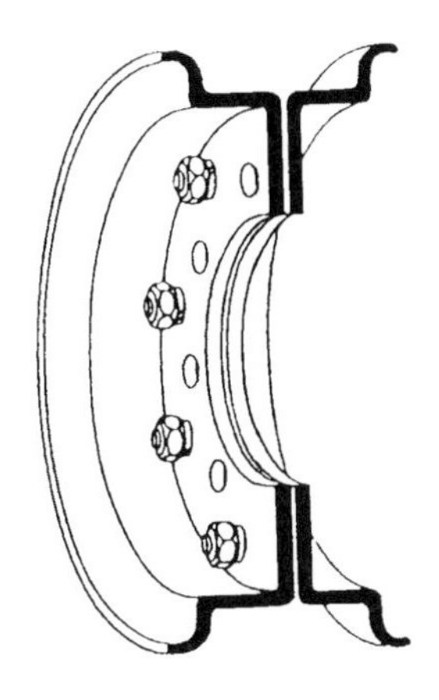

Divided or 'split' rims are used on many military vehicles and certain commercial vehicles where large, single, heavy-duty cross-country type tyres are fitted. Tyre fitting and removal are carried out by dividing the two halves of the wheel which are bolted together.

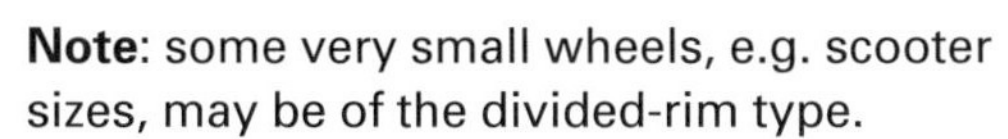

Note: some very small wheels, e.g. scooter sizes, may be of the divided-rim type.

Safety (divided rims): list safety precautions associated with the use of divided rims.

WHEEL ATTACHMENT AND LOCATION

The road wheel assembly is usually secured to the hub flange by studs and wheel nuts or by set bolts. Radial location of the wheel can be achieved by providing a conical seating on the wheel and using taper faced nuts, bolts or taper washers (see *Vehicle Mechanical and Electronic Systems (Levels 2 and 3)* book).

Describe the type of road wheel location shown below and name the important parts:

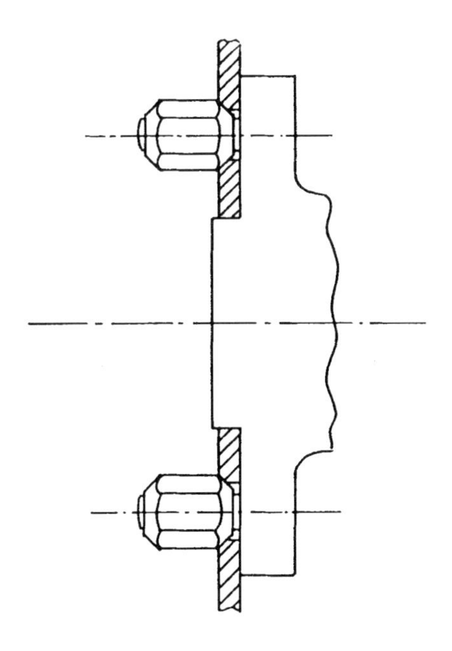

Give an example and reason for the use of left-hand screw threads for wheel attachment:

State the purpose of and describe, briefly, SPACESAVER wheel assemblies:

Name two vehicles using spacesaver spare wheels:

Make .... Model ....

Repair and maintenance

Routine tyre maintenance helps to ensure safe, efficient operation and maintains performance and tyre life. List the routine checks and adjustments for tyres and wheels:

Tyre fitting tools and equipment range from simple tyre levers to expensive purpose-built machines. However, the basic technique is the same irrespective of the equipment being used. Complete the sketch below to show how the type of wheel shown facilitates tyre removal:

Over which rim flange (a or b) shown below would the tyre be removed and refitted and why?

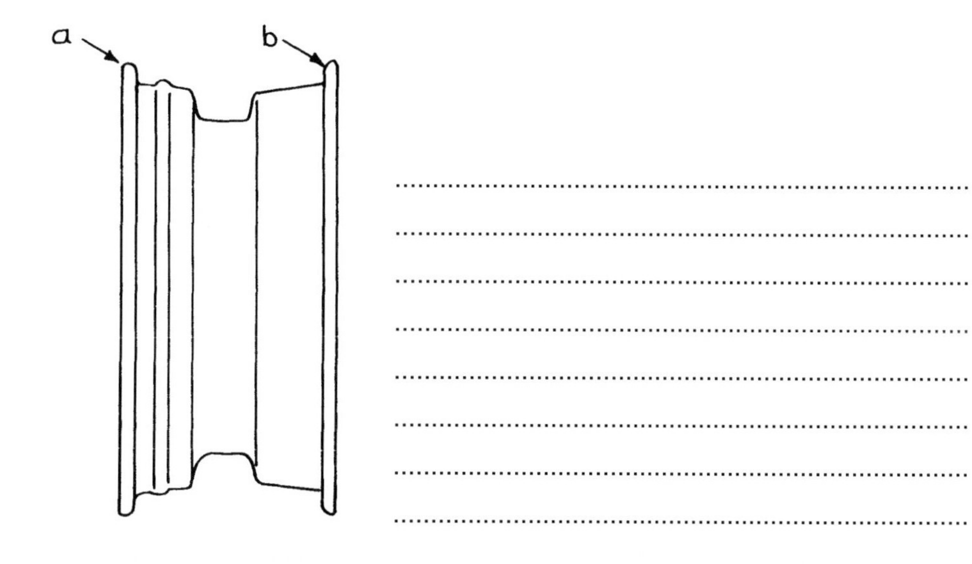

For what purpose is a bead-breaker used?

Other than safety precautions, list some important points to observe when tyre fitting:

Describe permissible tyre and inner tube repairs:

Describe with the aid of a sketch, how tyre 'tread depth' is measured:

....

....

....

State the purpose of 'tyre rotation' and show on the layout below, a typical rotation pattern:

....

....

....

....

Note: 'rotating' wheels can mask uneven wear caused by such problems as misalignment and worn steering or suspension components.

SAFETY PRECAUTIONS

There are many hazards associated with tyre servicing; it is essential therefore to be aware of these and to observe certain personal safety precautions when removing a wheel from the vehicle, and removing, refitting and inflating a tyre.

List the main safety points for servicing, removing and fitting tyres and wheels:

....

....

....

....

....

....

....

....

....

....

....

....

....

....

....

....

....

....

....

....

....

WHEEL-BALANCING

Unbalanced wheel and tyre assemblies adversely affect vehicle handling, stability and tyre wear.

These are two types of road-wheel imbalance:

.... and ....

A tyre and wheel assembly is in static balance when the mass of the assembly is uniformly distributed about its centre so that when mounted on a free bearing the assembly will come to rest in any position. If it always comes to rest in the same position, thus indicating a heavy spot, it is out of balance statically.
Describe the procedure for statically balancing a wheel:

© GOODYNA

....

....

....

....

....

....

Dynamic Balance

If the wheel shown statically balanced below is rotated at speed, certain out-of-balance forces will be produced, that is, the wheel is dynamically unbalanced.

Explain the reason for this and state the effect of the forces. Add arrows to the drawing to show the direction of the forces.

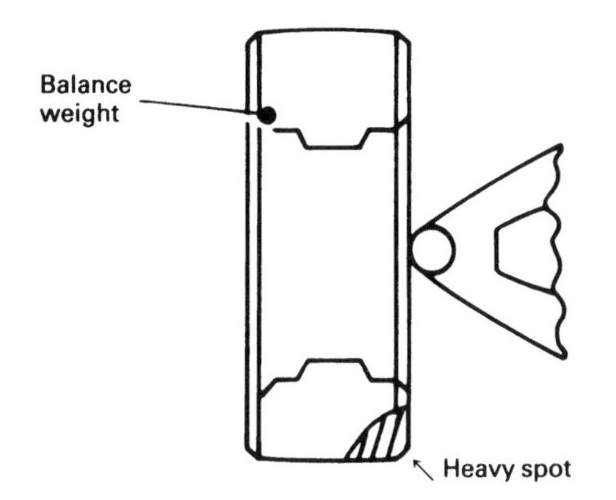

....

....

....

....

....

....

....

....

....

....

....

....

To dynamically balance the wheel shown below, additional weights are added to counteract the forces rocking the wheel. Show on the drawing where weights would be positioned to dynamically balance the wheel assembly:

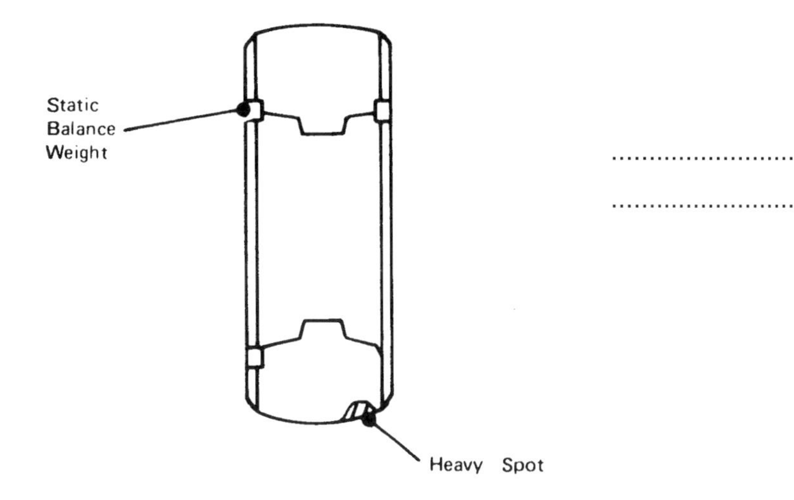

....

....

Two systems of wheel-balancing are used:

1. ....

....

2. ....

....

One advantage of balancing the wheel on the vehicle is that the hub and brake drum or disc are also taken into consideration.

Describe, by referring to equipment with which you are familiar, the procedure for dynamically balancing a wheel.

Wheel-balancing Procedure

....

....

....

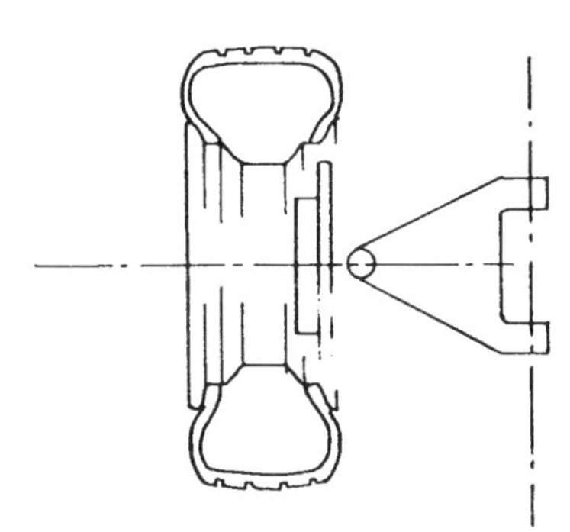

....

....

....

....

....

....

....

State the effects of road-wheel imbalance on vehicle handling:

....

....

....

....

....

TYRE PRESSURE

During driving, tyre pressure varies; this is due to:

....

If the vehicle travels at high speed over a long distance the temperature of the air in the tyre will ....

This will cause the tyre pressure to ....

State a possible increase in tyre pressure after a long journey ....

GAS LAW CALCULATIONS

It is possible to calculate the actual change in pressure (or temperature) of the air in tyres by using the 'gas equation' from Boyle's and Charles's laws, this is:

Assuming the volume of air in a tyre remains constant, $V_1 = V_2$; therefore the only factors to consider are pressure and temperature, thus:

$$\frac{P_1}{T_1} = \frac{P_2}{T_2}$$

When using the gas equation the pressure and temperature are 'absolute' values, that is

pressure = gauge pressure + ...

temperature K = °C + ...

EXAMPLES

1. At the start of a journey the gauge pressure in a tyre is 196 kN/m^2 at a temperature of 15°C. Calculate the increase in tyre pressure at the end of the journey if the temperature of the air in the tyre had increased to 35°C.

2. In cold conditions the tyres of a vehicle are at a temperature of 7°C and a gauge pressure of 2 bar. What will be their pressure reading if after standing in the hot sun their temperature rises to 28°C?

Tyre pressures may be set correctly, but can rise substantially during a journey. What causes this increase in pressure and what should be done about it?

....

....

....

Chapter 9

Braking

Brakes	164
Hydraulic brakes	164
Drum brakes	165
Wheel cylinders	166
Disc brakes	166
Parking brake	168
Automatic brake adjusters	168
Master cylinder	169
Pressure-limiting valves (brake force apportioning)	171
Brake vacuum servo units	172
Indicating devices	174
Brake testing	174
Power hydraulic braking systems	176
Anti-lock braking systems	177
Traction control (Ford/Bendix)	179
System check	179
Centre of gravity and load transfer	180
Stopping distance	181
Work done by the braking system	181
Vehicle speed and stopping distance	184
Investigation	185
Compressed air braking systems	186
System components	194
Trailer brakes	195
Compressed-air/hydraulic brakes	197
Jack-knifing	198
Anti-jack-knife system	198
Anti-lock braking	198
ABS (anti-lock braking system) HGV	199
Load sensing valves	200
Vehicle retarders (auxiliary brakes)	201
Braking system maintenance	203
Diagnostics	205

BRAKES

The purpose of a braking system on a vehicle is to:

1. *Stop the vehicle* ....

2. ....

3. ....

FRICTION is used to reduce the speed of a vehicle and bring it to rest. Friction material forced into rubbing contact with a rotating drum or disc, forming part of the wheel/hub assembly, generates HEAT. The energy of motion, that is, kinetic energy, is therefore converted into heat energy during the braking process. The heat energy is dissipated into the atmosphere. In addition to frictional contact at the braking surfaces, the ability to retard or stop a vehicle depends also on frictional contact at the

....

One of the functional requirements of a braking system is that the applying force, with which the brake surfaces are pressed together, should not require excess effort by the driver.

By what means is the driver's effort multiplied and transmitted to the braking surfaces?

....

....

....

....

HYDRAULIC BRAKES

System layout (hydraulic single line)
The 'single line' braking system is now only to be found on older cars as it has been superseded by safety systems such as 'split line'. Complete the drawing top right to show the piping layout for the system.

Hydraulic braking system (single line)

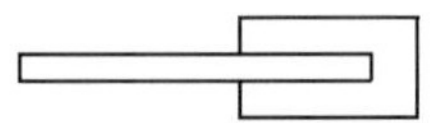

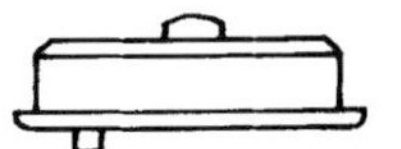

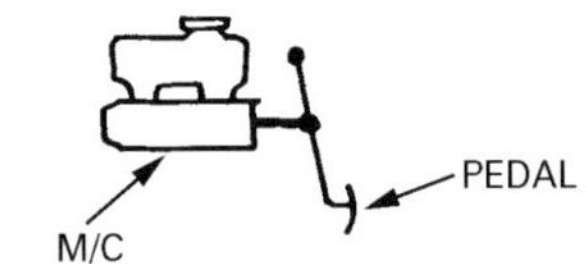

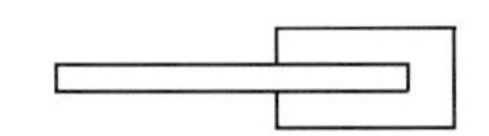

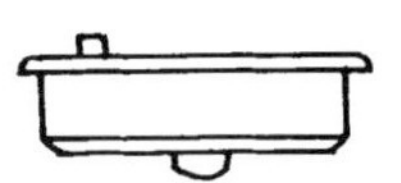

When the brake pedal is depressed, hydraulic fluid is displaced from the master cylinder into the system. The resulting increase of fluid in the wheel cylinders moves pistons which actuate the brake shoes and pads and force the friction material and calipers into contact with the rotating drums and discs.

The parking brake on most cars operates on two wheels only and is normally mechanically actuated by cables. Examine a vehicle and sketch the handbrake cable layout below (show any cable guides).

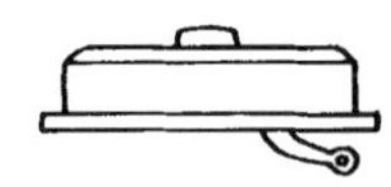

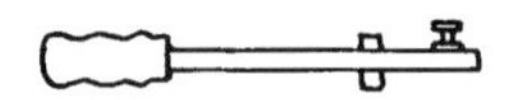

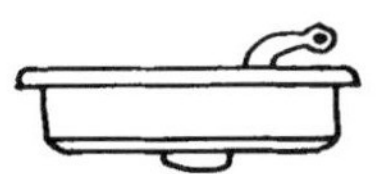

DRUM BRAKES

The 'Duo-Servo' brake shown below incorporates a double-piston wheel cylinder and a floating adjuster mechanism.

State the reason why this type of brake is referred to as 'Duo-Servo' and describe the action of the brake:

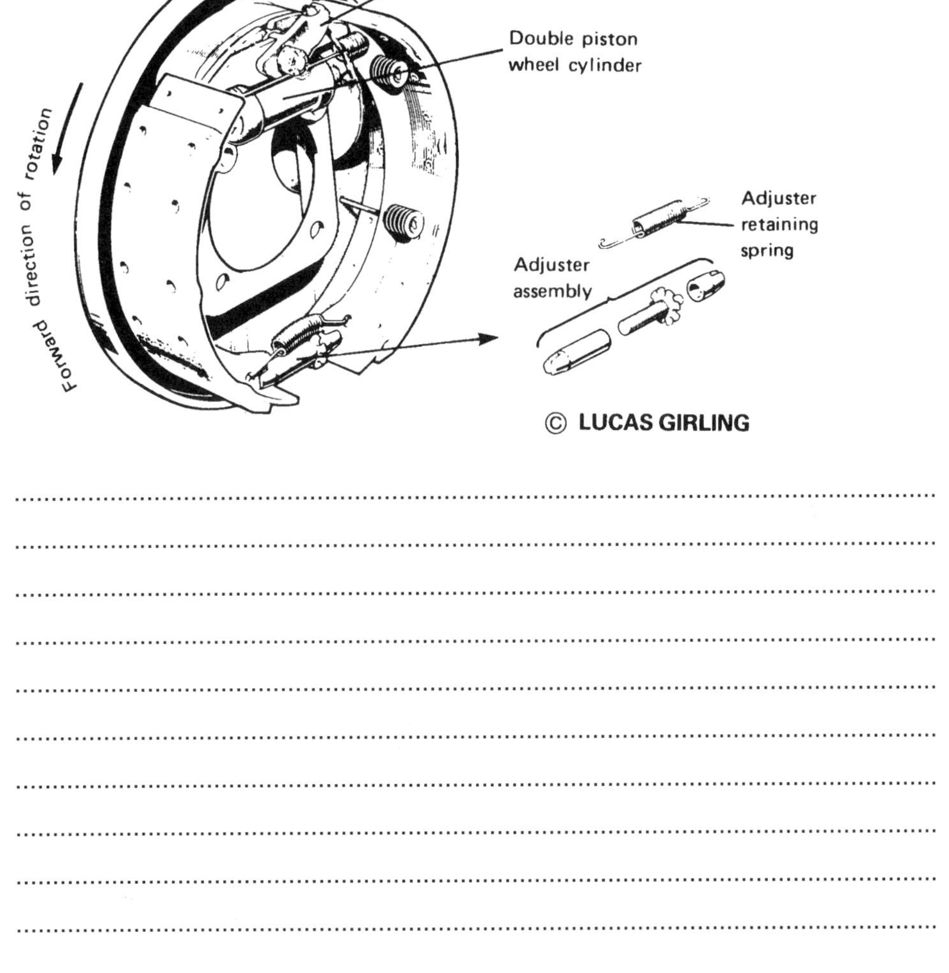

© LUCAS GIRLING

....

....

....

....

....

....

....

....

....

....

Name the brake lining/shoe fitments shown below:

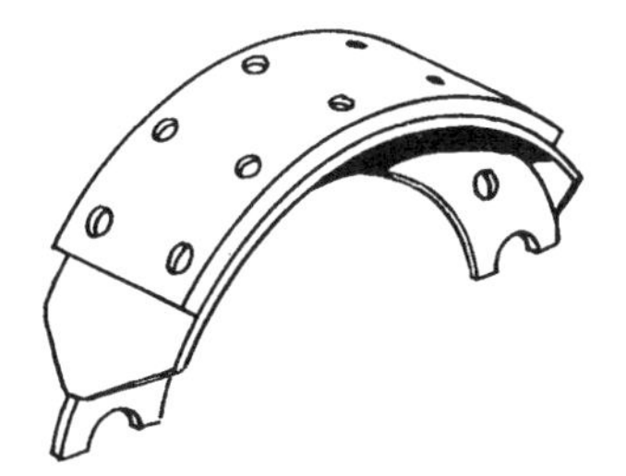

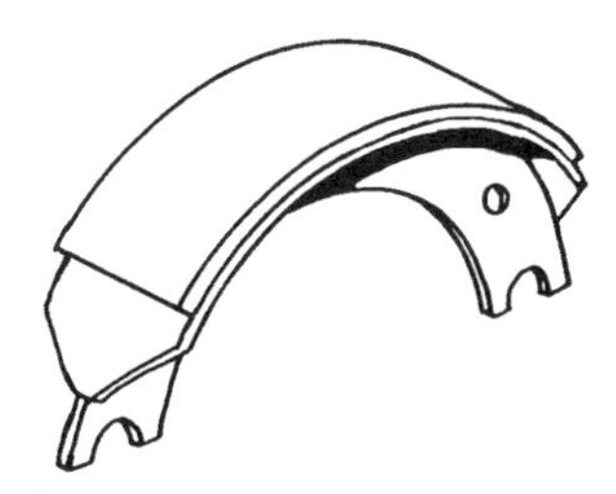

....

Friction materials

Materials used for friction linings must meet certain operational requirements, for example:

1. *Good resistance to wear.*
2. ....

   ....
3. ....

Resin-bonded asbestos has been the basic friction material used in braking systems; however, 'asbestos-free' materials are now superseding asbestos.

List some asbestos-free friction materials currently used in vehicle braking systems:

....

....

....

State a typical coefficient of friction for a brake friction material:

....

WHEEL CYLINDERS

The wheel cylinder consists basically of a plain cylinder and pistons which, energised by fluid pressure, actuate the brake shoes. A double-piston type is shown below. Label the drawing.

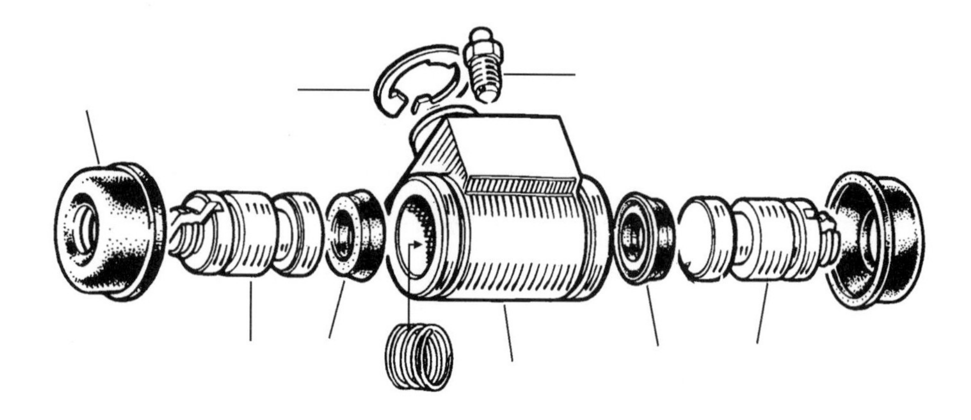

DISC BRAKES

The braking surfaces on a disc brake are more easily cooled than in a drum brake. They are self cleaning and less prone to BRAKE FADE.

In the 'FIXED CALIPER' disc brake shown here a common fluid pressure applied to the two pistons forces the friction pads against the disc. Label the drawing.

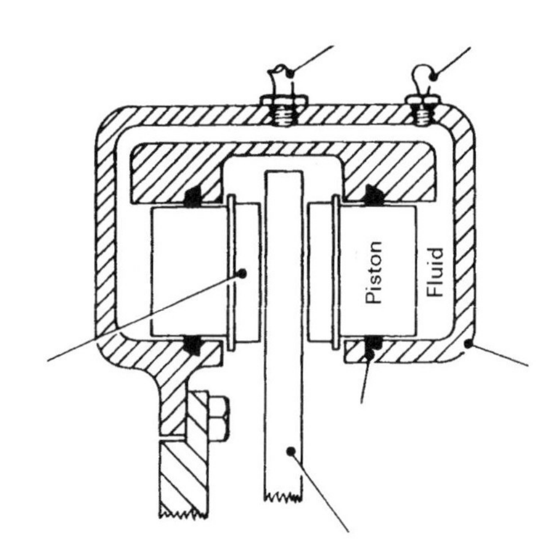

A two-piston fixed caliper assembly is shown below. What is the purpose of component 'A'?

WIPER SEAL RETAINER
PRESSURE SEAL
A
WIPER SEAL

© DELPHI LOCKHEED AUTOMOTIVE

State briefly what is meant by 'brake fade':

An alternative design of brake caliper is the floating or sliding caliper. With a single piston on one side only it needs less space on the wheel side, it operates cooler and can incorporate a handbrake mechanism.

Name the component parts of the floating-caliper disc brake assembly shown below:

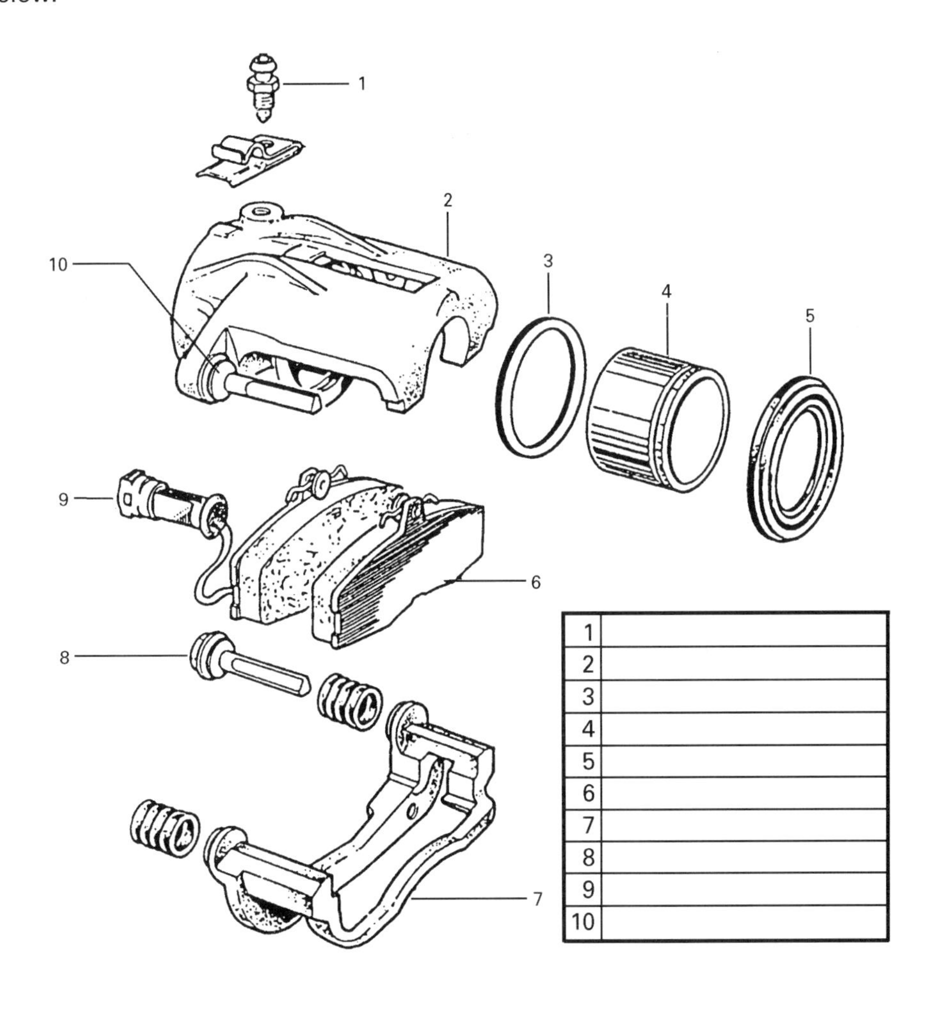

1	
2	
3	
4	
5	
6	
7	
8	
9	
10	

In the disc brake caliper assembly shown above, which is the 'sliding' component and which is the 'fixed' component?

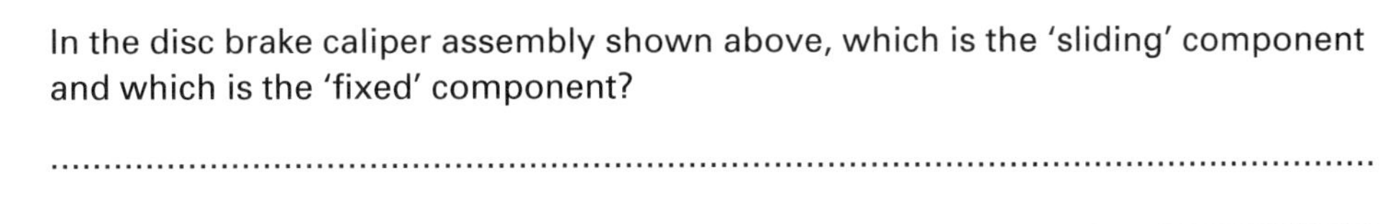

....

....

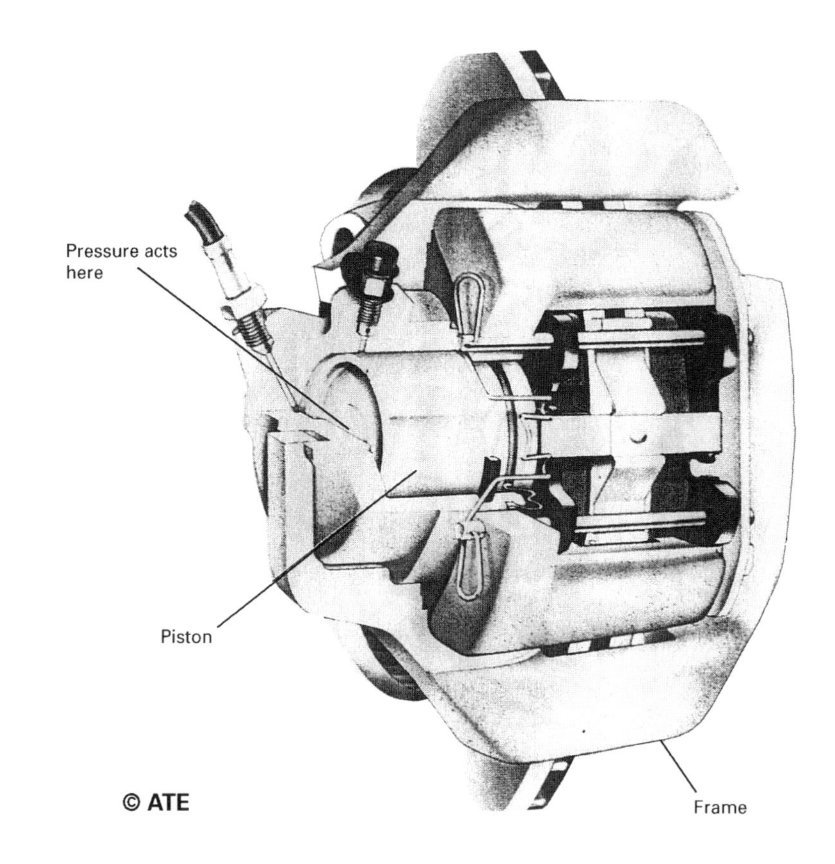

Describe the main differences in action of the disc brake caliper above to the one on the left. Label the 'fixed' component.

....

....

....

....

....

....

PARKING BRAKE

Name and state the purpose of components A, B and C:

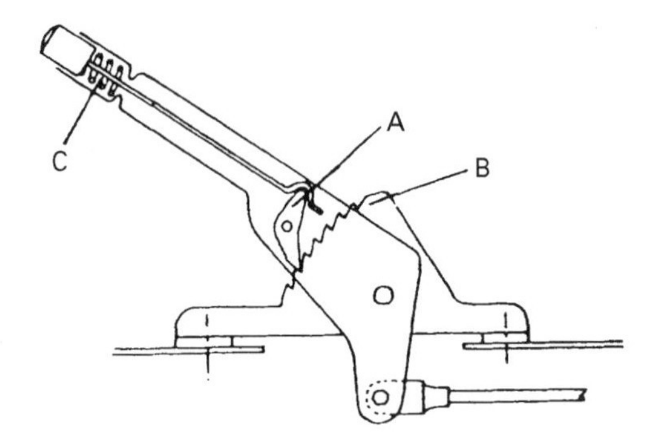

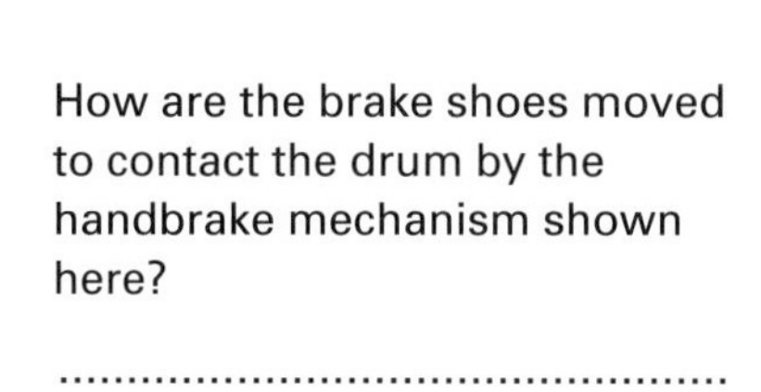

How are the brake shoes moved to contact the drum by the handbrake mechanism shown here?

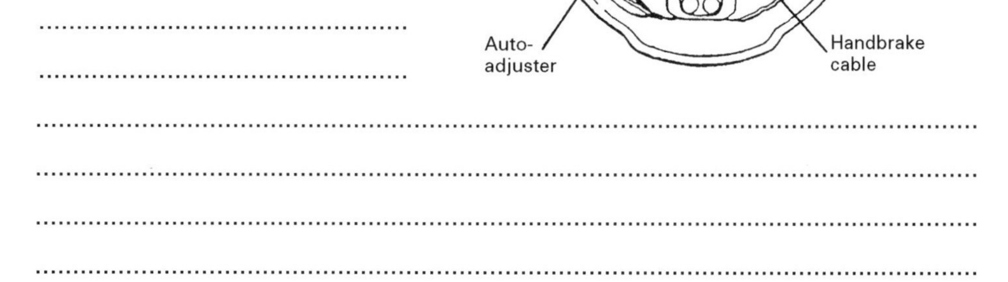

AUTOMATIC BRAKE ADJUSTERS

As the friction material (brake shoes and pads) wears down, some form of adjustment is required to maintain a minimum running clearance between the braking surfaces. Why is this necessary?

Describe the action of the automatic adjuster shown below:

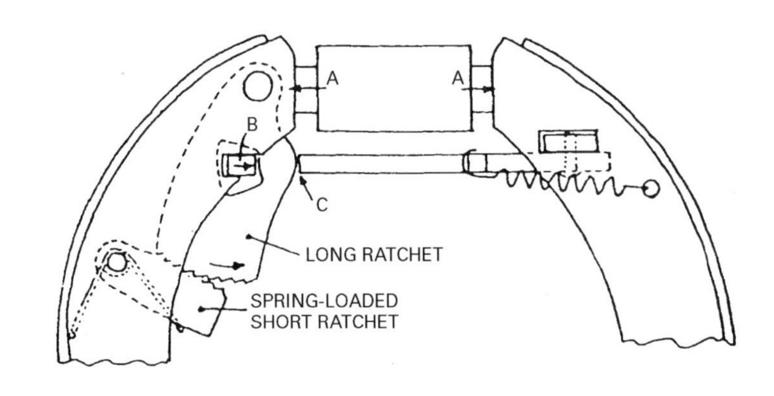

Disc brake adjuster

In most disc brake arrangements it is the rubber seal around the piston that maintains the correct clearance between pad and disc; it therefore serves two purposes, that is, it is a fluid seal and an automatic brake adjuster. Complete the simple sketch below and describe the action of the rubber seal in a disc brake caliper assembly, in particular how it acts as an automatic adjuster.

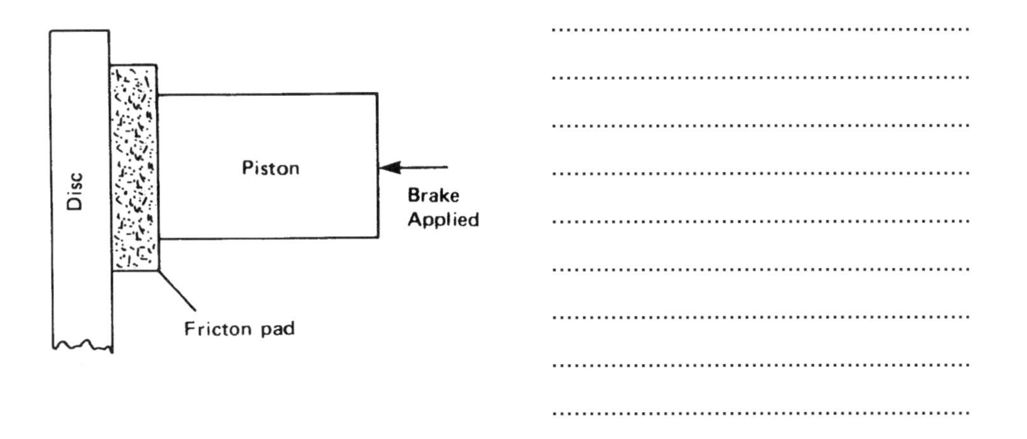

Brake fluid

This is in most cases a vegetable oil which, together with certain additives, gives the advantages of a low freezing-point and a high boiling-point, and maintains a constant viscosity across a wide temperature range. Brake fluid must also be compatible both with the metallic parts and with the rubber seals used in the system. State the precautions necessary with brake fluid:

....

....

Brake fluids must conform to SAEJ1703, DOT 3/4 specification relating to boiling point and performance at low temperature. Specially developed fluids for racing have higher boiling points.

MASTER CYLINDER

The master cylinder is a hydraulic cylinder with fluid reservoir which is operated by the brake pedal lever and push rod. Irrespective of the type of master cylinder employed there must be provision for:

(a) fluid compensation within the system; the reason for this is ....

....

....

....

(b) isolation of the fluid reservoir from the main cylinder pressure chamber; the reason for this is ....

....

....

....

Single Master Cylinder

Describe briefly the action of the valve (centre valve) during brake application and release.

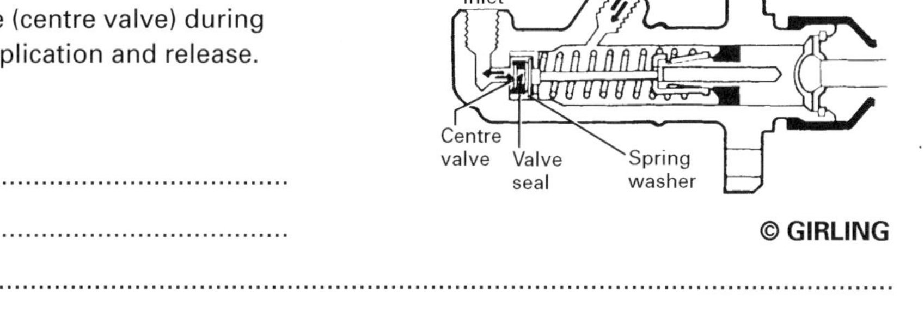

© GIRLING

....

....

....

....

....

....

Note: owing to the use of 'split braking systems single piston master cylinder line' are now obsolete on almost all new cars.

Tandem Master Cyclinder

A tandem master cylinder, as used in 'split line' braking systems, is shown below. Name the parts indicated.

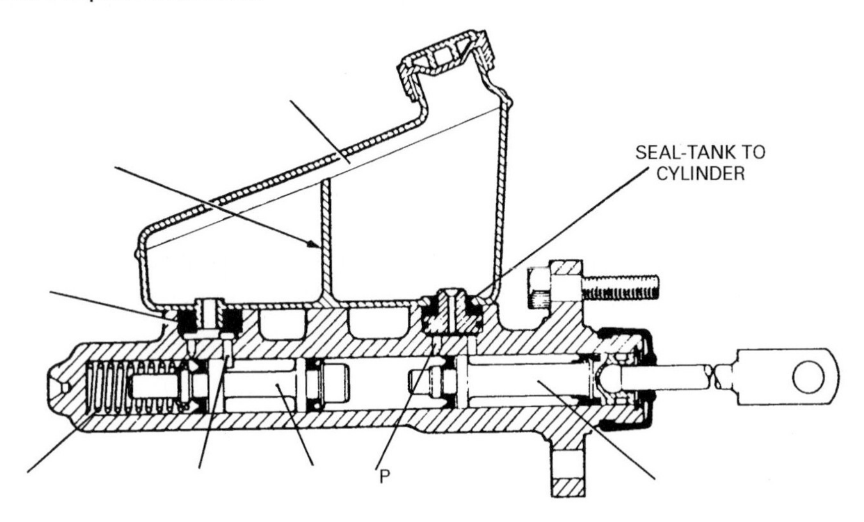

The tandem master cylinder is so constructed that the front and rear brake chambers within the cylinder each have their own fluid reservoir. When the footbrake is applied, pressure generated in the primary piston chamber acts on the secondary piston which, in turn, generates a pressure in the secondary chamber. In this way the brakes are actuated in each half of the braking system. Describe the effects of a brake failure (for example, fluid loss) in:

(a) front brake system and (b) rear brake system.

What role do the 'compensating' or 'by-pass ports' (P) play in the operation of the master cylinder?

Complete the drawings below to show SPLIT and DUAL line braking systems.

SPLIT LINE

DUAL LINE

PRESSURE-LIMITING VALVES (BRAKE FORCE APPORTIONING)

When the brakes are applied, 'weight transfer' from the rear wheels onto the front could cause the rear wheels to lock. This is more likely during rapid deceleration, particularly with front wheel drive vehicles. To overcome this problem, rear brake line pressure is reduced accordingly. This can be done in several ways.

1. Pressure limiting valves in the rear brake lines.
2. Pressure reducing valves; these reduce rear brake line pressure in proportion to front line pressure.
3. Load sensing valves: a link between a chassis mounted valve and rear axle monitors the load on the axle (axle to chassis height) and the valve apportions braking force according to load (See air brake section).

PRESSURE REDUCING VALVE

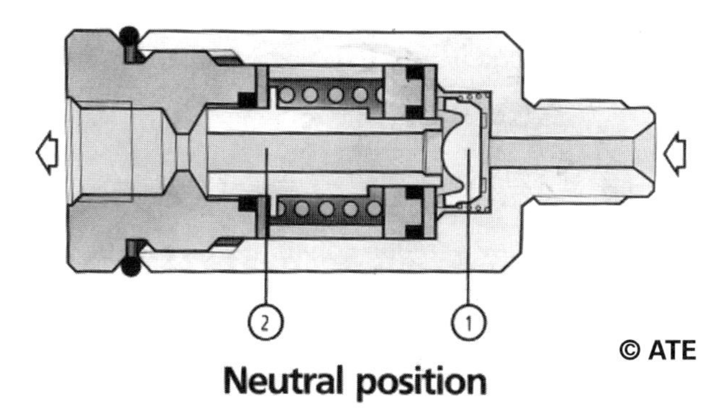

Neutral position

© ATE

The pressure reducing valve above is shown in the 'neutral' position. Fluid flows in the direction of the arrows from front to rear via open valve 1 and piston bore 2. Pressures front to rear are balanced. (Normal braking.)

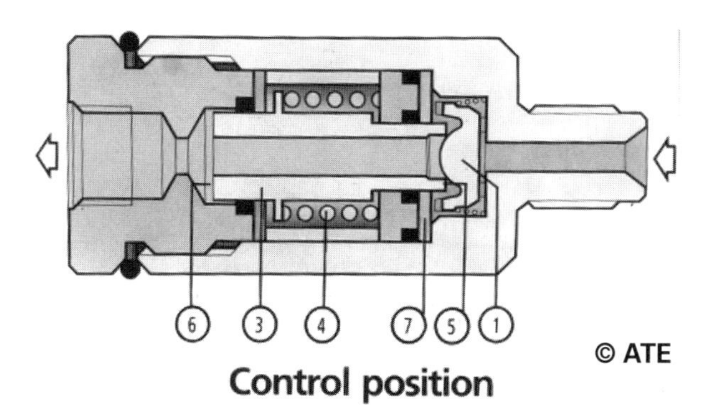

Control position

© ATE

In the control position above increasing brake line pressure acting at 6 on control piston 3 moves the piston against the spring to close the valve 1 and 'freeze' rear brake line pressure. How does the valve 'apportion' front and rear brake line pressure?

(Line pressures acting at 6 and 5 on the control piston are the key to understanding the action of this valve, i.e. FORCE = PRESSURE × AREA.)

....

....

....

....

....

....

....

....

Many ABS (anti-lock braking systems), incorporate electronic brake distribution. ABS wheel sensors and an electronic control unit measure small front and rear wheel speed differences and apportion rear wheel braking effort.

BRAKE VACUUM SERVO UNITS

The function of a brake servo unit is to augment the driver's pedal effort, thereby keeping the leverage required (and hence pedal travel) to a minimum.

General principle

The necessary pedal assistance is obtained by creating a 'pressure difference' across a large-diameter diaphragm or piston; the force of which, when applied to a hydraulic piston, increases the brake line pressure.

To obtain an air pressure difference a vacuum is created within the servo.

How is this achieved on a spark-ignition engine?

....

....

Describe the operation of the brake servo shown opposite:

....

....

....

....

....

....

....

....

....

....

Why is the servo known as a 'suspended vacuum' type?

....

....

....

Direct-acting brake servo

The drawing below shows a direct-acting suspended vacuum servo working in conjunction with a tandem master cylinder.

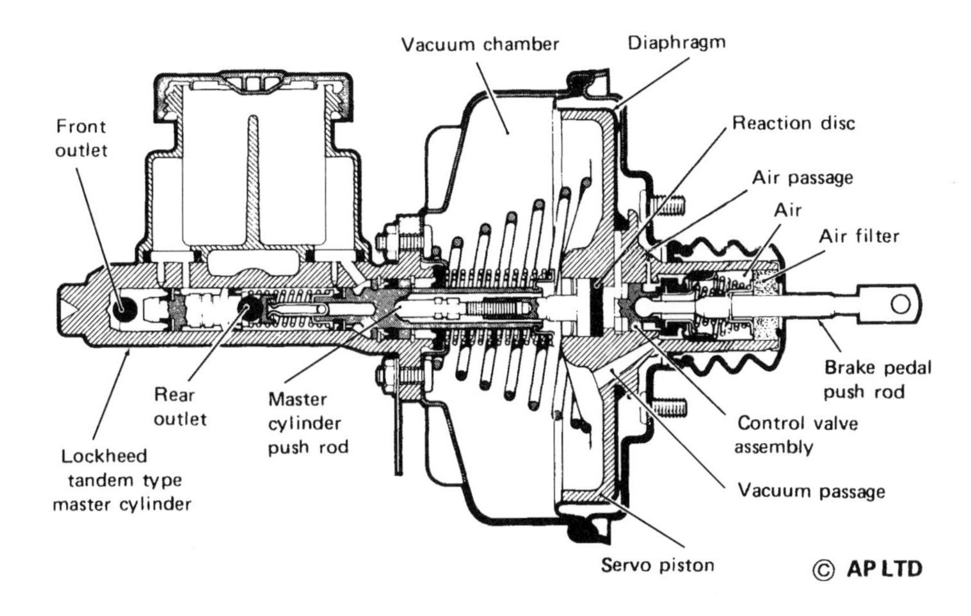

Complete the labelling on the enlarged control valve section shown below.

Brakes fully applied

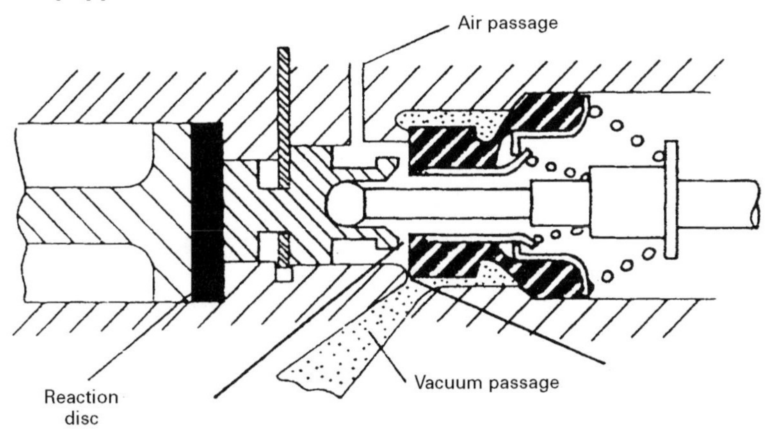

Many commercial vehicles employ larger-type suspended vacuum servo units which are activated from the brake pedal via a hydraulic master cylinder or by mechanical means; also included in certain systems are an engine-driven 'exhauster' and a 'vacuum reservoir'.

Complete the drawing below of a vacuum-assisted hydraulic braking system by adding the pipes and labelling:

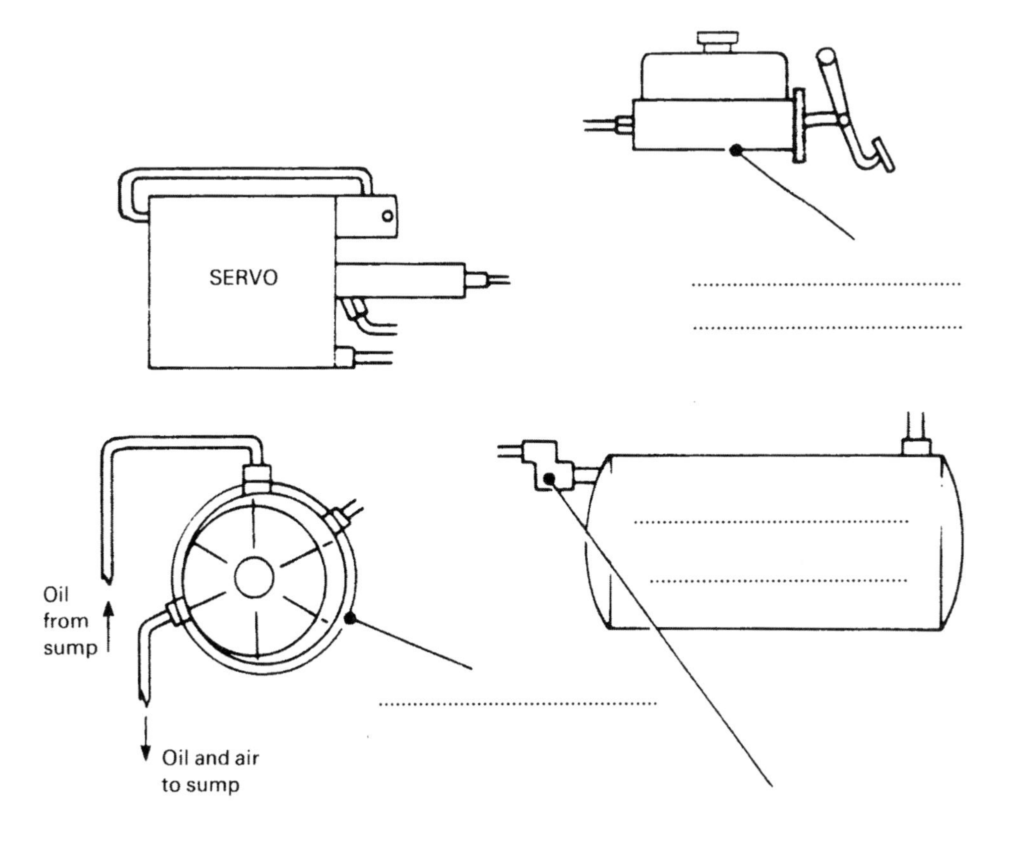

Give a reason for using, and describe the operation of, an exhauster:

....

Name the type of exhauster shown opposite:

....

State how an exhauster is usually driven:

....

State the purpose of the vacuum reservoir:

....

Describe the purpose of the non-return valve in the system:

....

In the arrangement shown opposite, oil is drawn into the exhauster unit. State the reason for this:

....

INDICATING DEVICES

Examine a vehicle and complete the drawings below to show the indicating devices and circuits.

STOP LAMP ACTUATION

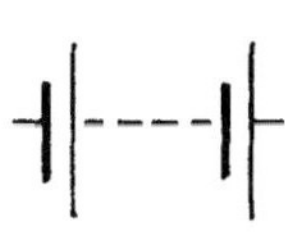

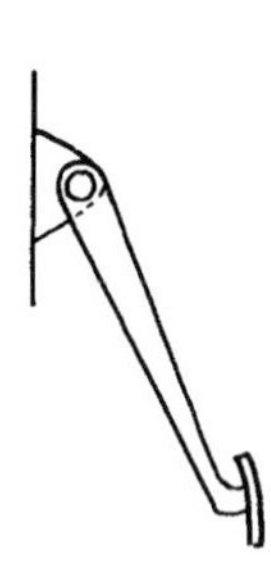

PAD/BRAKE SHOE WEAR INDICATION

FLUID LEVEL

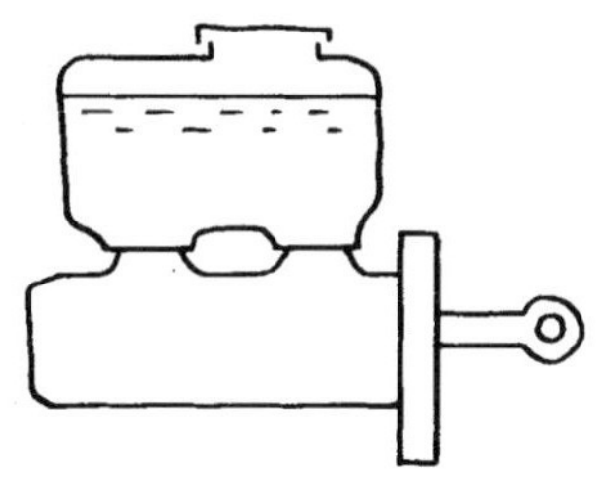

BRAKE TESTING

A brake test is carried out on a vehicle to determine the 'brake efficiency'. The efficiency is an indication of the ability of the braking system to stop the vehicle.

The brake efficiency of a vehicle is usually determined by:

(1) *Measuring the rate of deceleration of a vehicle, or* ....

(2) ....

Method (1) above involves the use of a ....

Method (2) above involves the use of a ....

List the main statutory requirements for braking systems relating to minimum efficiencies and MOT test requirements:

....

....

....

....

....

....

....

....

....

....

....

....

....

....

Describe the DECELEROMETER method and the ROLLER BRAKE TESTER method of determining the brake efficiencies for a vehicle.

DECELEROMETER METHOD (TAPLEYMETER)

....

....

....

....

....

....

....

....

....

....

....

ROLLER BRAKE TESTER

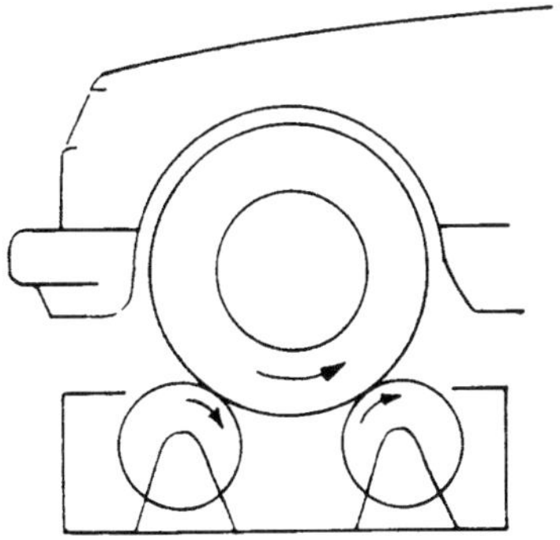

....

....

....

....

....

....

....

....

....

....

....

....

The somewhat obsolete decelerometer is still used in brake efficiency testing of certain vehicles. For which vehicles is this form of testing used and why?

....

....

....

....

....

In addition to normal brake efficiency tests, vehicles with ABS require rather more sophisticated test procedures to ensure the operation of the anti-lock braking system.

This calls for the use of dedicated or specific test equipment. Describe such equipment and testing carried out on anti-lock braking systems:

....

....

....

....

....

....

....

....

....

....

....

....

POWER HYDRAULIC BRAKING SYSTEMS

Hydraulic power, rather than compressed air or vacuum assistance, can be utilised to provide an applying force rather greater than that available by normal pedal effort.

Hydraulic booster system

This system entails the use of:

Fluid reservoir
Hydraulic pump (mechanically or electrically driven)
Accumulator
Valve block

Complete the labelling and describe the operation of the simplified layout below:

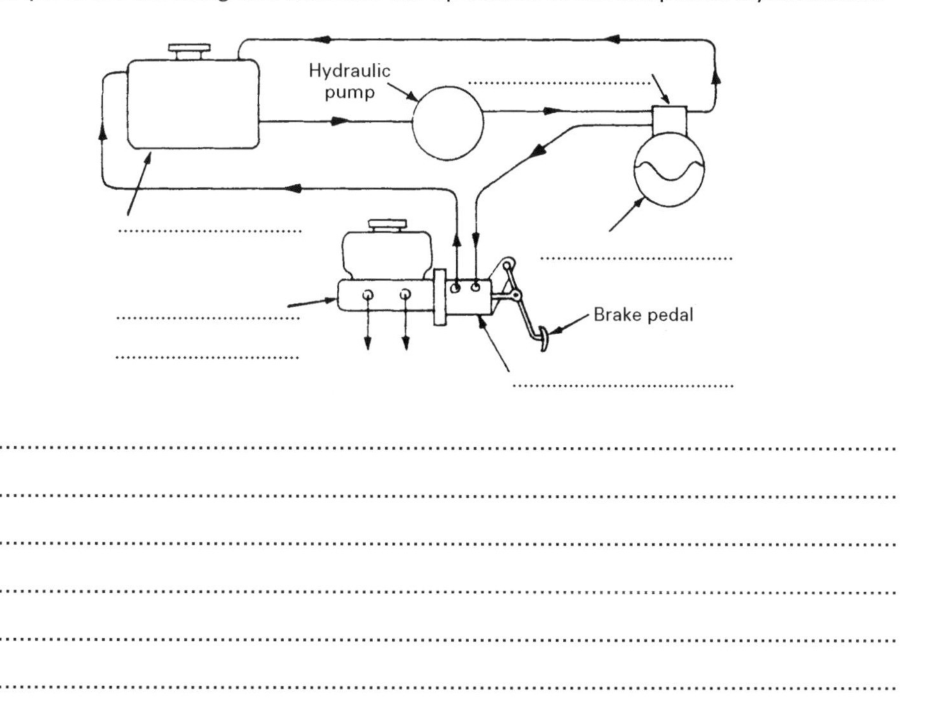

....
....
....
....
....
....
....
....

Alternatively the fluid can flow directly from the accumulator to the braking system (dynamic flow) via a driver's foot valve.

Describe the operation of the HGV 'full power' hydraulic system below; include in the description the construction and action of the hydraulic accumulator.

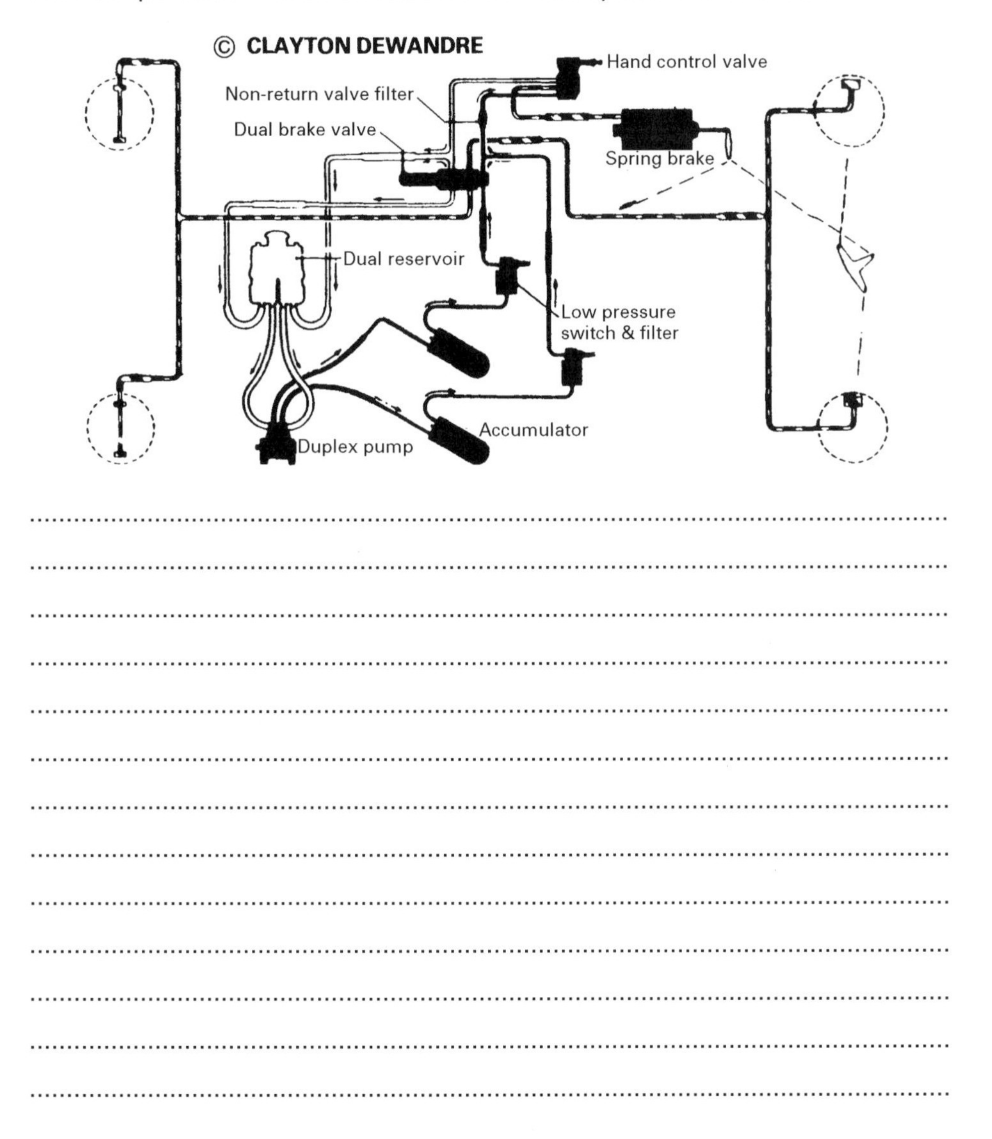

....
....
....
....
....
....
....
....
....
....
....
....

ANTI-LOCK BRAKING SYSTEMS

To achieve maximum retardation and maintain steering control during braking, the road wheels must be on the point of locking or skidding. If skidding occurs, adhesion is reduced and the steering becomes ineffective, that is, control is lost. Anti-lock braking systems control the braking torque at the road wheels to a level which is acceptable according to the tyre/road adhesion, that is, the braking torque applied is the maximum permissible without locking-up the wheels. The main components in an anti-lock system are:

Hydraulic pump (mechanically or electrically driven).

In anti-lock systems, sensors respond to imminent wheel lock (related to deceleration rate) and cause pressure modulating valves to 'freeze', reducing and increasing line pressure several times per second and thereby never allowing the pressure to increase to a point where wheel lock could occur.

The system shown opposite is a relatively inexpensive arrangement incorporating mechanical sensors.

Label the drawing and describe the operation of the system.

OPERATION

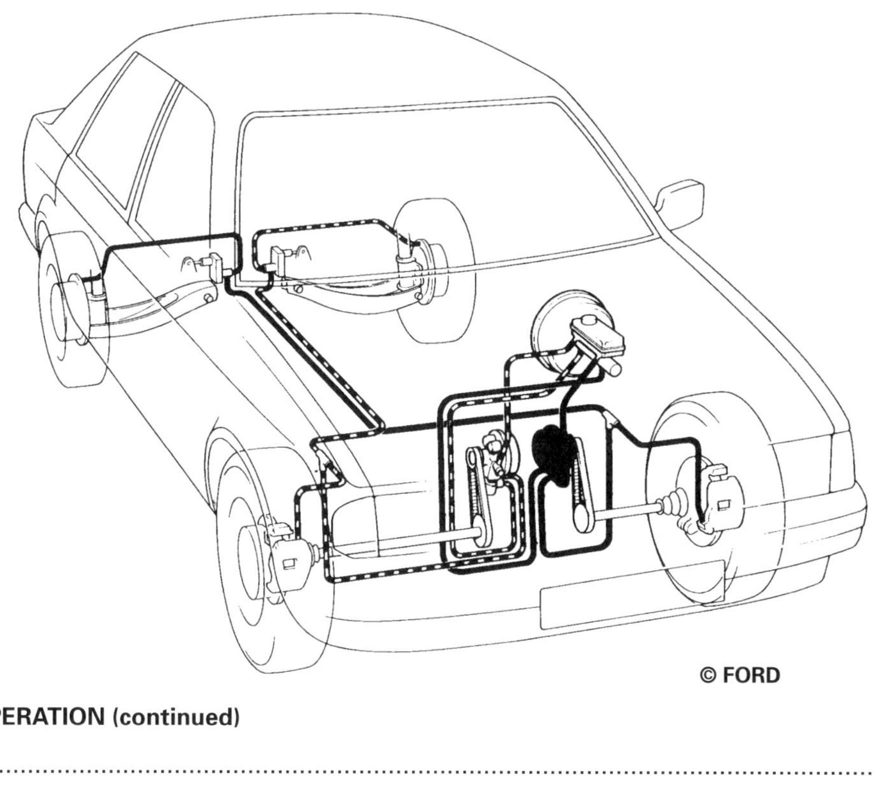

© FORD

OPERATION (continued)

The ABS system shown opposite incorporates a power hydraulic booster. The valve block contains an input and an output valve for the rear brakes, separate input and output valves for each front brake, that is, the front brakes have independent anti-lock control, and a master valve.

OPERATION (normal braking)

The master cylinder actuates (via the driver's pedal) the front brakes and simultaneous operation of a spool valve puts the accumulator in direct communication with the rear brakes and master cylinder booster. Describe the operation of the system during ABS regulation:

....

....

....

....

....

....

....

....

....

....

....

....

....

....

....

1. .... 2. ....

3. .... 4. ....

5. .... 6. ....

7. .... 8. ....

Complete the labelling on the drawing:

A ——— ....

B ------- ....

C ....

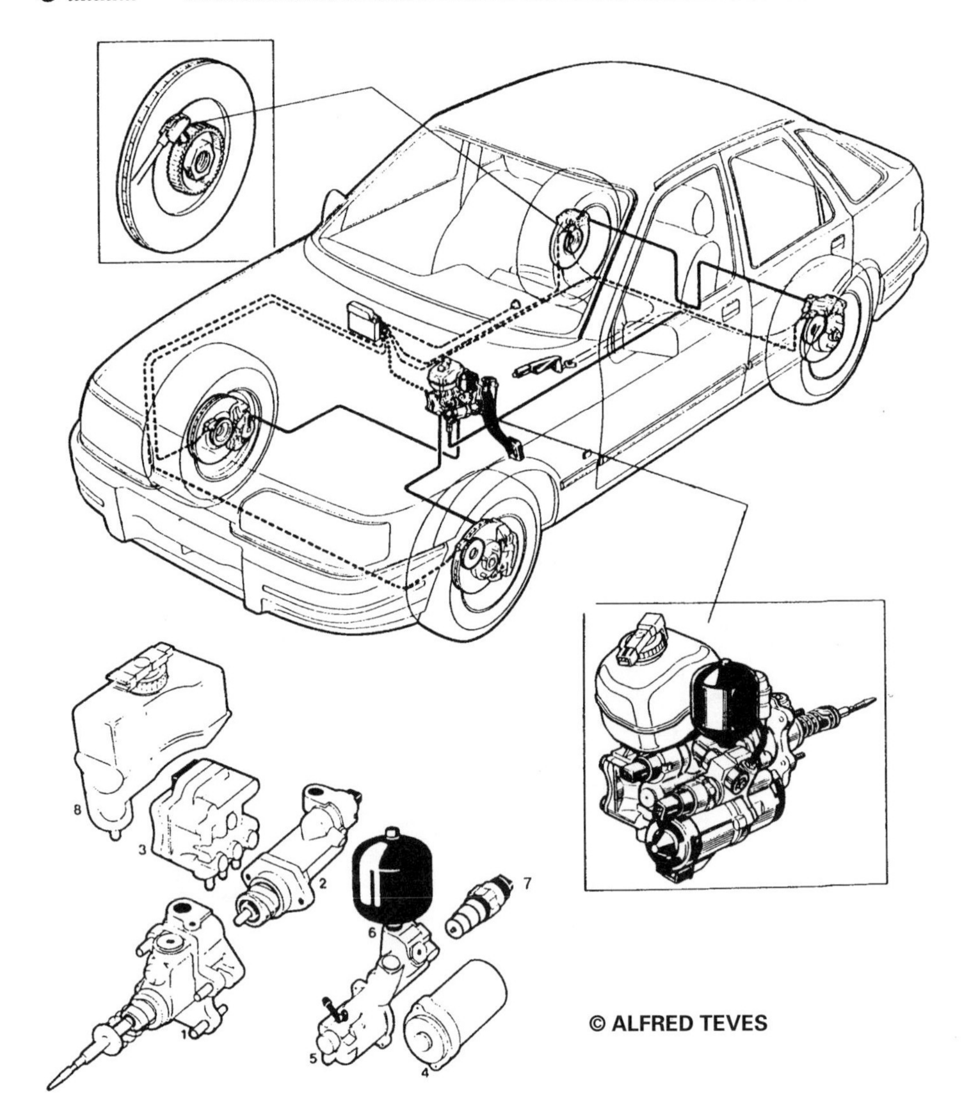

TRACTION CONTROL (FORD/BENDIX)

Controlling *wheelspin* during a standing start or when cornering etc. allows maximum usage of the traction available to the tyres.

The TCS shown opposite is a development from ABS and is an alternative to the limited slip differential or torque-sensitive differential as a means of controlling wheelspin.

In the system shown, wheelspin is controlled by a combination of brake and throttle control. The two key components in the system are a combined *ECU and hydraulic modulator and a throttle control intervention motor.*

Indicate and name the major parts on the drawing.

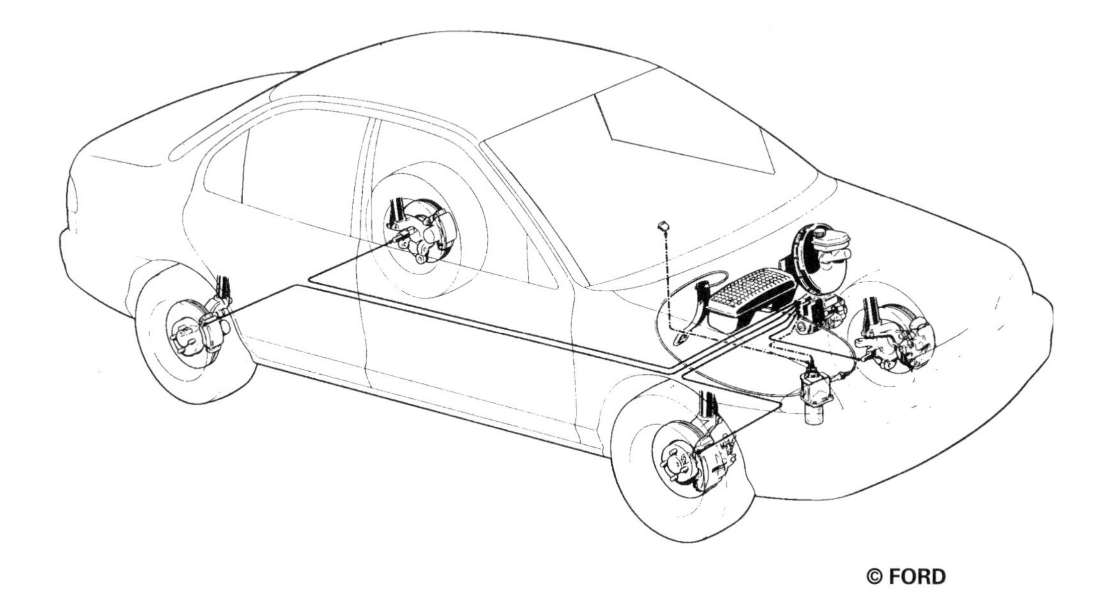

© FORD

SYSTEM CHECK

The ECU carries out a complete system check every time the ignition is turned on, and constantly monitors its own performance from then on. Any fault found causes a warning light to illuminate.

Workshop tests would normally be carried out using a diagnostic computer. This virtually replaces equipment such as multi-meter, star tester and the break-out box. A computer is plugged into the vehicle to carry out a sequence of diagnostic checks, for example, ECU memory, sensors, etc. The system will also analyse and detect faults in the car's engine management module, ABS, adaptive suspension, auto-transmission, car security, etc.

CENTRE OF GRAVITY AND LOAD TRANSFER

On the vehicle shown below the cross marks the *centre of gravity*. With a front engine, front-wheel-drive vehicle the C of G would be nearer to the front. However, with a rear engine, rear-wheel-drive vehicle the C of G would be nearer to the rear of the vehicle.

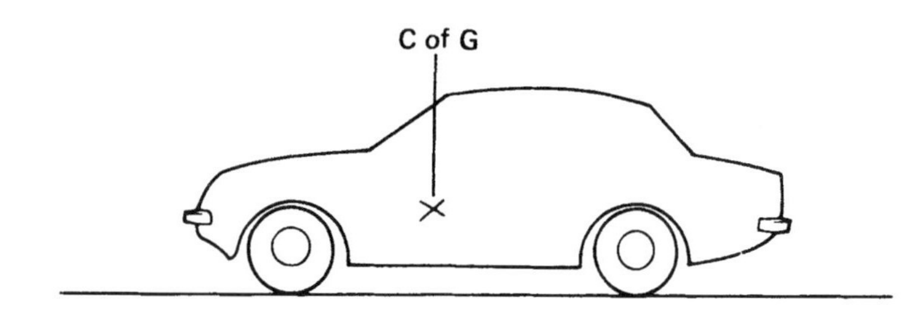

Briefly describe what is meant by the term 'centre of gravity':

....

....

....

....

....

....

The stability of a vehicle when braking (or cornering) is greatly affected by the actual position of the vehicle's C of G.

Give examples of vehicle types which have:

1. Low C of G ....

2. High C of G ....

What other vehicle factors can adversely affect braking stability?

....

....

Consider the vehicle shown below. During braking the force *F* acting at the C of G tends to tilt the vehicle about the front wheels, thus putting more load on to the front wheels, and reducing the load on the rear wheels. 'Load transfer' takes place. State TWO brake system features that are designed to overcome problems created by this:

1. .... 2. ....

C of G

F

Wheelbase

The degree of load transfer is dependent upon a number of factors. How do the following affect load transfer during braking?

1. Height of centre of gravity:

....

....

2. Wheelbase:

....

....

3. Rate of retardation:

....

....

4. Ratio of C of G height to wheelbase:

....

....

STOPPING DISTANCE

The stopping distance for a particular vehicle is dependent on a number of factors other than the design and condition of the vehicle's braking system.

List FIVE factors, external to the vehicle, which can affect stopping distance.

1. ....

2. ....

3. ....

4. ....

5. ....

Maximum retardation of a vehicle is achieved just before wheel lock (skidding) occurs. This is because, when sliding or skidding of the tyres on the road surface occurs, friction is reduced. This is known as *kinetic* or *sliding* friction.

Why is the braking force higher if the wheels continue to roll?

....

....

....

....

....

WORK DONE BY THE BRAKING SYSTEM

When a vehicle is brought to a halt, the 'work done' by the braking system is a product of:

1. ....

2. ....

....

The formula used to calculate work done is:

The retarding force produced by the brakes when stopping a vehicle is 5000 N. If this force is applied over a distance of 25 m, calculate the work done by the brakes.

If the work done by the brakes in stopping a vehicle is 200 kJ, calculate the retarding force if the distance travelled during brake application is 40 m.

The work done by the brakes in stopping a vehicle is 600 kJ. If the retarding force produced by the brakes is 20 kN, the distance travelled by the vehicle during braking is:

(a) 3 m

(b) 30 m

(c) 60 m

(d) 90 m?

Answer (   )

WORK DONE BY THE BRAKING SYSTEM

Braking performance

Many factors affect the actual distance in which a vehicle can be brought to rest by the brakes.

State the effect on stopping distance of:

Vehicle speed

....

....

Vehicle mass

....

....

Road conditions

....

....

VELOCITY/ACCELERATION/DECELERATION

Velocity is the rate at which a body moves in a given direction. The rate of movement is expressed in m/s^2.

The speed of a body is also the rate of movement; the difference between speed and velocity is that speed does not involve direction.

Define:

Acceleration

....

....

Deceleration

....

....

Acceleration due to gravity

The acceleration of a free-falling body due to the force of gravity is

....

In theory the maximum deceleration of a vehicle can only equal the acceleration due to gravity, since gravity maintains contact between tyres and road.

Braking efficiency

State the meaning of braking efficiency:

....

....

....

For the braking force to be equal to the vehicle weight the coefficient of friction between the tyre and the road would have to be ....

The braking efficiency of a vehicle can be determined by:

(a) expressing the deceleration as a percentage of 9.81 m/s^2

or

(b) expressing the braking force as a percentage of the

....

If a vehicle decelerates at 9.81 m/s^2, the braking efficiency is said to be .... %.

Using method (a) above:

Braking efficiency =

Using method (b) above:

Braking efficiency =

During a brake test the maximum rate of deceleration for a vehicle was 7.5 m/s^2. Calculate the braking efficiency.

A vehicle has a mass of 2000 kg. Calculate the braking efficiency if the total retarding force during brake application is 14 kN.

A vehicle is uniformly decelerated at 4 m/s^2. Calculate the retarding force and the braking efficiency if the vehicle has a mass of 1000 kg.

When testing the parking brake on the rear axle of a vehicle, the efficiency was found to be 27%.

Calculate the load on the rear wheels if the recorded braking force was 4 kN.

Calculate the rate of deceleration of a vehicle during braking if the total retarding force is 11.5 kN and the vehicle mass is 1600 kg.

Complete the table:

% Efficiency	Vehicle mass (kg)	Braking force (kN)	Decel. (m/s^2)
25		5	
	1500	12	
	2300		6.5
72	3000		

VEHICLE SPEED AND STOPPING DISTANCE

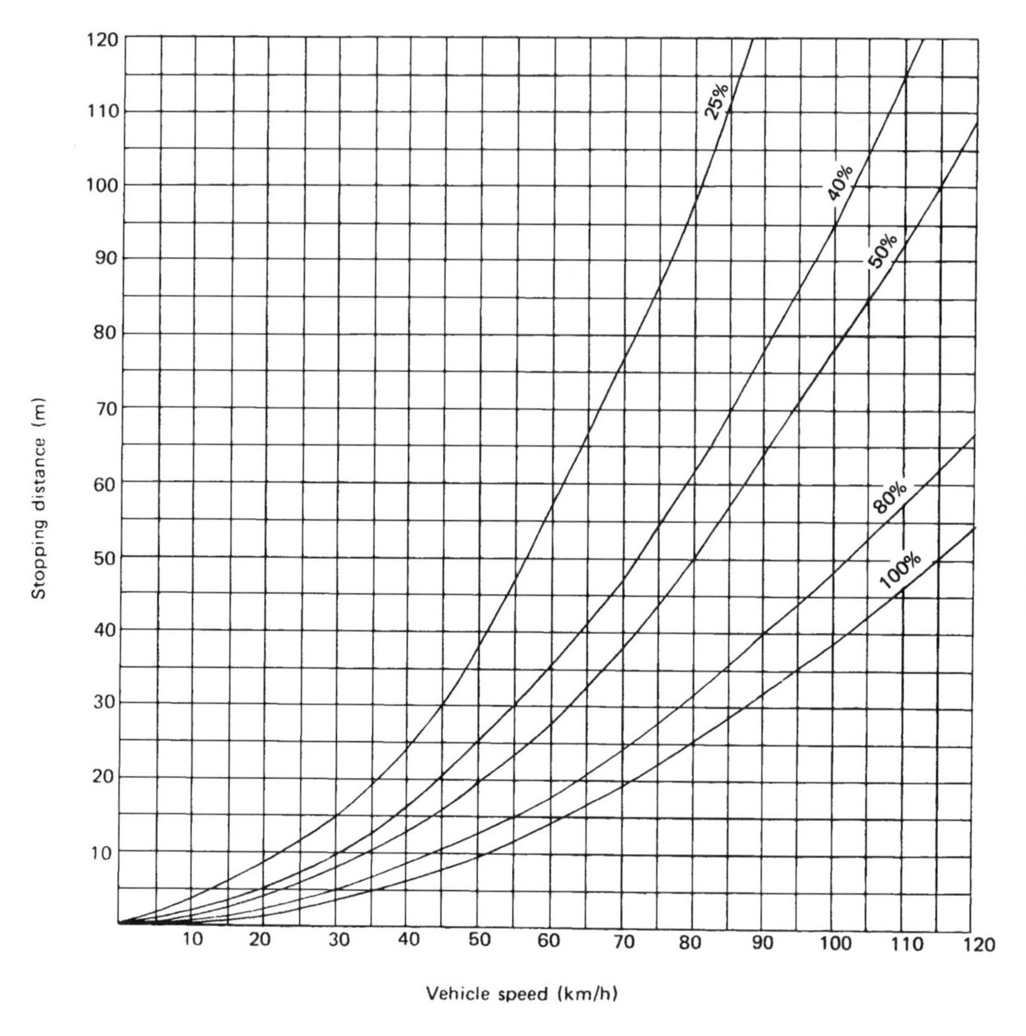

It can be seen from the graph that, if the speed doubles, the stopping distance is .... greater. That is, the stopping distance increases as the .... of the speed.

The graph opposite illustrates the effect of vehicle speed on stopping distance.

Determine from the graph:

1. The braking efficiency if the stopping distance is 48 m from a speed of 100 km/h.

   ....

2. The stopping distance from a speed of 75 km/h if the braking efficiency is 40%.

   ....

From the figures given below, plot a curve on the graph opposite of stopping distance against vehicle speed and, from information already on the graph, state the brake efficiency for the vehicle (see note below).

Speed (km/h)	36	54	72	90	108
Stopping dist. (m)	7.7	17.3	30.8	48	69.25

It can be seen from the graph that the stopping distance is proportional to the efficiency, that is, if the braking efficiency is halved the stopping distance is doubled.

Therefore for a given speed, if the stopping distance with 100% efficiency is known, it is possible to establish the efficiency for a vehicle using this graph if its stopping distance is known, for example, from a graph.

Stopping distance at 95 km/h at 100% efficiency is 35 m.

Stopping distance at 95 km/h at 40% efficiency is 87.5 m.

$$\text{Efficiency} = \frac{35}{87.5} \times \frac{100}{1} = 40\%$$

Braking force

When the brakes are applied on a vehicle, the hydraulic or air pressure acting on pistons or diaphragms forces the friction material into contact with the rotating drum or disc to produce a braking force and braking torque. It is important to appreciate the relationship between the pressure in the system, the area of the pistons or diaphragms and the force exerted by the pistons or diaphragms. State the relationship between PRESSURE, FORCE and AREA:

....

....

....

....

....

....

State the two factors that contribute to the ratio of movements and forces between the brake pedal and wheel cylinder pistons in a hydraulic braking system:

1. ....

2. ....

State the effect the inclusion of air in a hydraulic braking system would have on:

(a) the 'movement ratio'

....

....

(b) the 'force ratio'

....

....

INVESTIGATION

To show the relationship between movement ratio and force ratio in a simple hydraulic system.

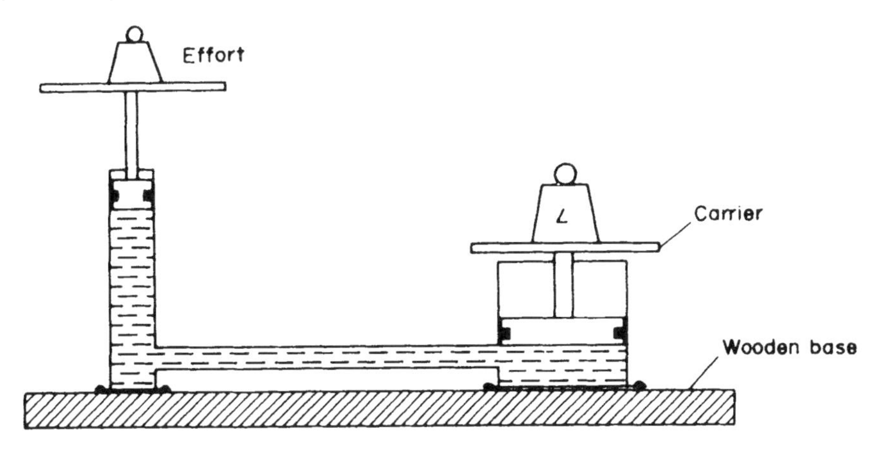

Use a simple hydraulic system similar to that shown above.

Force ratio

1. Place a load on the large piston carrier.
2. Determine, by experimenting, the effort required to raise the load at a uniform speed.

Load =

effort =

force ratio $= \frac{\text{load}}{\text{effort}} = \frac{\qquad}{\qquad} =$ ....

Movement ratio

Allow the load to rise a measured amount and determine the corresponding movement of the effort piston.

$$\text{Movement ratio} = \frac{\text{distance moved by effort}}{\text{distance moved by load}}$$

Movement ratio $= \frac{\qquad}{\qquad} =$ ....

COMPRESSED AIR BRAKING SYSTEMS

In its simplest form a full air braking system consists of the following basic components: air compressor, reservoir, unloader valve, foot valve and brake chambers or actuators.

Outline the basic operation of the simplified single-line system shown opposite with particular reference to the function of the compressor, unloader valve and foot valve:

....

....

....

....

....

....

....

....

....

....

Shown below is an S-type cam with roller followers. Indicate on the drawing the direction of cam and shoe movement during brake application:

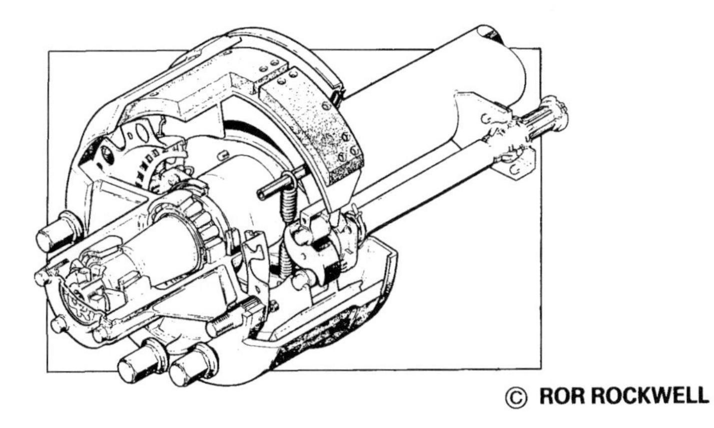

© ROR ROCKWELL

Single-line air brake system

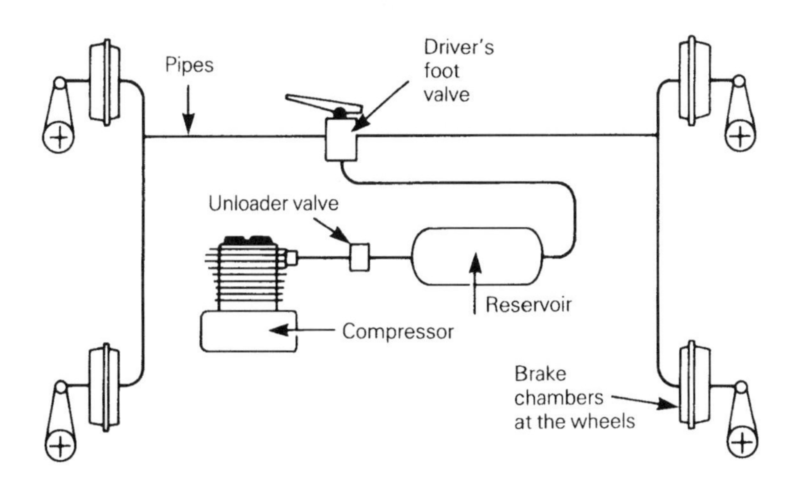

Complete the drawing on the right below to show the action of a single-diaphragm actuator when the footbrake is applied.

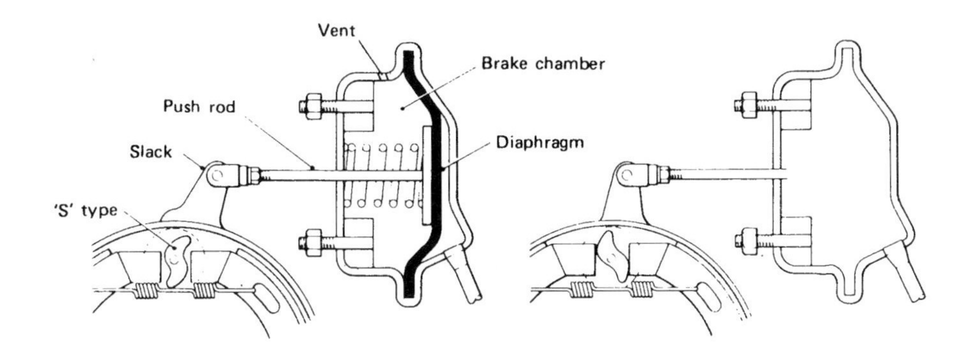

State the reason why the S-type cam is used in preference to the simple cam:

....

....

Split-line system

In the split-line system the front and rear brakes are independent systems operated from separate reservoirs via a dual foot valve. This system provides a secondary brake, operative on 50% of the wheels, in the event of a failure in one part of the system. Complete the labelling and add the pipes to the drawing below:

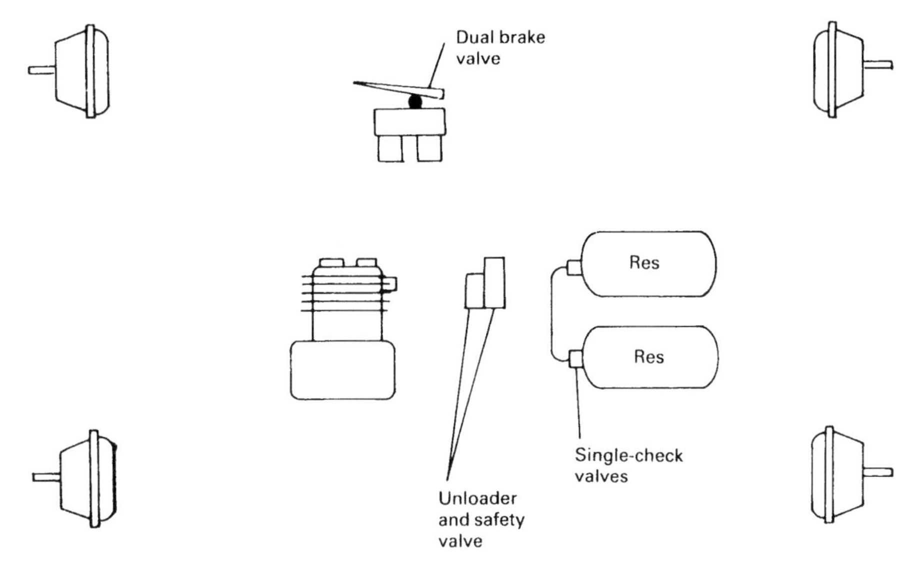

State the purpose of the:

Single check valves

....

....

....

Safety valve

....

....

....

State the causes and effects of water accumulation in a compressed-air braking system:

....

....

....

....

....

State the purpose and describe the action of an alcohol evaporator in an air braking system, and add one to the system opposite:

....

....

....

....

....

....

Some systems employ an ALCOHOL INJECTOR which automatically and frequently injects a quantity of alcohol into the system. Its function is to:

....

....

....

One method of overcoming the problem of water and other contaminants entering the main braking system is to employ an AIR DRIER. This unit is located between the compressor and reservoir, warm air entering the unit is cooled and the water content is absorbed into crystal-type material.

When the governor unloads the compressor, the water is automatically expelled from the air drier.

COMPRESSED AIR BRAKING SYSTEMS

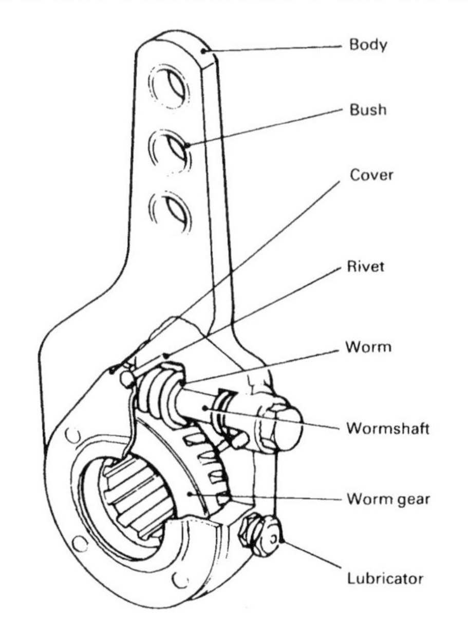

© WESTINGHOUSE CVB

The SLACK ADJUSTER lever shown left has a manual adjuster (worm and worm gear).

In the automatic version, the worm gear is on a ratchet and automatically takes up slack if the lever movement is excessive. State the purpose of the component:

....

....

....

....

....

....

....

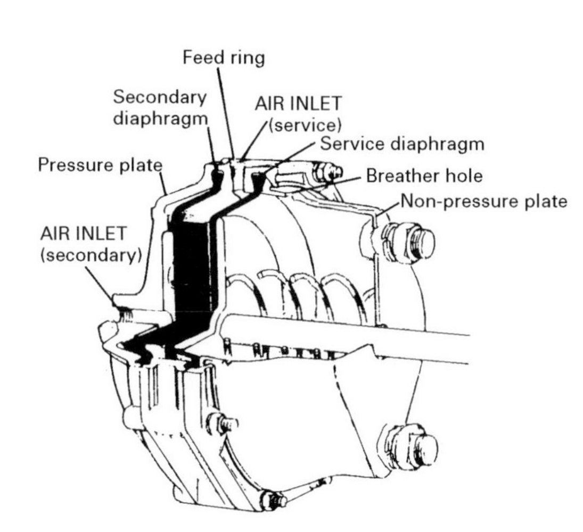

Name and describe the operation of the brake chamber shown on the left:

....

....

....

....

....

....

....

....

Dual-line system

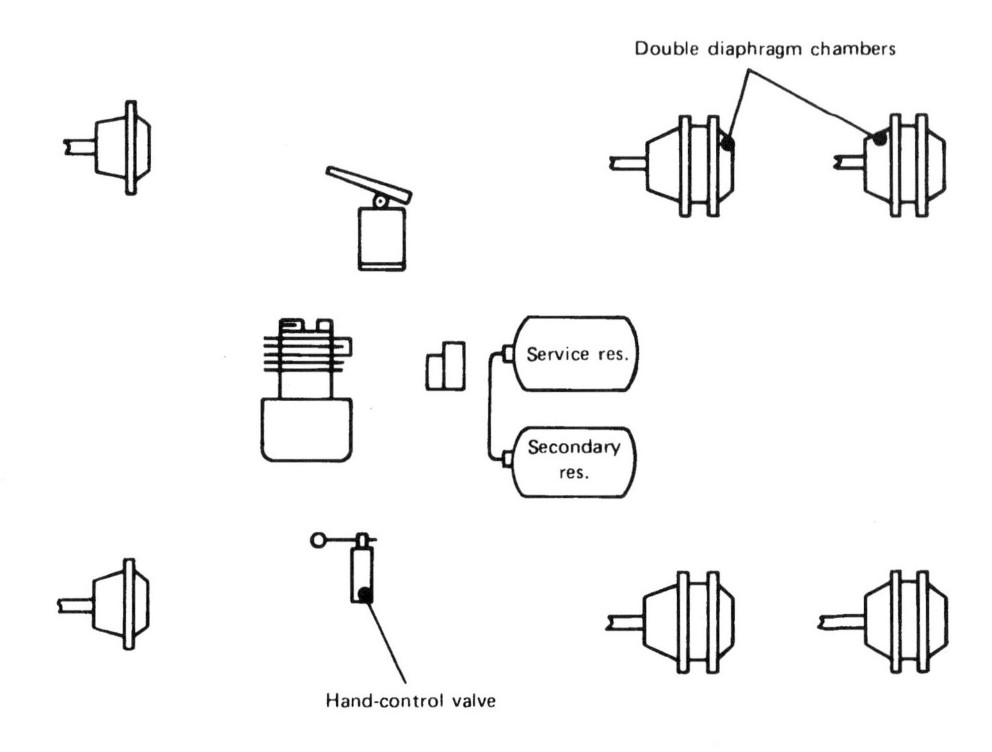

A dual-line layout for a six-wheeled vehicle is shown above; add the pipes to the drawing and describe the operation of the system:

....

....

....

....

....

....

....

....

Quick release valve

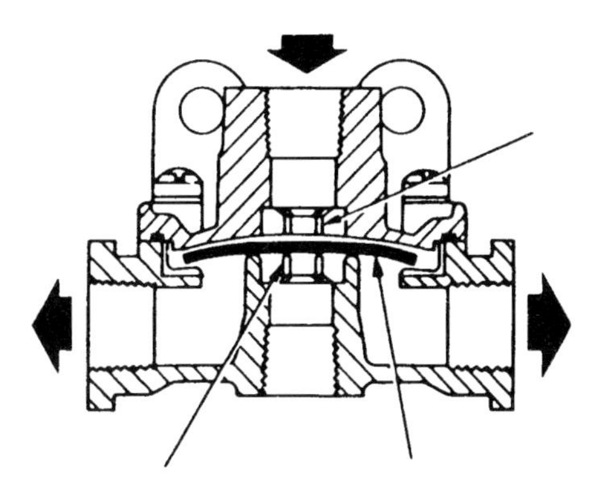

State the purpose and describe the operation of the valve shown above:

....

....

....

....

....

....

....

....

DRAIN VALVES in the system may be manual or automatic in operation. Why are these valves necessary?

....

....

....

....

Relay valves

Relay valves provide a means of admitting and releasing air to and from the brake chambers, in accordance with brake valve operation, without passing the air for the brake chambers through the driver's foot valve.

This gives much quicker braking responses and also permits the use of smaller control valves in modern cab designs.

Add the pipes to the arrangement shown below to show how a relay valve would be incorporated into the system:

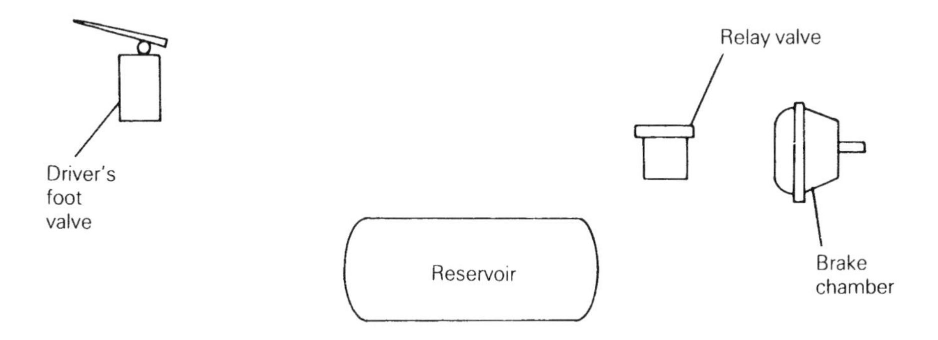

Pressure-regulating valve

The drawing below shows the location of a pressure-regulating valve in an air brake system; state the purpose of this valve.

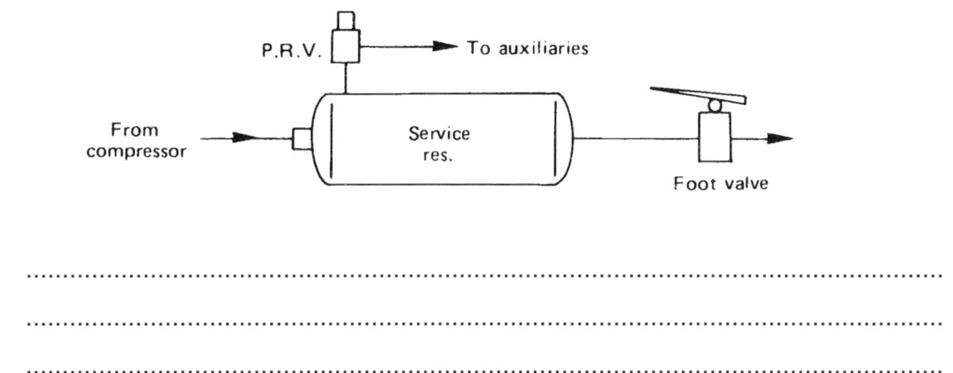

....

....

....

COMPRESSED AIR BRAKING SYSTEMS

Spring-brake system

The spring-brake chamber is a single-diaphragm brake chamber with an extension cylinder on the rear containing a spring-loaded piston. Label the spring-brake chamber shown below:

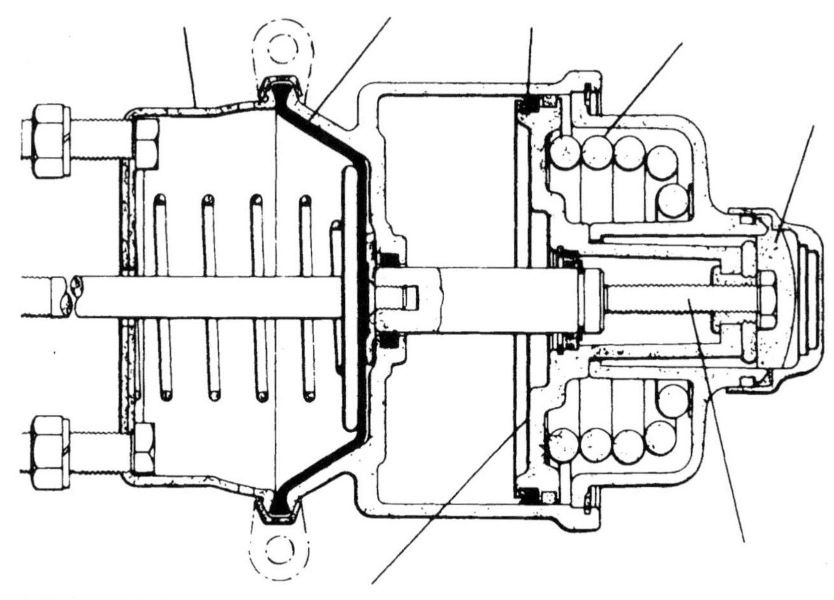

© **WESTINGHOUSE CVB**

The spring-brake chamber fulfils three functions:

1. *It provides a service brake.*
2.
3.

Give three advantages of spring brakes:

1.
2.
3.

The drawings below show different operating conditions. Explain each one briefly:

Normal driving

Service brake

Service brake portion acts in similar manner to standard brake chamber. Air pressure supplied (normally) via foot valve. Spring is held compressed by a steady air pressure in spring chamber.

Secondary and parking

Manual release

Three-line System

The braking layout shown is a THREE-LINE system as used on many articulated vehicles and drawbar trailer combinations. Spring-brake chambers are employed on the tractor and double diaphragm units on the trailer.

Add the three-line trailer braking system to the drawing below and indicate the colour coding for the tractor/trailer connections.

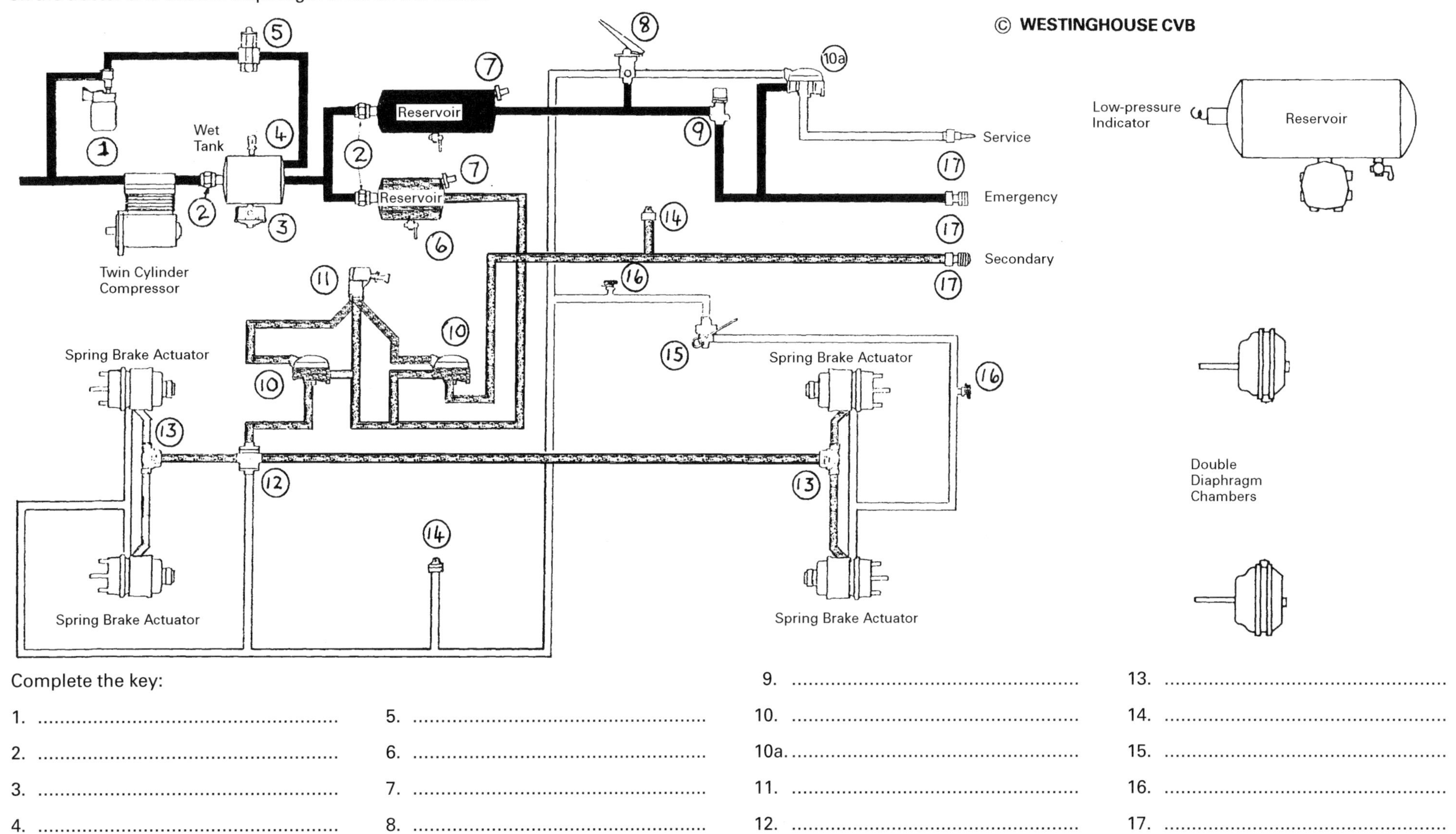

Complete the key:

1. ....
2. ....
3. ....
4. ....
5. ....
6. ....
7. ....
8. ....
9. ....
10. ....
10a. ....
11. ....
12. ....
13. ....
14. ....
15. ....
16. ....
17. ....

Two-line System

The EU Type Approved layout shown below is a TWO-LINE system for articulated vehicles and drawbar trailer combinations. The system on the tractor shown can be used with either two or three tree-line trailer systems.

Complete the drawing by adding a two-line trailer system and indicate the colour coding for the tractor/trailer connections.

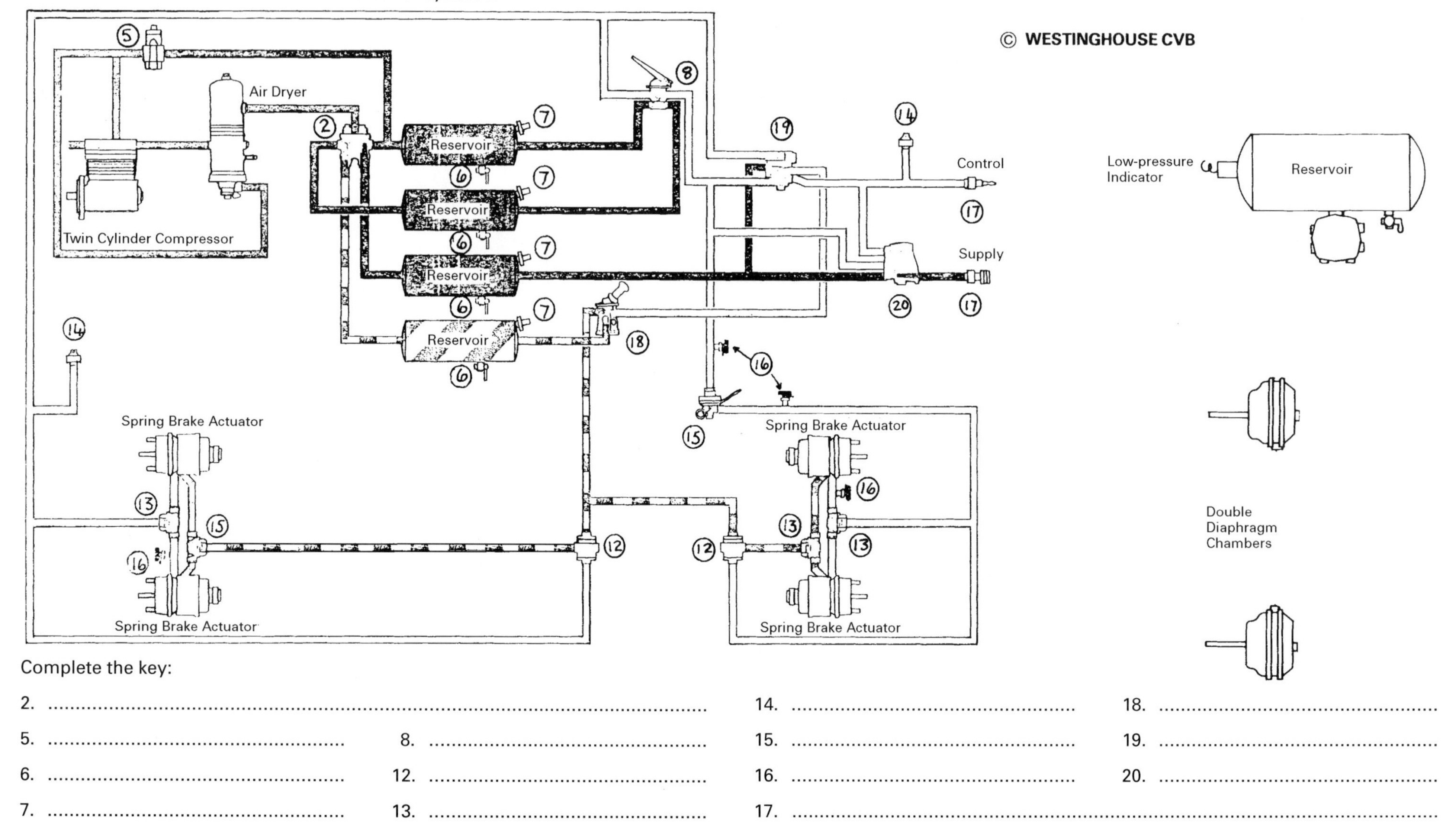

Complete the key:

2. ....

5. ....

6. ....

7. ....

8. ....

12. ....

13. ....

14. ....

15. ....

16. ....

17. ....

18. ....

19. ....

20. ....

Refer to system on page 191.

Describe the operation of the three-line system.

Service brake

Air flows from the service reservoir, through the foot valve to the service diaphragms on the tractor. Relay 10a is actuated to admit air through trailer service line which operates the trailer relay valve to admit air from the trailer reservoir to the trailer service diaphragms.

Secondary/auxiliary brake

Parking brake

Emergency system

Refer to system on page 192.

Describe the operation of the two-line system.

Service brake

Air flows through the dual foot valve to each half of the split tractor system and to the multi-relay valve (19) to operate the trailer service brake. Should one half of the tractor service brake fail, the multi-relay valve will still apply the trailer brakes.

Secondary/auxiliary brake

Parking brake

Emergency system

SYSTEM COMPONENTS

State the purpose of the following:

Governor valve

....

....

....

Tractor protection valve

....

....

....

....

....

Differential protection valve

....

....

....

....

....

Describe the methods used to identify trailer air line connectors:

....

....

....

....

....

Double-check valve

This valve permits the air from two sources to flow down a common line when either system is operative.

When both systems are operated at the same time the double-check valve responds to the system registering the highest pressure.

The differential protection valve is a 'biased' type of double-check valve.

Complete the drawing below to show a simplified double-check valve:

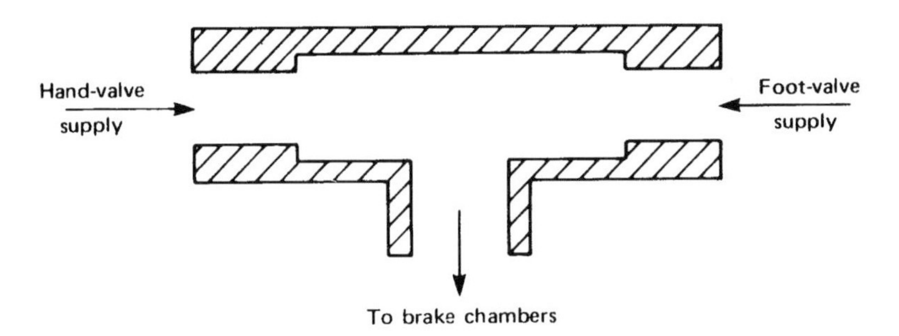

On the simple diagram below, add the necessary pipes and show where the valve above would be used in a brake system:

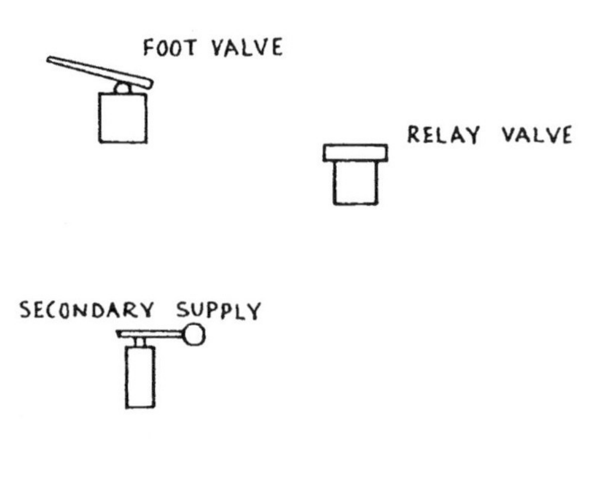

TRAILER BRAKES

The drawings on this page illustrate the use of SPRING BRAKES on a trailer. State the purpose of:

1. Shunt valve

....

....

....

2. Park valve

....

....

....

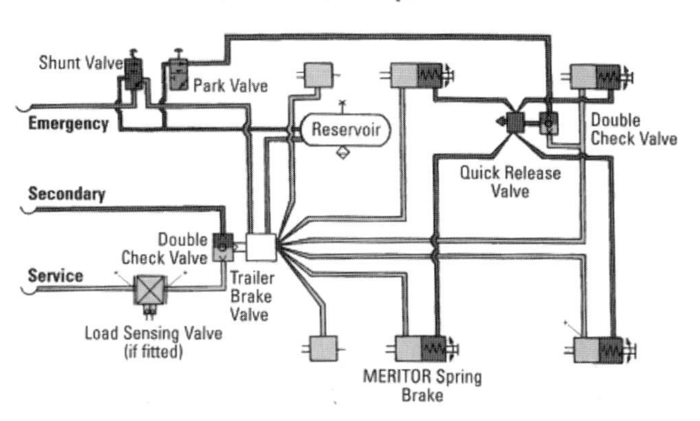

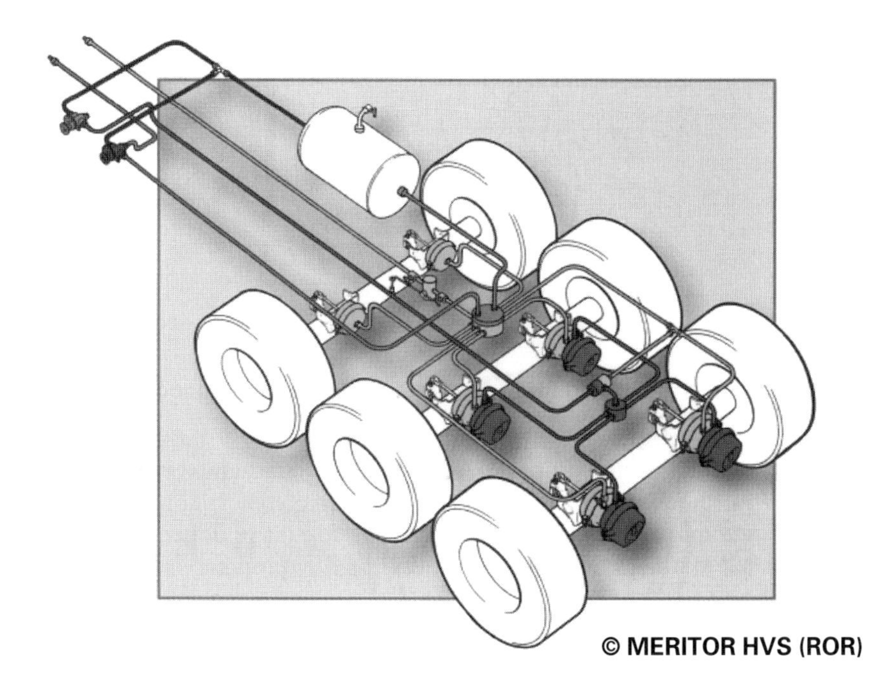

© MERITOR HVS (ROR)

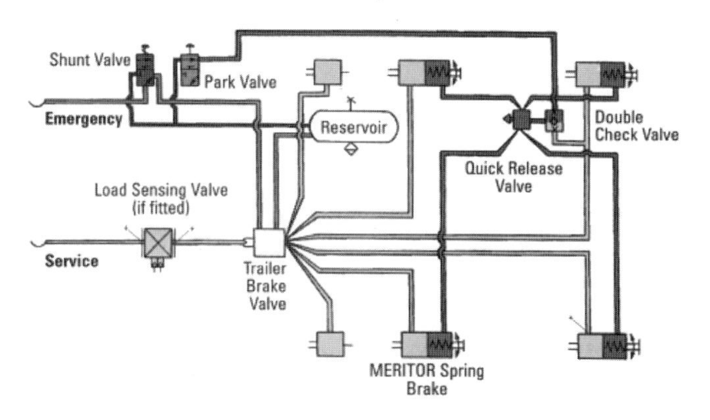

© MERITOR HVS (ROR)

TRAILER BRAKES

Lock actuators

The lock actuator is a double-diaphragm or diaphragm/piston brake chamber fulfilling the function of service and secondary braking. In addition the unit incorporates a device which enables the brakes to be 'mechanically' locked in the applied position when parking the vehicle, thus eliminating the conventional handbrake linkage.

Complete the labelling on the lock actuator shown below:

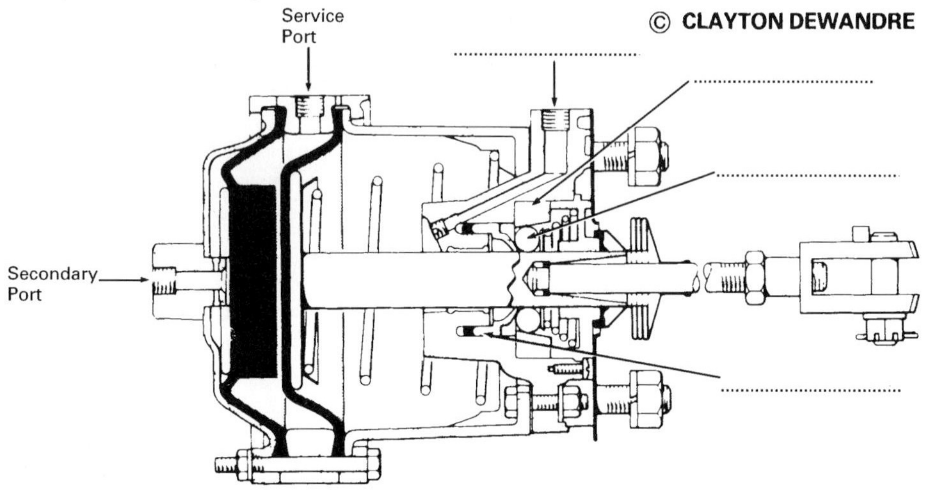

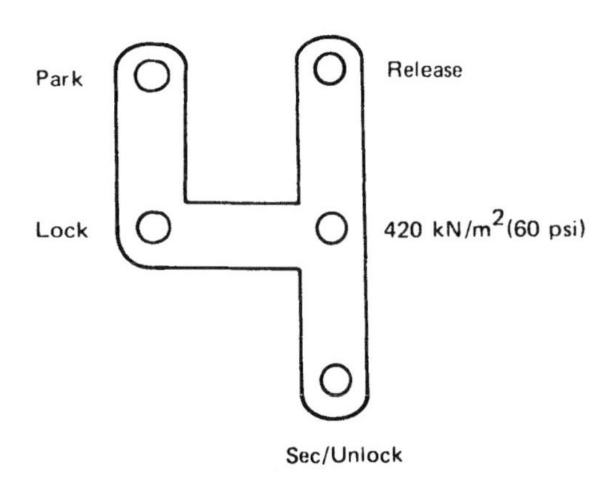

The gate for the full-flow secondary/park valve, which is used in the lock actuator system, is shown left; describe, alongside the headings opposite, the operation of the lock actuator system.

Service brake

With the control lever in the release position the lock line is pressurised to keep the rollers out of engagement with the push rod. When the footbrake is applied, air enters the service chamber to apply the brakes.

Secondary brake

Parking brake applied

Parking brake released

With the lever held in the unlock position, secondary line pressure is applied to the unloader or governor valve; this allows the secondary brake line pressure to build up beyond normal operating pressure (provided the engine is running) to free the locks under abnormal conditions.

COMPRESSED-AIR/HYDRAULIC BRAKES

Compressed-air/hydraulic brakes are used on many light/medium-range commercial vehicles.

Basically the system is hydraulic with some form of compressed-air unit providing assistance.

Two main types of system are in use:

1. A compressed-air actuator, controlled from a foot valve, operating directly on a hydraulic master cylinder piston.
2. ....

....

Complete the layout below to show a tandem air actuator operating the master cylinder shown:

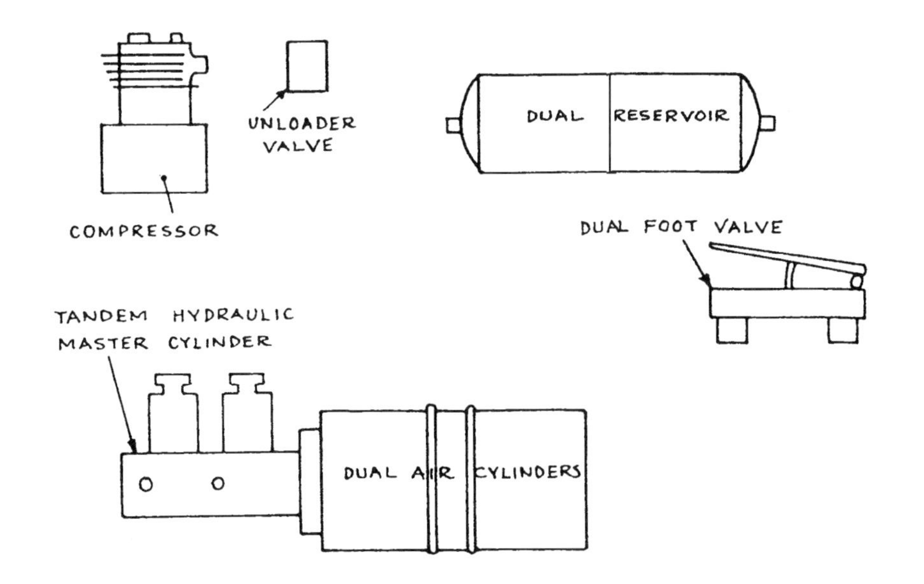

The layout below shows simplified a compressed-air servo unit operated mechanically from the footbrake.

Study the drawing and describe the operation of this system:

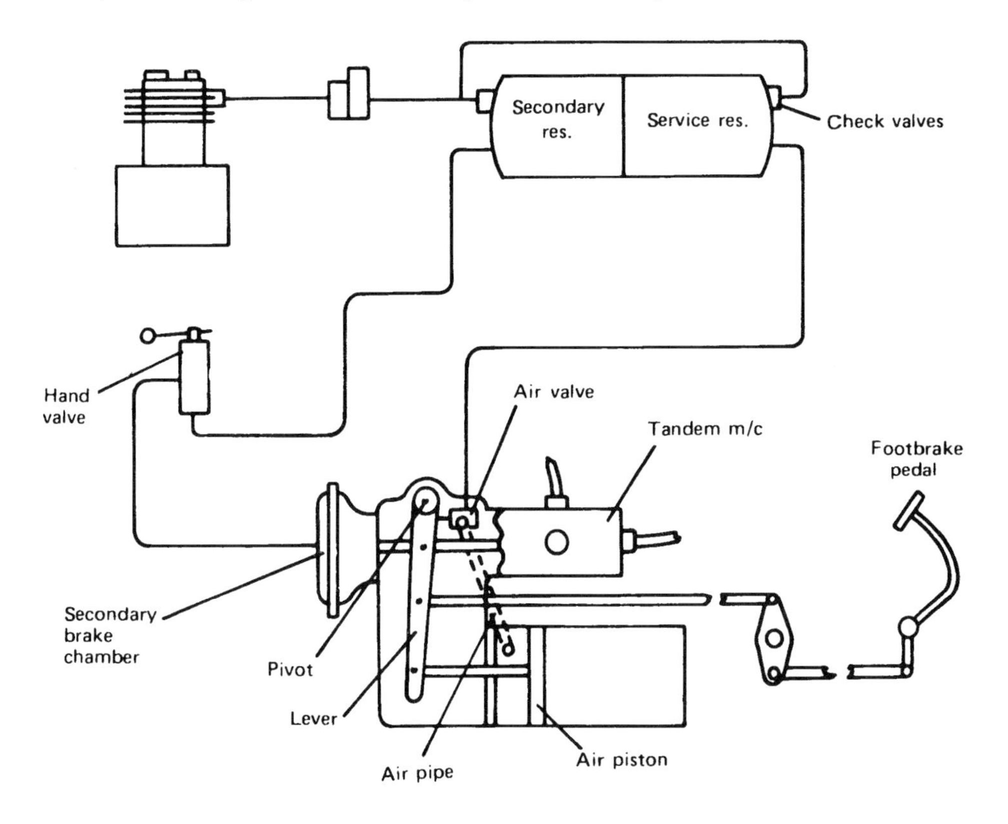

OPERATION

....

....

....

....

....

....

JACK-KNIFING

Jack-knifing is a problem associated with articulated vehicles. The drawing below illustrates what happens when jack-knifing occurs; that is, the tractor unit folds round on to the semi-trailer.

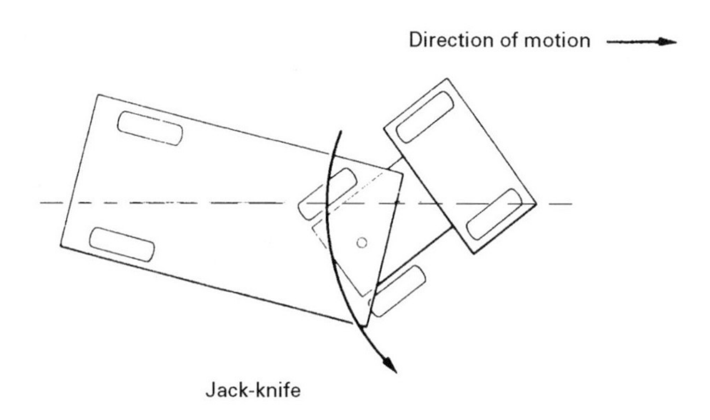

When is jack-knifing most likely to occur and what are the main causes of the problem?

....

....

....

....

ANTI-JACK-KNIFE SYSTEM

1. *A multi-disc brake can be incorporated into the fifth wheel coupling so that when the brakes are applied the fifth wheel coupling is effectively locked to the trailer king pin. This is normally added as a modification to a standard vehicle.*
2. ....
3. ....

ANTI-LOCK BRAKING

The system shown below is very popular on articulated tractor units; describe briefly how it prevents wheel lock during braking.

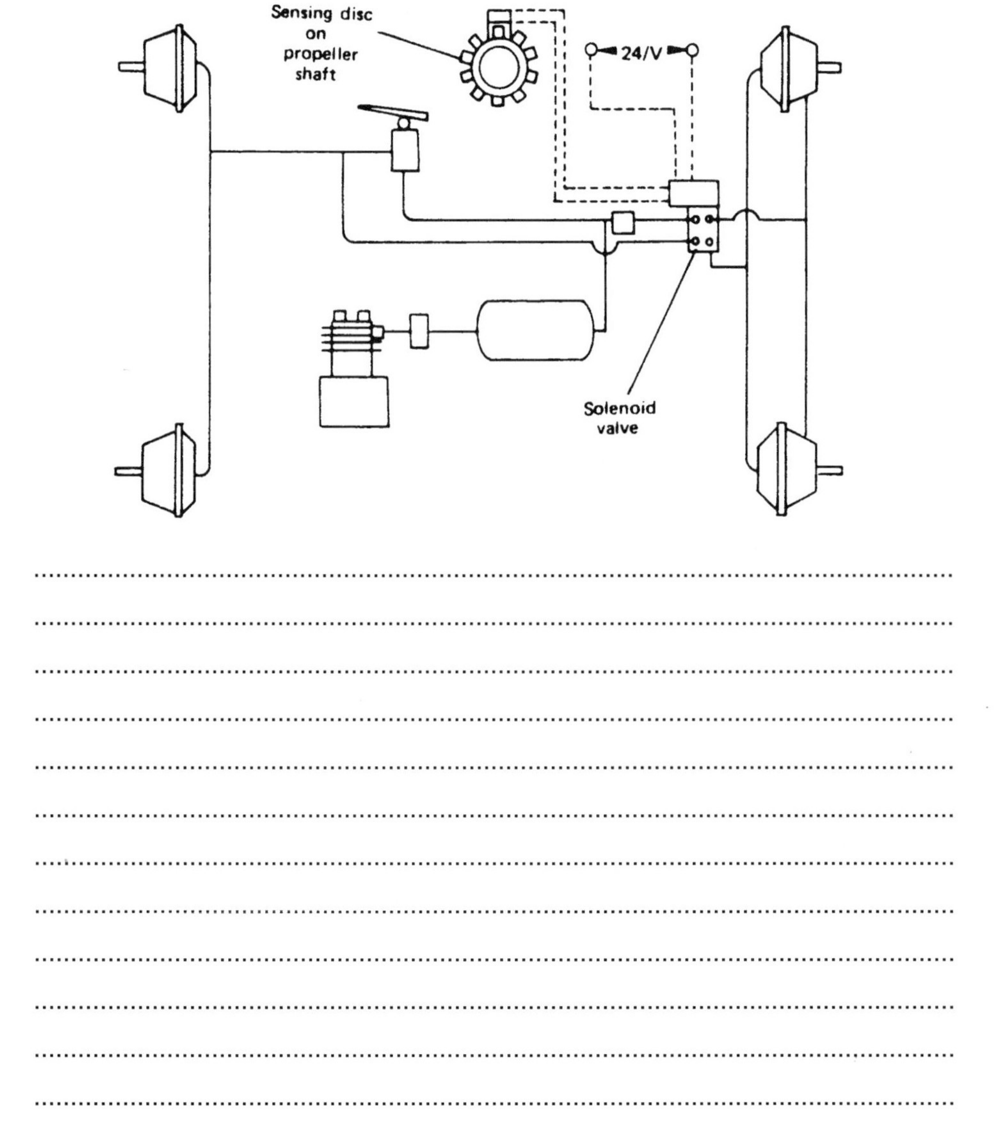

....

....

....

....

....

....

....

....

....

....

....

....

ABS (ANTI-LOCK BRAKING SYSTEM) HGV

The drawing opposite shows the basic layout of a full air ABS system for a truck. This system, which is employed on rigid and articulated HGVs, prevents wheel lock during braking and automatically utilises the maximum road surface friction available. This shortens the stopping distance and maintains vehicle stability.

Describe the operation of the system shown:

....

....

....

....

....

....

....

....

....

....

....

....

....

....

....

....

....

....

System layout

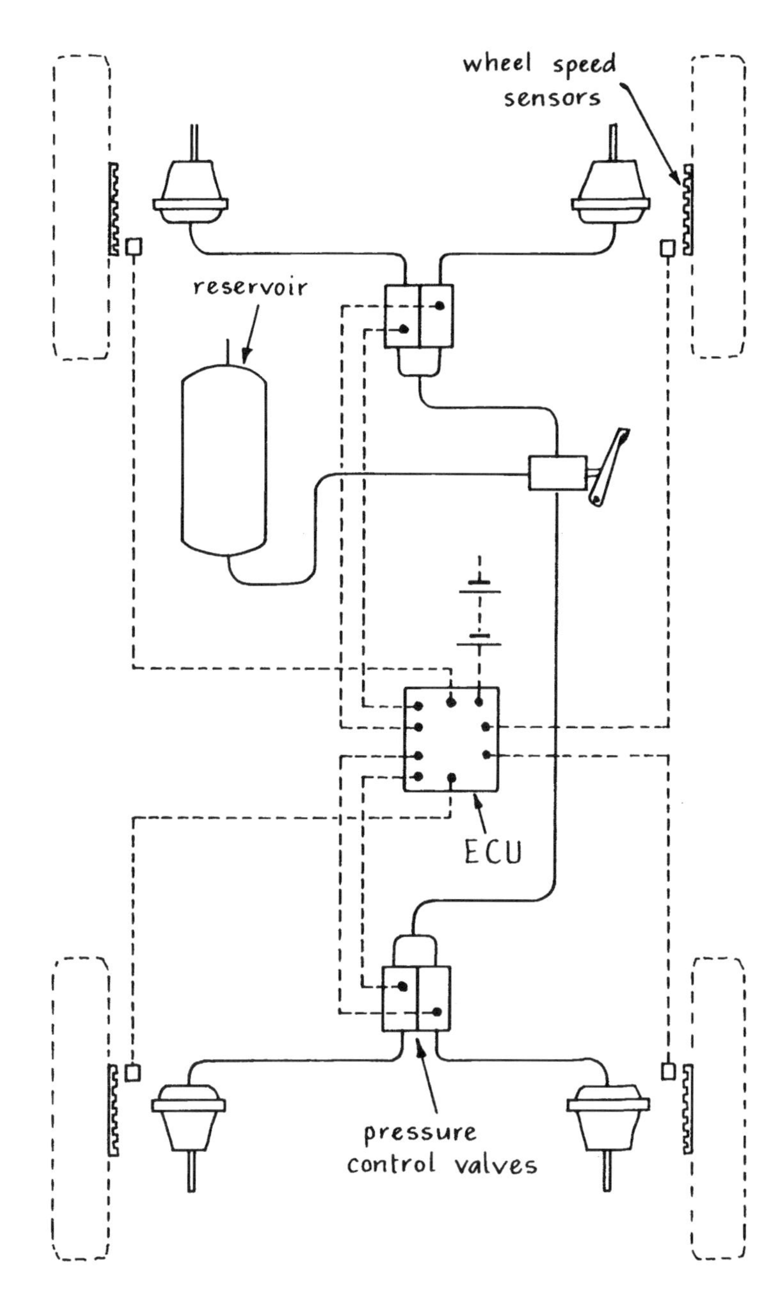

LOAD SENSING VALVES

Load sensing valves are used in the braking systems of cars and HGVs. The valves apportion the braking force at an axle according to the loading on the axle. It is necessary to reduce the braking force as the load on the axle reduces. Why is this?

Load sensing valves are used on axles where the load is likely to vary considerably due to payload, passengers or weight transfer. Examples include: at the rear axle of front wheel drive cars, such valves are fitted to balance braking according to the weight of passengers and luggage carried; on HGV trailer axles, these valves balance braking between fully laden and unladen operation; and on some tractor driving axles, such valves help prevent wheel lock and hence jack-knifing.

Add a load sensing valve to the drawing below:

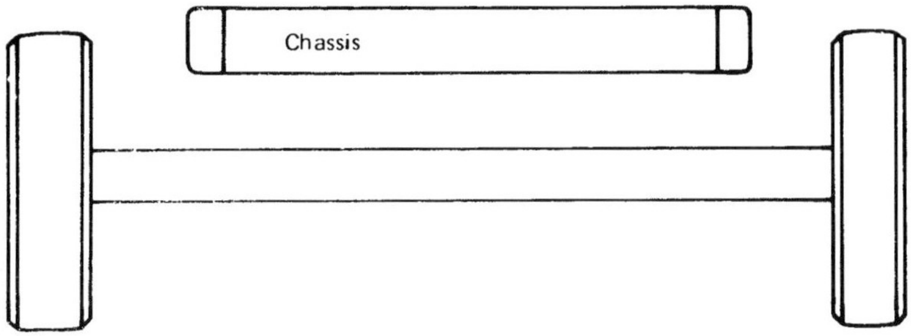

Describe how a load sensing valve can be checked and adjusted:

OPERATION

A schematic view of a load sensing valve is shown below; describe the operations of the valve.

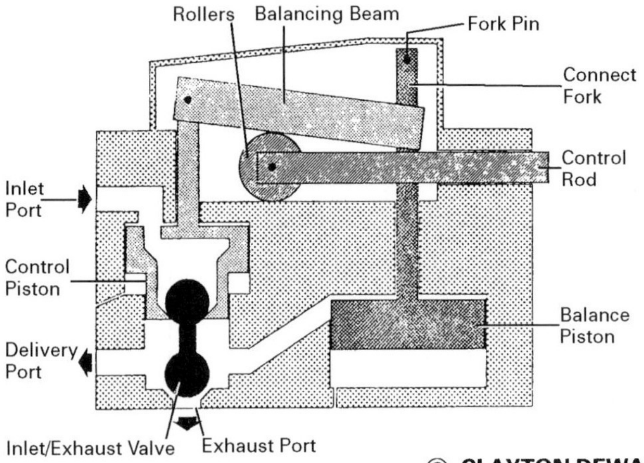

© CLAYTON DEWANDRE

VEHICLE RETARDERS (AUXILIARY BRAKES)

A vehicle retarder is a device that slows down a vehicle without the driver having to operate the normal braking system.

Retarders were originally used on heavy vehicles and coaches operating on journeys which involved long descents (for example, mountainous roads on the Continent). Using a retarder a driver can control the speed of a vehicle during a long descent without the need for continuous brake application which can cause brake fade.

However, owing to improvements in retarders, increased vehicle speeds and motorway operation, many more advantages are to be gained by using vehicle retarders and their fitment has become widespread.

List the advantages of vehicle retarders:

....

....

....

....

Name three types of vehicle retarder:

....

....

....

Vehicle retarders usually operate on the transmission system to slow the vehicle and may be manual or automatic in operation; they can also provide varying degrees of retarding force.

Exhaust Brake

The exhaust brake is a popular type of vehicle retarder; complete the drawing below to show the arrangement and describe briefly the principle of operation.

Exhaust frontpipe

....

....

....

....

....

....

....

....

....

Name alternative types of shut-off valves:

....

....

....

Electro-magnetic Retarder

The type of retarder shown is located in the transmission, usually directly behind the gearbox.

Name this type, label the drawing and describe briefly how it operates:

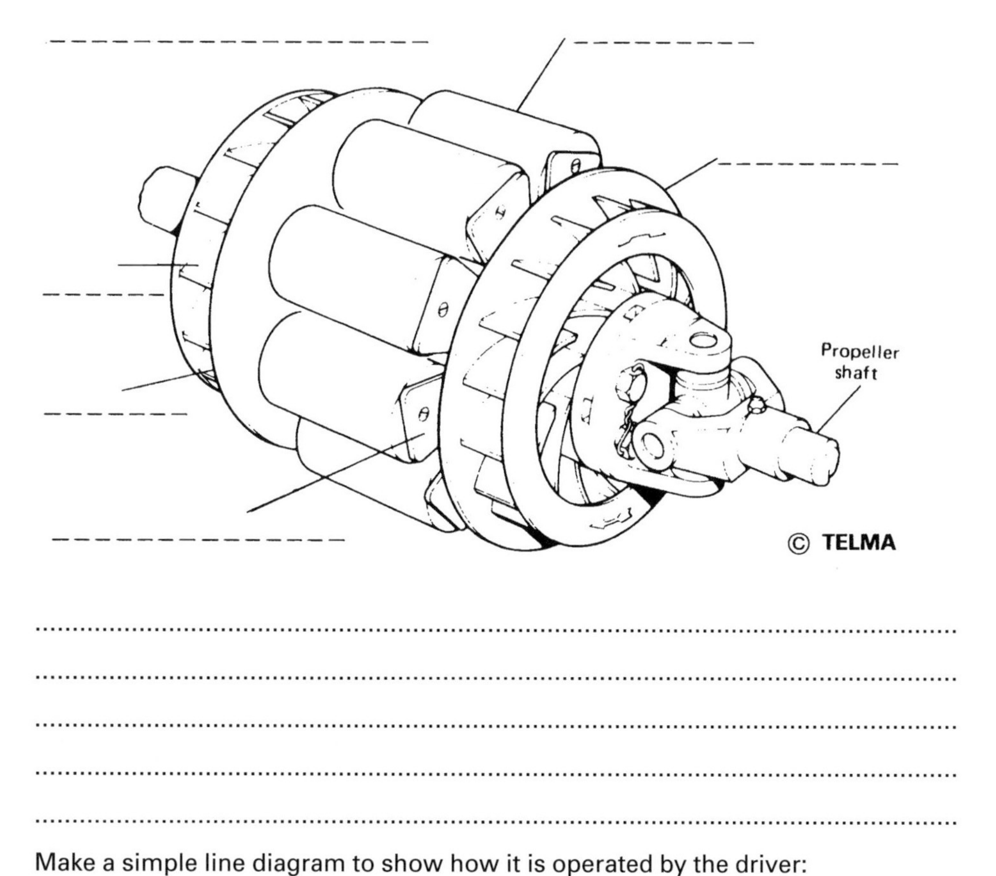

Make a simple line diagram to show how it is operated by the driver:

Hydraulic Retarders

The retarder shown below is usually situated in the drive line between the gearbox and final drive. In operation it is like a fluid coupling under 'stall' conditions, that is, the fluid is accelerated on to a stationary turbine (stator). The mechanical energy is therefore converted to heat energy. The degree of retardation is dependent on the amount of fluid in the retarder, this being controlled by a driver operated air-valve. On some vehicles the retarder is incorporated into the automatic gearbox.

Complete the drawing below to show the control system and heat exchanger:

It is necessary to protect the braking system from hazards arising during repair or use.

List such hazards and protective measures necessary:

BRAKING SYSTEM MAINTENANCE

A preventive maintenance system will maximise braking system reliability, efficiency, service life and vehicle safety.

List preventive maintenance/MOT checks and tasks associated with the braking system.

Check:

Describe any special equipment used for bleeding a hydraulic braking system:

....

In relation to the braking system, for what purpose would the following equipment be used:

HYGROMETER ....

MULTIMETER ....

State the reason for and describe the process of skimming brake drums and discs:

....

List the general rules for efficiency and any special precautions to be observed when maintaining and repairing braking systems:

....

DIAGNOSTICS: BRAKING SYSTEM – SYMPTOMS, FAULTS AND CAUSES

State a likely cause for each symptom/system fault listed below. Each cause will suggest any corrective action required.

SYMPTOMS	FAULTS	PROBABLE CAUSES
Excessive pedal travel; low efficiency	Worn friction material	....
Pulling to one side when braking; low efficiency	Seized wheel cylinder pistons	....
Pulling to one side when braking; low efficiency	Oil on brake linings	....
Excessive pedal/handbrake travel	Incorrect adjustment	....
Spongy pedal	Air in system	....
Faulty seals; corroded pipe	Fluid leakage	....
Inoperative parking brake	Seized parking brake cable	....
Pulling to one side under severe braking	Faulty ABS sensor	....
Vibration noise when braking; low efficiency	Worn brake drum or disc	....

Chapter 10

Bodywork

Vehicle body design	207
Shape of body members	208
Construction and use regulations	209
Body maintenance and protection	209

VEHICLE BODY DESIGN

Factors affecting body design
The intended use of a vehicle and cost constraints are two main factors affecting the design of a vehicle body.

Size and basic dimensions
The dimensions for body, occupants and load are indicated below.

Name the dimensions against the letters below.

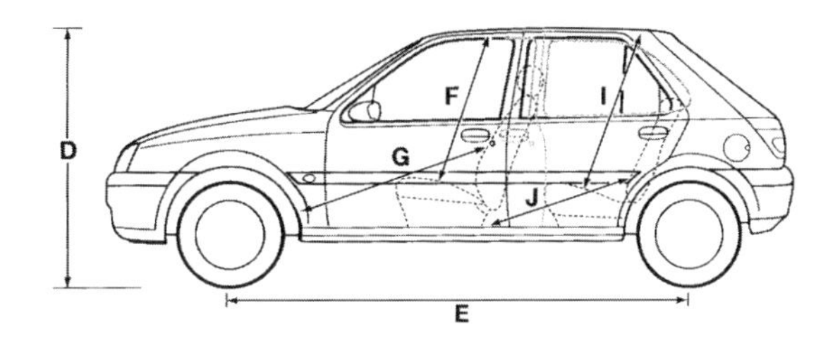

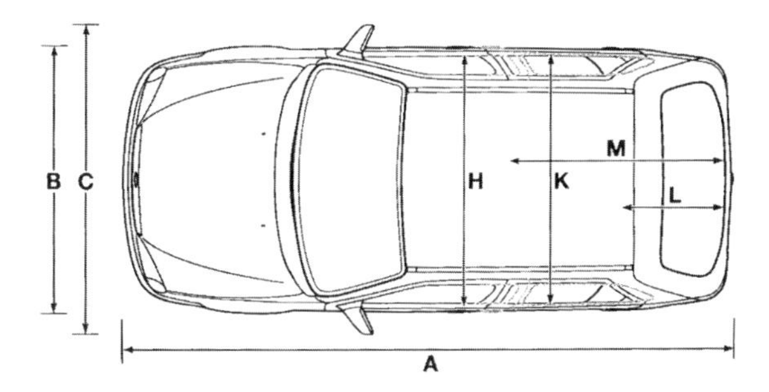

© FORD

A ....

B ....

C ....

D ....

E ....

F ....

G ....

H ....

I ....

J ....

K ....

L ....

M ....

Explain the difference between GROSS VEHICLE WEIGHT (GVW), PAYLOAD and CAPACITY in relation to the vehicle and body.

GVW ....

....

PAYLOAD ....

....

CAPACITY..

....

Strength
What forces is a body subjected to during normal operation?

....

....

....

....

What do the darker parts of the structure represent on the vehicle shown below?

© PEUGEOT

....

....

SHAPE OF BODY MEMBERS

Cross-sectional shape of key structural body members is an important factor with regard to body strength and rigidity.

Examine various body sections (or refer to a manual) and make sketches below to illustrate the cross-sectional shape of the body members listed.

INNER AND OUTER SHELL	**DOOR PILLAR**
WINDOW PILLAR	**CANTRAIL**
CHASSIS MEMBER	**FLOOR SECTION**

State the differences between a STRESSED body panel and an UNSTRESSED body panel, and give two examples of each:

....

....

....

....

....

....

Give examples of where on a vehicle body the following methods of joining/ attachment are used:

Welding ....

Adhesive joining ....

Mechanical fixing ....

Resilient (rubber) mountings ....

....

The CENTRE OF GRAVITY is the centre of the entire mass of the vehicle. What major factors determine the position of the C of G and how does its position affect vehicle operation?

....

....

....

....

....

....

....

See page 180 for more details of C of G.

CONSTRUCTION AND USE REGULATIONS

The purpose of the Construction and Use Regulations is to control the manner in which motor vehicles (and their equipment) are constructed, adapted and used. This is to ensure that legal standards regarding weight, construction and use, dimensions and design for vehicles and trailers are complied with. The Construction and Use Regulations do therefore contribute to the safety of vehicles on the road.

What is the purpose of the Type Approval Regulations?

List the areas of vehicle body construction and condition controlled by the Construction and Use Regulations:

Describe the procedure for installing a component on the body of a vehicle which involves drilling or cutting, for example, radio aerial, light, etc. Sketch any special cutting tool used in the operation.

BODY MAINTENANCE AND PROTECTION

The benefits of routine maintenance and running adjustments on the vehicle body are:

Maintenance of appearance

List routine maintenance tasks associated with the vehicle body:

Describe the procedure for:

1. Removing corrosive residues (salt/mud/tar, etc.) from paint finishes and underbodies.

2. Touching up minor paintwork damage.

© FORD

Methods used to protect the body/chassis against corrosion and damage arising from use or repair include the use of such as:

(a) galvanised panels (doors, bonnet, boot lid, etc.)

(b) cathodic phosphate electrocoating

(c multi-layer painting (primer, synthetic enamels, etc.)

(d) wax injection

(e) stone chip protection

(f) flange sealers

(g) anti-scuff strips

(h) plastic wing liners

(i) transit protective coating

Indicate on the drawing, top right, where the protective measures listed would be likely to be applied.

How is the protection maintained during vehicle use and repair?

1.

2.

Chapter 11

Electrical and Electronic Systems

Basic electrical circuits	212
Electrical symbols	213
Circuits	214
Cables	215
Wiring harness	216
Circuit protection	216
Switches and relays	217
Wiring diagrams	218

BASIC ELECTRICAL CIRCUITS

The electrical/electronic system in a motor vehicle is very complex, therefore to allow an understanding of how the system works, it must be broken down into individual circuits. The individual electrical circuits fitted to light, heavy and public service vehicles may use, in some form, the items listed in the table below. For each of the items listed state its basic working principle:

ITEM	FUNCTIONAL REQUIREMENT	WORKING PRINCIPLE TO ACHIEVE FUNCTIONAL REQUIREMENT
Generation of electricity	To convert mechanical energy into electrical energy	
Storage batteries	To store electrical energy in chemical form	
Motor and drive assemblies	To convert electrical energy into mechanical energy for use in linear or rotary form	
Lights	To convert electrical energy into light for illumination and indication purposes	
Control components ECUs	To provide a means of accurately controlling electrical and electro-mechanical systems	
Switches and relays	To provide a means of activating, controlling and isolating the electrical supply	
Circuit protection devices	To provide a means of protecting circuits against overload, surges and reversed polarity	
Driver information circuits	To provide driver information on speed, temperature, pressure, operation, contents, correct function, condition, etc.	

ELECTRICAL SYMBOLS

Sketch the appropriate symbol which is normally used to represent each of the components listed:

BATTERY		**DIODE**		**GAUGE**	
Switch		Zenor diode		Horn	
Relay		Thermistor		Alternator	
Fuse		Thyristor		Starter motor	
Bulb		Plug and socket connector		Voltmeter	
Motor		Radio		Ammeter	
Resistance		Speaker		Ohm-meter	
Capacitor		Transistor		Variable resistor	

CIRCUITS

The electrical circuits in this book are generally, as in most vehicles, EARTH RETURN in which the earth path for the circuit is provided by the body/chassis. This means that the current passes from the battery through switchgear to a component, such as a light bulb, and then through the chassis frame back to the battery to complete the circuit.

Complete the drawings below to show (a) an EARTH RETURN circuit and (b) an INSULATED EARTH RETURN circuit:

(a)

(b)

Give examples of the use of insulated earth return and state the reason for its use:

....

....

....

Complete the electrical circuits opposite by adding the wiring layout and components such as: switches, relays, circuit breakers, resistors, etc.

Include on the drawings the colour coding for the circuits.

Charging circuit

Heated rear-window circuit

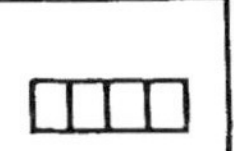

CABLES

Components within an electrical circuit are interconnected by flexible copper cables which provide a path for the electrical current. When selecting a cable for a particular application, a number of factors must be considered, such as current to be carried, length of cable required, routing protection/insulation, identification, etc.

Cable size and current rating
The size of a cable is designated by the number of strands of wire and the thickness of each strand.

A typical cable size used on vehicles is ....

The first number represents ....

and the second number is ....

The actual 'current rating' for such a cable is ....

The size of cable is increased as the current carrying requirement increases. Why is this necessary?

....

....

What influence does the length of cable have on its size? ....

....

Complete the table below to give the cable sizes and ratings for the applications listed:

APPLICATION	RATING (AMPS)	SIZE
Side and tail lamps		
Head lamps		
Alternator		

Wire size can also be designated by a number which relates to the cable diameter in millimetres, for example, No. 1 – 0.7 mm, No. 2 – 0.9 mm, No. 3 – 1.0 mm, No. 4 – 1.2 mm etc.

State one other system of designating wire size:

....

Colour coding
The system of identifying cables and components on wiring diagrams varies from one manufacturer to another.

On black and white diagrams, cable colours are identified by a code letter. Complete the tables below to show different coding systems:

Vehice make ....

CODE	COLOUR	FUNCTION
		Earth wires
		Battery feed to ignition
		Sidelamps
		Headlamps
		Permanent battery supply
		Ignition
		Fused supply from ignition

Vehice make ....

CODE	COLOUR	FUNCTION
		Earth wires
		Battery feed to ignition
		Sidelamps
		Headlamps
		Permanent battery supply
		Ignition
		Fused supply from ignition

Why are the code letters used?

....

WIRING HARNESS

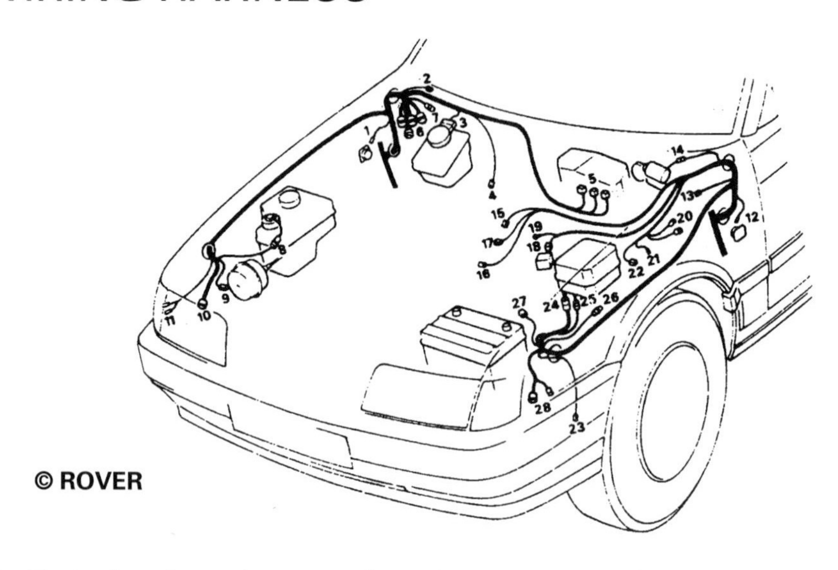

© ROVER

The illustration above shows a main engine compartment harness or loom. A number of wiring harnesses are used throughout a vehicle to bind together wires servicing a common area. Why is this system used?

Examine main harness connectors such as the type shown below and make simple sketches on the right to illustrate (1) the connector locking device, (2) the method of securing cables in the connector.

© ROVER

CIRCUIT PROTECTION

State the function of a fuse or circuit breaker:

The glass type fuse shown at (a) contains a paper indicating the amperage rating. The more recent type (b) is colour coded for rating purposes. Name the fuse types at (c) and (d).

(a) (b) (c) (d)

Explain the difference between PEAK CURRENT rating and CONTINUOUS RATING of fuses:

How does a FUSIBLE LINK compare with the fuses shown and where would it be used?

Bi-metal circuit breakers, in which the heating effect of excessive current would bend the bi-metal strip and cause contacts to break, is another form of circuit breaker. These are very often the 'manual reset' type. Describe this method of circuit breaking and give an example of its use:

SWITCHES AND RELAYS

A switch in a circuit is used to make or break the flow of current in the circuit. Depending on the application, switches can be simple on/off spring-loaded toggle types or complicated multi-function switches such as the stalk type steering column arrangements.

In many circuits on a motor vehicle the manually operated switches energise RELAYS which in turn carry the main current load for the circuit.

Give two reasons for the use of relays:

....

....

....

....

Relay operation: (Note: also see horn section)

A typical relay panel is shown opposite.

Complete the list opposite with names of typical relays.

Some interior light circuits employ a DELAY relay and some heated rear window circuits employ a TIMER relay. Describe how these relays operate within their circuit:

....

....

....

....

....

....

....

Give an example of an INTERMITTENT RELAY:

....

....

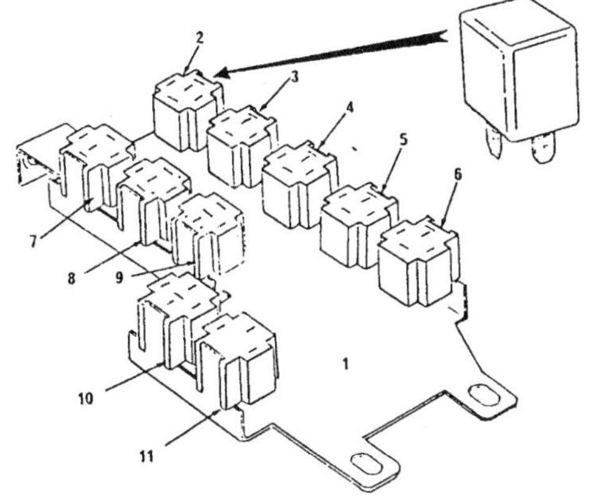

1. *Relay panel*
2. ....
3. *Interior light delay unit*
4. ....
5. ....
6. ....
7. ....
8. ....
9. ....
10. *Heated rear-window timer*
11. ....

Describe the procedure for testing the relay shown below:

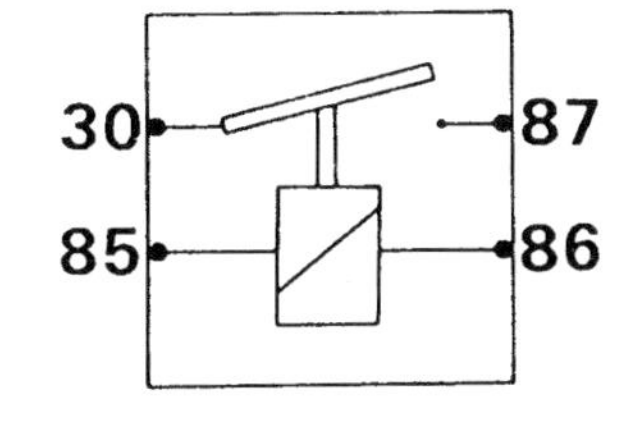

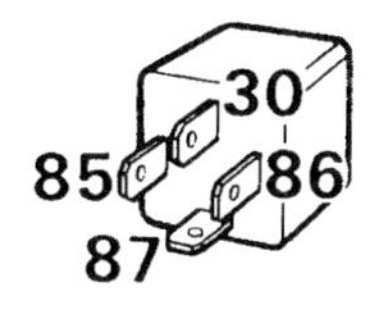

....

....

....

....

....

....

....

The ignition switch assembly consists of:

1. An electrical switch for

   (a) ....

   (b) ....

   (c) ....

2. A spring-loaded plunger for locking the steering column.

3. A locking barrel – when the key aligns the wards, the steering lock release cam is rotated and the electrical switch can be operated.

Inertia switches
Give examples of the use of inertia switches and describe briefly how they are activated:

....

....

....

....

....

....

....

....

WIRING DIAGRAMS

Fault-finding on the electrical system of a vehicle very often involves tracing electrical circuits on wiring diagrams. The wiring diagram for a vehicle is normally shown on a number of sheets with components grouped together for quick and easy reference. The diagrams show power supplies, earth connections, multi-plugs, branch points, etc.

In the type of diagram (DIN) shown on the next page, the electrical components are given a code number and a current path or track number (see sample index opposite).

Comp.	Description	Current Track
H4	Oil pressure warning light	373
H5	Brake warning light	370
H6	Hazard warning flasher repeater light	249
H7	No charge (ignition) warning light	372
H8	Main beam warning light	378
H9	LH stop-lamp	243
H10	RH stop-lamp	244
H11	LH front direction indicator	250
H12	LH rear direction indicator	251
H13	RH front direction indicator	254
H14	RH rear direction indicator	253
H15	Fuel level warning light	388
H16	Glow plug preheating warning light	380
H18	Horn	332, 333
H19	Headlight warning buzzer	264, 265
H20	Choke warning light	380
H21	Handbrake warning light	382
H30	Engine management warning lamp	380
H33	LH direction indicator repeater	249
H34	RH direction indicator repeater	255
K1	Heated rear window relay	279, 280
K2	Flasher unit	247
K8	Wiper delay relay	310 to 313
K15	Timing control unit	484 to 495
K20	Ignition module and coil	138 to 140, 435 to 437, 464, 465
K30	Rear wiper delay relay	325 to 327
K37	Central locking relay	337 to 343
K57	Injection control unit	178 to 196
K58	Fuel pump relay	198, 199, 461, 462
K59	Day running lamp relay	227 to 233
K62	Dim-dip control unit	236 to 240
K63	Horn relays	332, 333
K68	Injection control unit relay	496 to 499, 527 to 530
K72	Engine speed relay	135 to 137
K76	Glow plug control unit (Diesel)	418 to 424
K77	Sensor resistor relay	424, 425
K78	Pre-resistor relay	427, 428
K79	Charging indicator relay	413 to 416
K80	Filter heating relay	431, 432
K84	Ignition control unit	469 to 482
K91	Engine management control unit	502 to 526
K97	Headlight washer delay relay	316 to 318
K100	Engine management control unit	441 to 459
S2	Light switch	
S2.1	Lighting switch	207, 210
S2.2	Interior light switch	261
S3	Heater blower/heated rear window switch	281 to 285
S5	Multi-switch	
S5.2	Dipswitch	216, 217
S5.3	Direction indicator switch	253, 254
S6	Distributor (contact breaker)	117
S7	Reversing light switch	277
S8	Stop-light switch	244
S9	Wash/wipe switch	

Comp.	Description	Current Track
E1	LH parking lamp	201
E2	LH tail lamp	202
E3	Number plate lamp	212
E4	RH parking lamp	210
E5	RH tail lamp	211
E6	Engine compartment lamp	319
E7	LH main beam	215
E8	RH main beam	217
E9	LH dipped beam	216
E10	RH dipped beam	218
E11	Instrument illumination	376, 377
E13	Luggage area light	257
E14	Courtesy light	261
E15	Glovebox light	276
E16	Cigarette lighter illumination	274
E17	LH reversing lamp	277
E19	Heated rear window	279
E24	Rear foglamp	224
E25	LH seat heater	358
E26	Light switch illumination	207
E30	RH seat heater	362
E32	Clock illumination	272
E34	Ashtray illumination	368, 370
E37	Driver's vanity mirror illumination	260
E40	Passenger's vanity mirror illumination	259
F1 to F11, F13, F14, F16 to F18	Fuses (in fuse box)	Various
F21	Fuse – headlamp washers	318
F27	Fuse – horn	333
F36	Fuse – filter heating (Diesel)	432
F41	Fuse – glow plugs (Diesel)	425
F43	Fuse – oxygen sensor	527
G1	Battery	101
G2	Alternator	109
G3	Battery – Diesel	401
G6	Alternator – Diesel	411 to 413
H1	Radio	269 to 271
H2	Horn	331
H3	Direction indicator warning light	375
S15	Luggage area light switch	257
S16	Driver's door switch	262
S17	Front passenger door switch	263
S18	Glovebox light switch	276
S22	Rear foglamp switch	222, 224

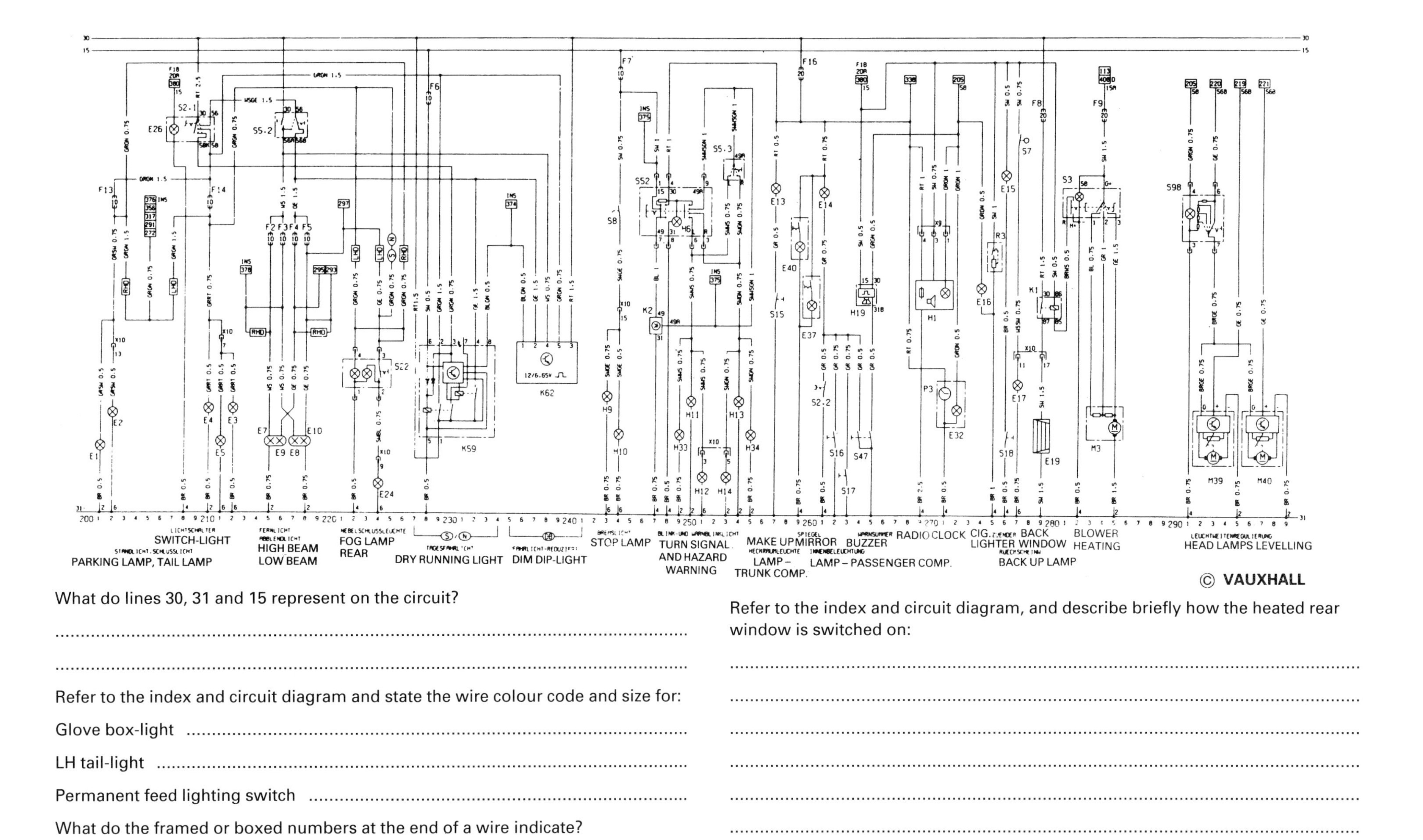

What do lines 30, 31 and 15 represent on the circuit?

....

....

Refer to the index and circuit diagram and state the wire colour code and size for:

Glove box-light ....

LH tail-light ....

Permanent feed lighting switch ....

What do the framed or boxed numbers at the end of a wire indicate?

....

....

Refer to the index and circuit diagram, and describe briefly how the heated rear window is switched on:

....

....

....

....

....

....

....

....

Chapter 12

Lighting Systems

Lighting regulations and circuits 221
PSV interior lighting 222
Bulbs and lamps 223
Headlamp alignment 224
Direction indicators/hazard warning 225
Lamp warning devices 225
Lighting system features 226
Multiplexing 226
Electronic architecture 227
Trailer/caravan electrics 228
Diagnostics 229
Volt-drop check 230
Current consumption 230
OHM-meter 230

LIGHTING REGULATIONS AND CIRCUITS

The purpose of the lighting system on a vehicle is to:

(a) *Identify the vehicle and indicate its position*

(b) ....

(c) ....

(d) ....

(e) ....

State the main statutory regulations relating to:

(a) Bulb rating

....

....

(b) Colour of light emission

....

....

(c) Lens and reflector condition

....

....

(d) Fog and driving lamps

....

....

(e) Reversing lamps

....

....

(f) Use of lamps

....

....

Lamps and reflectors fitted to road vehicle lighting systems include:

1	....	7.	....
2.	....	8.	....
3.	....	9.	....
4.	....	10.	....
6.	....	11.	....
7.	....	12.	....

Complete the wiring diagrams for the circuits 1 to 5; include any fuses, relays, etc. in the circuits.

1. Side and tail lamp circuit

2. Headlamp circuit

3. Fog and driving lamps

⊗

⊗

The fog lamp gives a low flat beam to pick out the kerb, whereas a driving lamp provides a long penetrating beam useful for fast driving.

4. Reversing lamps

⊗

⊗

The stop lights and indicator lamps can be operated via resistors switched into the circuits by a relay. What is the reason for this?

....

....

5. Interior lights

⊗

⊗

⊗

PSV INTERIOR LIGHTING

The interior lighting load for a PSV is obviously considerable. It is therefore advantageous to employ fluorescent lighting for this purpose. State two advantages of fluorescent lighting:

1. ....

2. ....

It is common practice to use three-foot fluorescent tubes operating in pairs in conjunction with an inverter and transformer. What is the purpose of these two components?

....

....

....

....

....

....

BULBS AND LAMPS

The drawings opposite illustrate typical front and rear lamp assemblies. Complete the labelling on the drawings and the table below:

BULB	TYPE/RATING DESCRIPTION	FUNCTION
A		
B		
C		
D		
E		

How does the SEALED BEAM headlamp unit differ from the type shown opposite?

....

....

....

Lenses

The lens in, for example, a headlamp distributes the light rays to provide correct illumination of the road ahead during main beam and dip operation.

State the purpose of the area (x) on the lens shown below.

....

....

....

....

Describe how the lens above can be modified for continental driving.

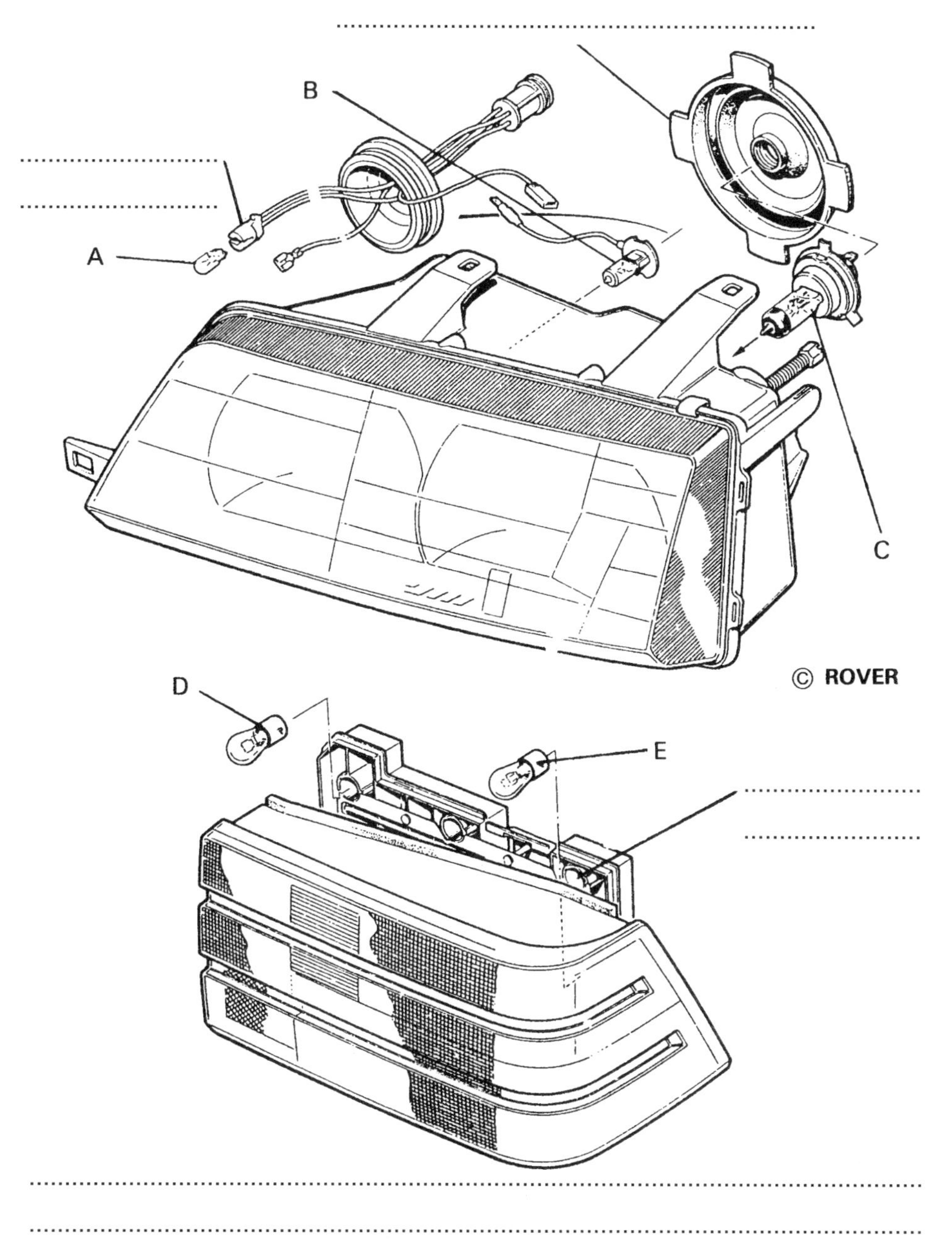

....

....

HEADLAMP ALIGNMENT

All vehicle headlamps in the UK must comply with the Department of Transport Road Vehicle's Headlamp Regulations, which state the position, or angle, of the dipped beams.

The specialist equipment used to align headlamps measures the angle of dipped beam and the beams' positions relative to one another when both on main and dipped beam.

Show a sketch of such equipment in its testing position:

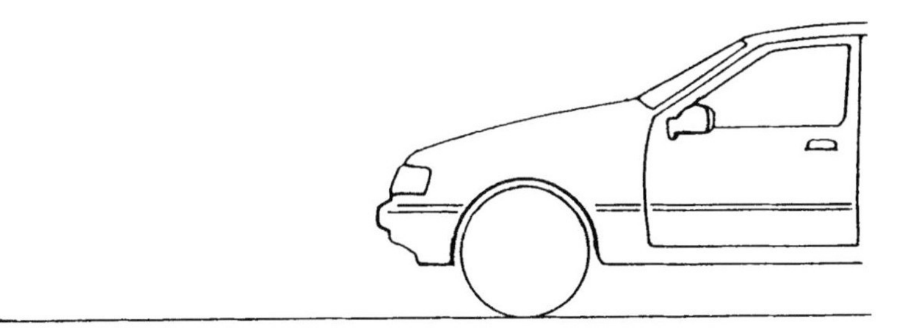

Type of gauge: ....

What pre-checks are necessary to ensure that headlamp alignment is accurately carried out? ....

....

Check the headlamp alignment of a vehicle, using available equipment, and describe the alignment procedure.

Vehicle Make .... Model ....

....

....

....

....

....

On the aiming screens shown, indicate correct main and dipped beam positions.

1. Typical main driving beam position all types

....

....

....

....

....

....

....

....

2. Symmetric dipped (passing) beam

....

....

....

....

....

....

....

3. Asymmetric dipped (passing) beam

....

....

....

....

....

....

DIRECTION INDICATORS/HAZARD WARNING

Amber coloured flashing lights are used to signal a hazard warning, or intent to change the direction of a vehicle, to other road users. In the indicator circuit a FLASHER UNIT is used to make and break the electrical supply to the lamps causing them to flash.

Complete the wiring diagram for the circuit shown and describe the operation of the system during normal turn indicating and hazard warning:

....

....

....

....

....

....

State the statutory flashing rate for direction indicators:

....

Electronic Flasher Unit

As is the case with many other electrical components, the use of electronic circuitry has improved the performance of the flasher unit.

The system comprises an integrated circuit, capacitor and resistors working in conjunction with a relay.

Intermittent current pulses operate the relay which controls the current to the lamps.

LAMP WARNING DEVICES

A typical lamp warning device is the headlamp main beam warning light. This generally consists of a bulb wired in parallel with the main beam lamps to warn the driver when the circuit is in use. Give another example of the use of this form of warning lamp:

....

For reasons of safety it is now common practice to warn the driver of a bulb failure in any part of the lighting or signalling system.

Magnetically operated reed switches can be used for this purpose. How does the system operate?

....

....

....

....

....

....

....

....

Give an example of the use of a 'switch off reminder':

....

LIGHTING SYSTEM FEATURES

Current features of vehicle lighting systems include:

1. The use of XENON lamps. The Xenon bulb emits a blue-tinged white light (close to natural light) which is produced by an arc between two electrodes (no filament). This lamp provides an increase in the level of light emitted.
2. Automatic headlight adjustment. Sensors at the front and rear of the vehicle calculate the ride height. A computer analyses a number of parameters and determines the appropriate headlight position. The system then energises electric actuators which adjust headlamp angles.
3. Environment illumination. In conditions of darkness a delay after 'switch off' allows the headlights to remain on (for about 1 minute) to illuminate a path or garage door, for example. Locking the vehicle via the remote control switches the lights off.

Describe one other current headlight feature.

....

....

....

....

....

....

....

....

....

....

Brake warnings

1. During very rapid deceleration, hazard flashers come on automatically.
2. A high level stop light at the base of the rear screen consisting of a row of LEDS behind a crystal lens.

MULTIPLEXING

In recent years the level of driver and passenger facilities or services has increased greatly. This has led to an increase in on-board electronics and electrical wiring. The wiring harness for a top-of-the-range vehicle, for instance, becomes unmanageably complex and bulky, while the number of pins at the individual ECUs also becomes excessive.

The principle of MULTIPLEXING is to link the various elements which must communicate with each other to the same 'databus' or communication line (BUS), providing one link and only two connections per element (one to the BUS and the other to the management system).

The Controller Area Network (CAN) system is used in automobiles.

Briefly describe CAN.

....

....

....

....

....

....

....

....

....

....

....

State the main benefits of using a multiplex network.

....

....

....

ELECTRONIC ARCHITECTURE

A typical multiplexing arrangement for a vehicle is shown here. The Intelligent Control Module (BSI) collects and processes all data from the MECHANICAL systems. The BSI is also linked to three other networks which are dedicated to: COMFORT, BODY and SAFETY.

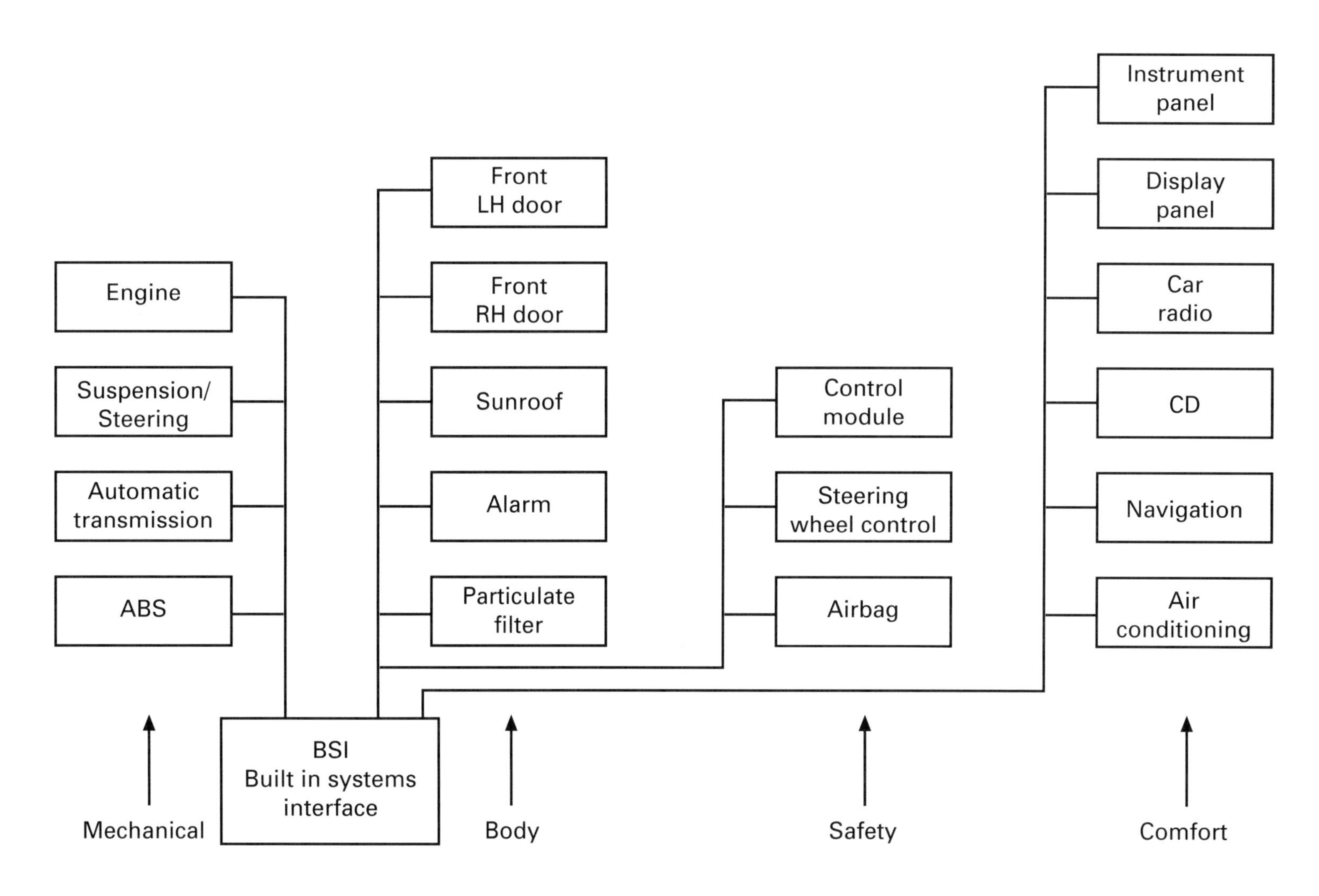

TRAILER/CARAVAN ELECTRICS

All types of trailer units (including caravans) must show the obligatory rear-facing lamps. These must be operated from the driving vehicle's battery and not a separate battery (that is, that could be fitted to the trailer to provide interior lighting).

In order to provide this supply safely, and allow it to be easily disconnected from the vehicle, 7-pin connectors are used. An example is shown below.

Identify each part:

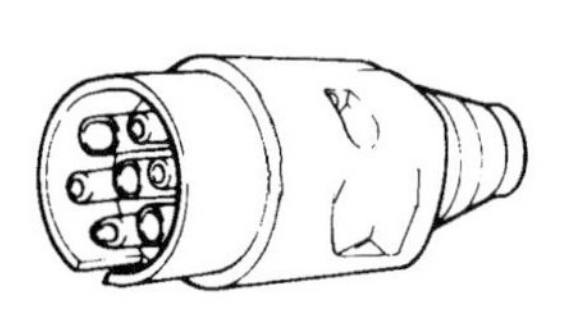

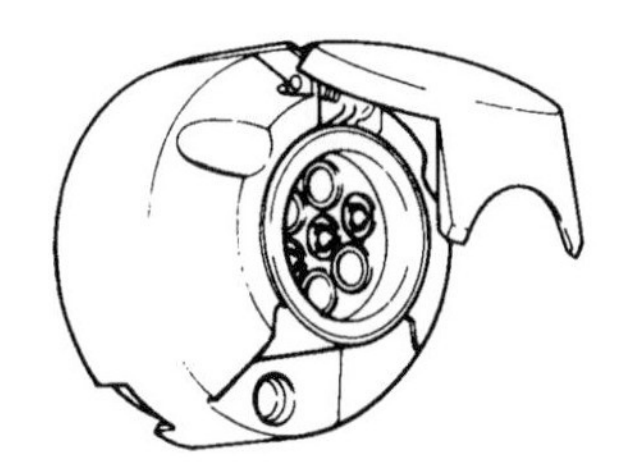

On older vehicles and caravans a single 7-pin connector is able to supply all the electrical needs. On modern caravans it is obligatory to fit rear fog lights. This extra item, together with the possible fitting of reversing lights and internal caravan electrical equipment, has made the fitting of two 7-pin connectors essential.

The first (or existing) connector is known as a ....

The second connector is known as a ....

Number each connection on the drawings and indicate how the sockets are arranged to prevent interchangeability:

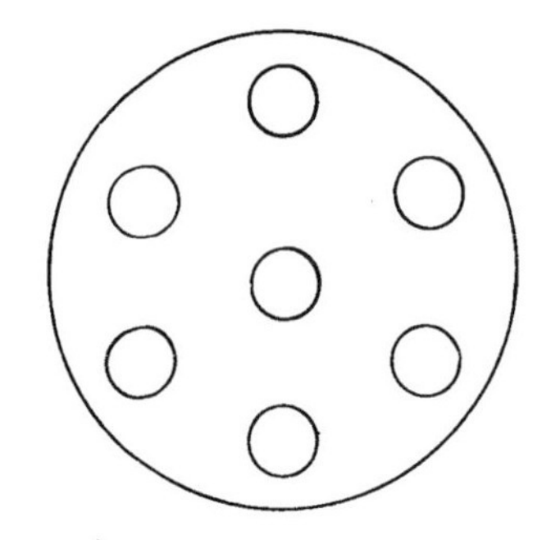

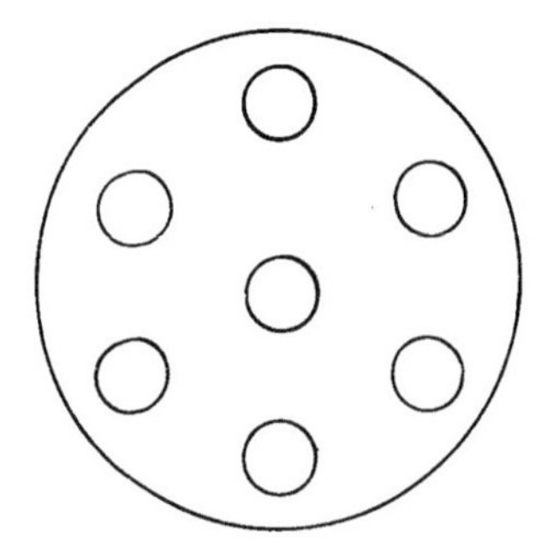

The vehicle's rear cable harness arrangement shows the items that should be connected to the 12 N socket. With the aid of the table below, complete the diagram to show the socket correctly wired.

Number the socket connections and indicate the cable colours.

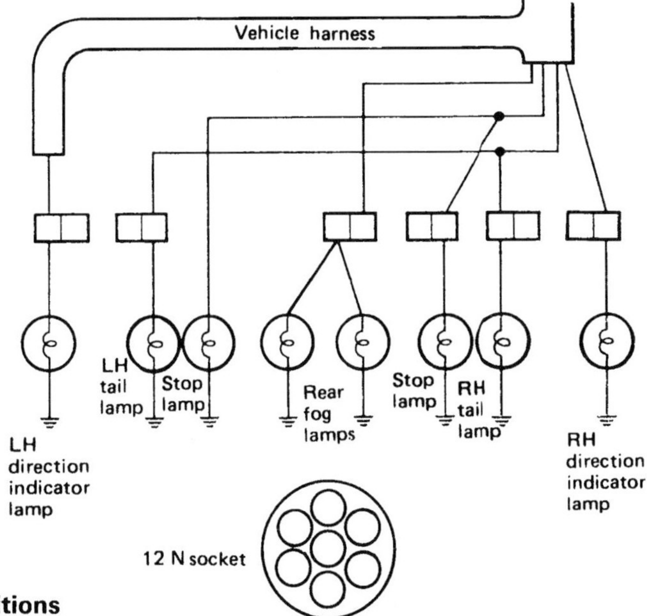

Pin dispositions

Pin no.	7-core cable colour	Circuit allocation: 12 N connector (ISO 1724)	Circuit allocation: 12 S connector ISO (3732)
1	Yellow	LH flashers	Reverse and/or mechanisms
2	Blue	Rear fog (auxiliary, older vehicles)	No allocation (additional power)
3	White	Common return (earth)	Common return (earth)
4	Green	RH flashers	Power supply (caravan interior)
5	Brown	RH side/tail/no. plate	Sensing device (warning light)
6	Red	Stop	Power supply (refrigerator)
7	Black	LH side/tail/no. plate	No allocation

DIAGNOSTICS: LIGHTING – SYMPTOMS, FAULTS AND CAUSES

State a likely cause for each symptom/system fault listed below. Each cause will suggest any corrective action required.

SYMPTOMS	FAULTS	PROBABLE CAUSES
Intermittent light operation; inoperative lamp	Loss of earth	....
Intermittent lamp operation	Corroded bulb holder	....
Beam too low or too high	Beam misalignment	....
Short circuits; ingress of water	Physical damage	....
Indicators inoperative	Faulty flasher unit	....
Low intensity light	Corroded or discoloured lamp reflector	....
Light inoperative; blown fuse	Short circuit	....
Intermittent light operation	Faulty switch	....
Inoperative light	Open circuit	....

VOLT-DROP CHECK

When checking for high resistance (bad connections), a volt-drop check should be made.

State what is being checked on the diagrams and carry out a similar check.

Give expected and actual values:

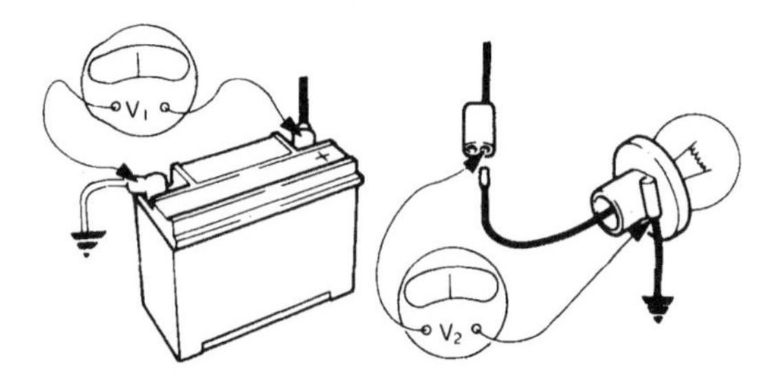

Readings		Voltage Expected	Voltage Actual
V_1			
V_2			
V_3			
V_4			
V_5			

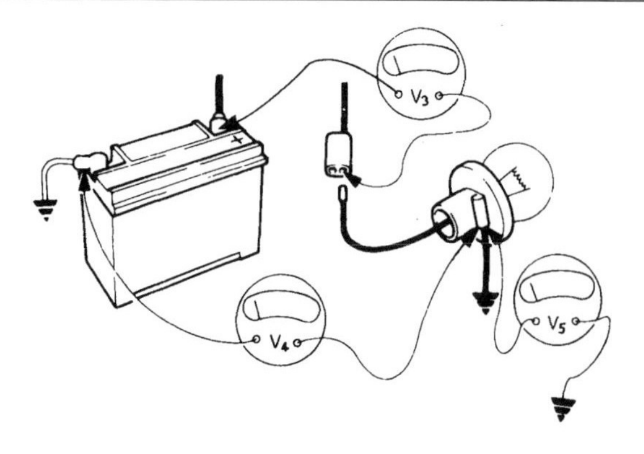

CURRENT CONSUMPTION

Describe how the wiper motor circuit shown would be checked for current consumption. Show on the circuit where the instrument would be connected.

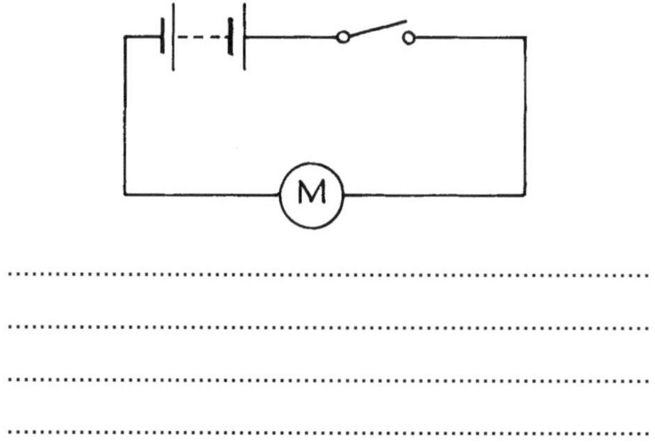

....

....

....

....

....

State typical values for the current flow in a wiper circuit:

....

OHM-METER

State one example of the use of an ohm-meter and show by sketching how the meter could be connected to check circuitry or components:

....

....

What are the advantages of a MULTIMETER as opposed to the use of individual gauges?

....

....

....

....

....

....

Calculate the current consumed by a 12 V, 21 W lamp bulb.

....

....

....

....

....

....

....

The current supplied to an HGV wiper motor during operation is 25 A and the voltage at the motor terminals is 12.5 V. Calculate the power consumed by the wiper.

....

....

....

....

....

....

Chapter 13

Auxiliary Electrical Systems

Auxiliary electrical systems	232
Windscreen wiper system	232
Horns	236
Bi-metal fuel and temperature gauges	238
Sensors	239
Speedometer/tachograph	240
Electrically operated windows	241
Central door locking	242
Diagnostics	242
Infra-red locking/unlocking	243
Electronic components	243
Diagnostics	244

AUXILIARY ELECTRICAL SYSTEMS

Auxiliary electrical systems fitted to road vehicles include the following:

1. *Windscreen and headlamp wipers and washers.*
2. ....
3. ....
4. ....
5. ....
6. ....
7. ....

WINDSCREEN WIPER SYSTEM

Statutory regulations require that a road vehicle should be equipped with one or more windscreen wipers and washer to give the driver a good view of the road ahead in all weather and driving conditions. Most windscreen wiper mechanisms are electrically operated. A relatively powerful motor is required to drive the mechanism and it is desirable that the motor and mechanism are quiet in operation.

Describe the operational features of a windscreen wiper/wash system:

....

....

....

....

....

....

....

....

Windscreen Wipers

The wipers are operated by a mechanism which is given a to-and-fro action by a suitably designed linkage which is connected to the armature shaft of a small electric motor.

Identify the type of drive assembly: ....

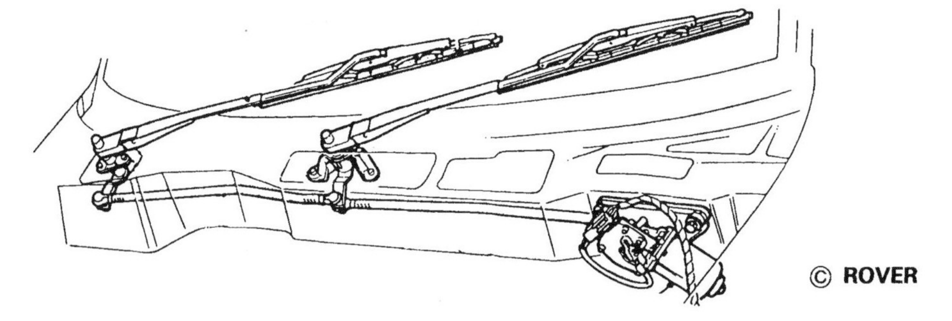

Name one other type of wiper drive mechanism:

....

The electric motor may be of a single-twin or variable-speed type. Almost all types have a limit-switch device incorporated into the drive assembly.

Below is shown the wiring diagram of a single-speed windscreen wiper motor. Explain the operation of the limit-switch when the main control switch is opened:

....

....

....

....

....

....

....

....

....

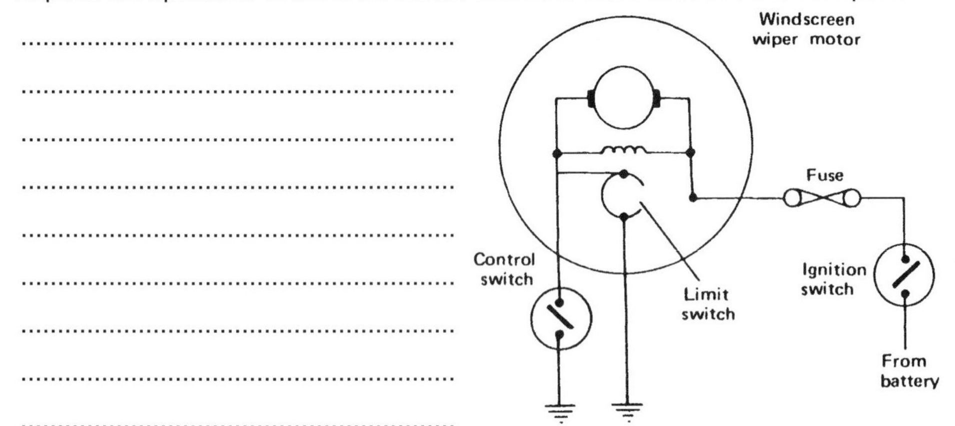

Single- and Two-speed Wiper Motors

The electric motors used for wiper operation are normally two-speed PERMANENT MAGNET type. How do permanent magnet motors differ from the WOUND FIELD type?

....

....

....

....

Describe how two-speed motor operation is achieved. Complete the sketch below to show the brush layout and circuit for two-speed operation.

....

....

....

....

....

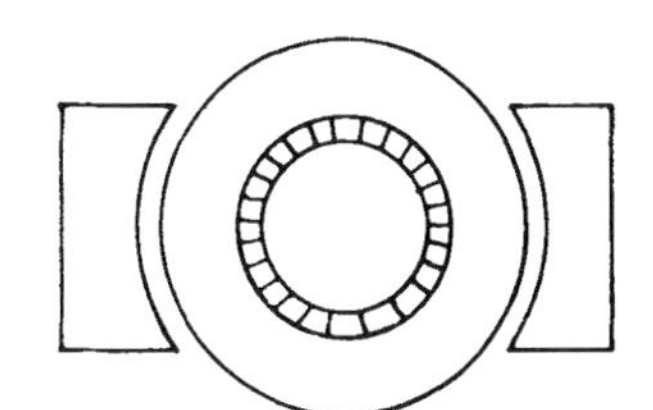

Describe the drive arrangement on both motors and state if single or twin speed, giving reason for choice:

....

....

....

....

....

....

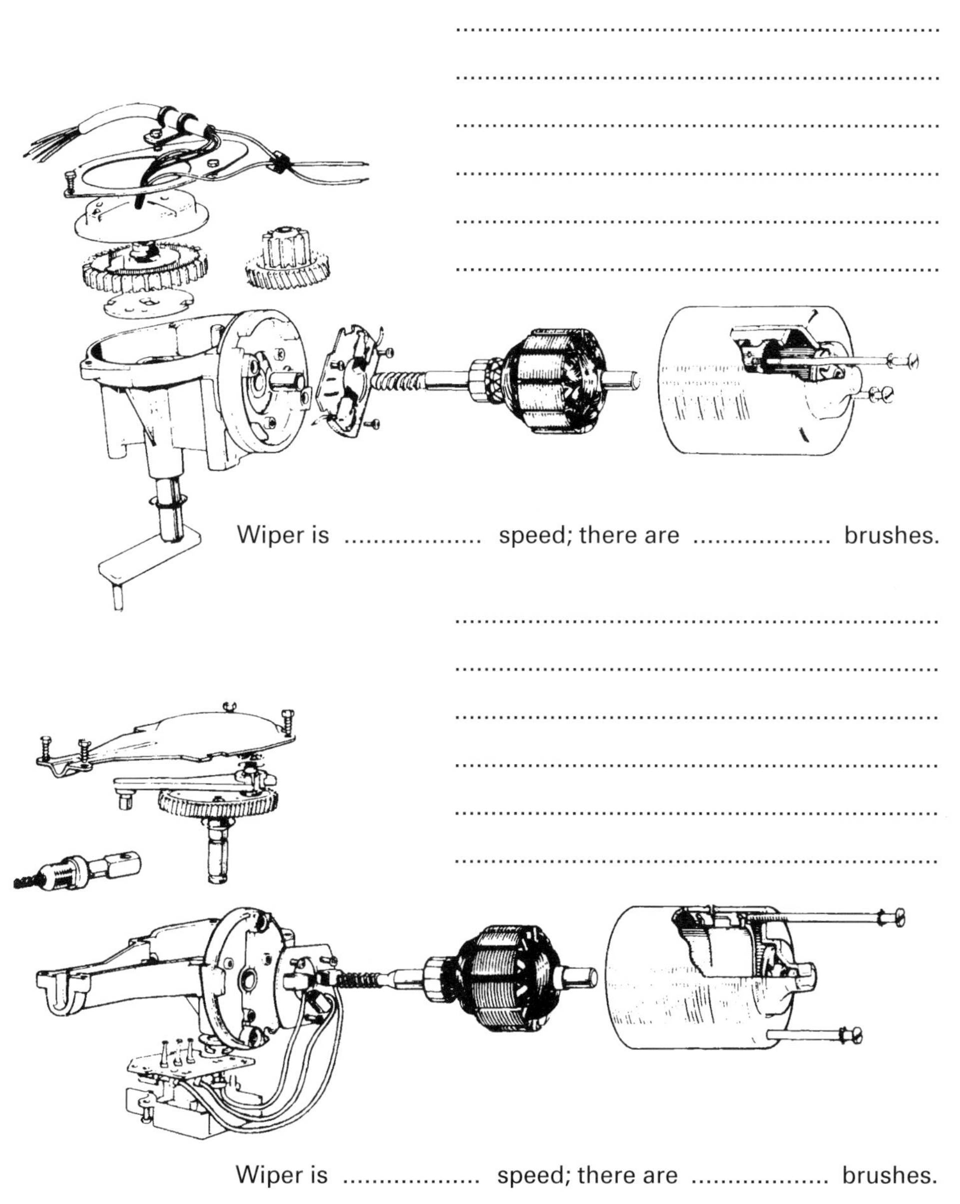

Wiper is .... speed; there are .... brushes.

....

....

....

....

....

....

Wiper is .... speed; there are .... brushes.

Intermittent Wipe System and Screen Wash

The key components in the wiper electrical circuit controlling the intermittent wipe operation are:

1. Wiper Relay 2. Park Switch 3. ECU.

The ECU provides a current pulse, for example 0.65 second every 6 seconds to energise the relay when intermittent wipe is selected.

Study the wiper circuit shown on the next page and describe the operation during intermittent wipe.

Describe also the operation during screen wash:

Label the components on the screenwash system shown at top right.

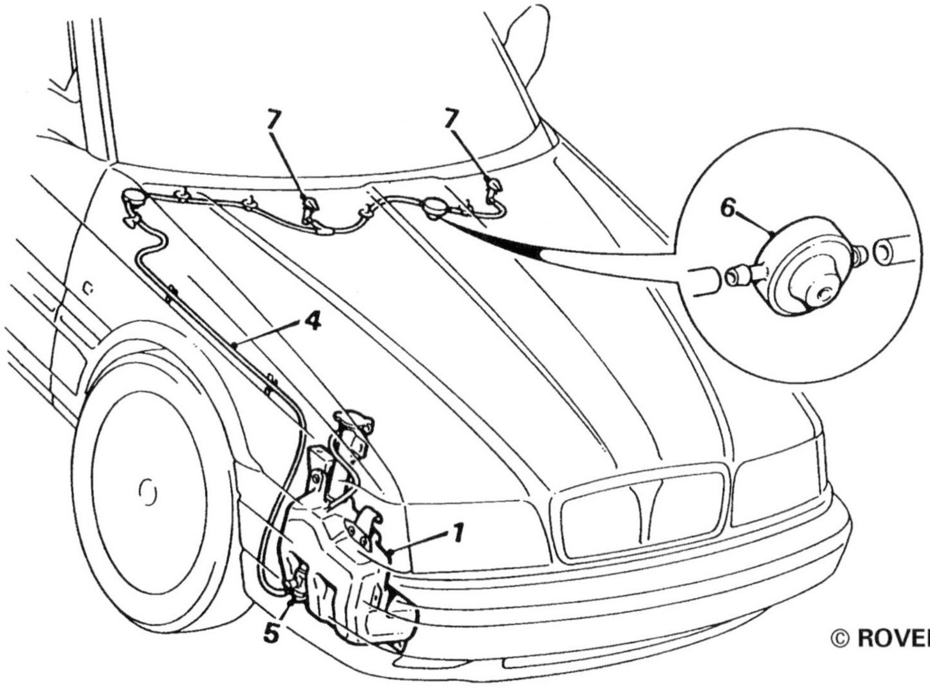

© ROVER

State the purpose of the one-way valves in the system:

A THERMAL PROTECTION device can be fitted in series with the electrical supply to the wiper motor. State the function of such a device and briefly describe one type:

Headlamp Wash/wipe

On the more expensive type of vehicles, headlamps are often fitted with a POWER WASH SYSTEM or a WASH/WIPE SYSTEM. The wash pump for the headlamps is normally a higher powered pump than the screen wash pump and is actuated by a timer.

Examine a vehicle and complete the drawing below to show the headlamp wash/wipe arrangement:

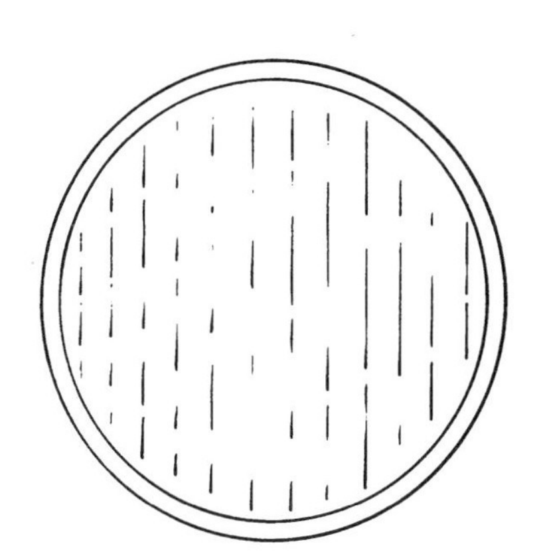

Circuit Diagram – Front wipers

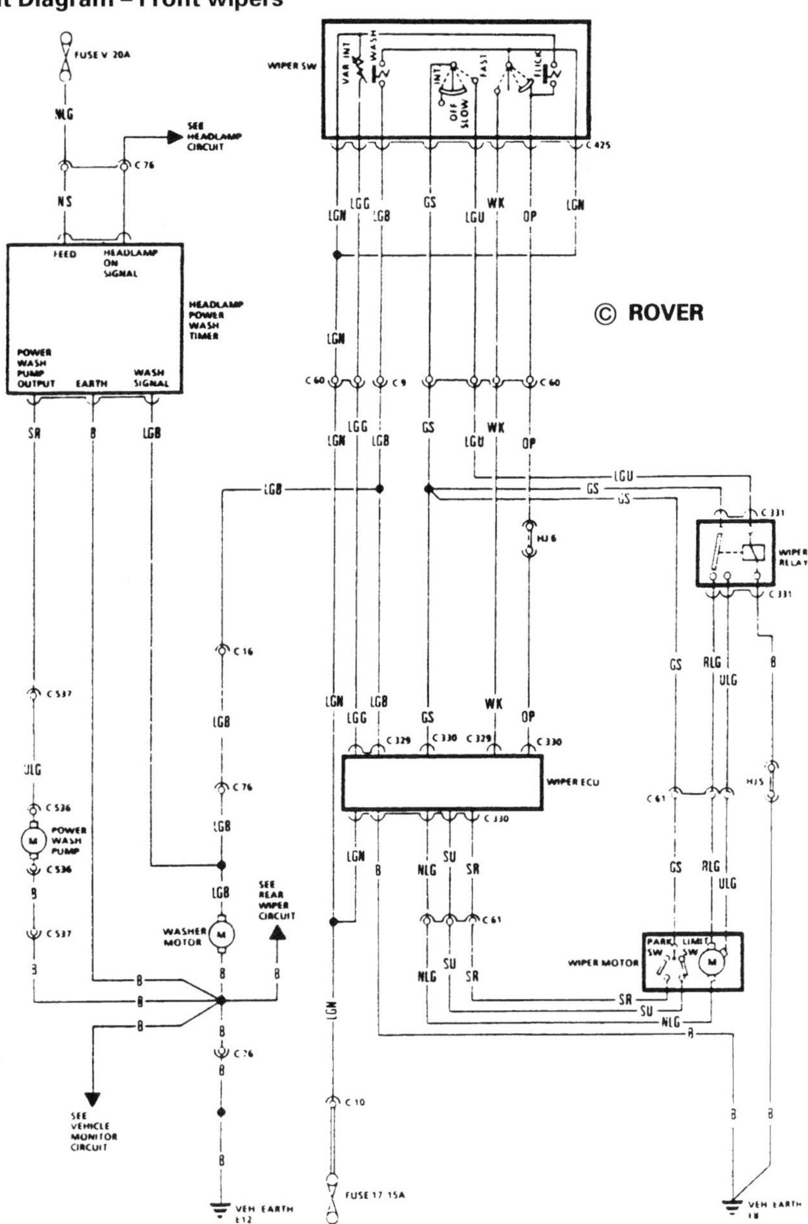

HORNS

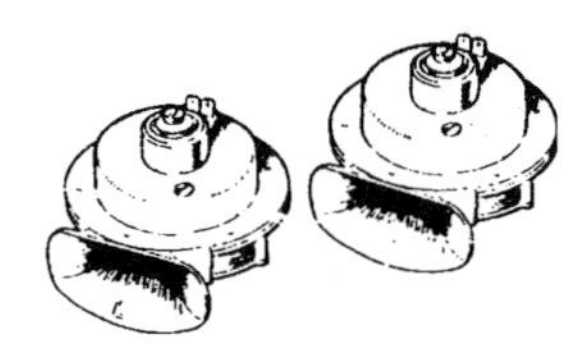

A requirement of the Construction and Use Regulations is that all vehicles must be equipped with an audible warning device.

Three types of horn used for this purpose are:

1. *High frequency horn.*
2.
3.

Complete the drawing below to show a 'high frequency' horn and explain how it operates.

Name the various parts:

Terminals

Windtone horns shown opposite are often used in pairs on a vehicle. Why is this?

Complete the drawing below to show a 'windtone' horn and describe its operation:

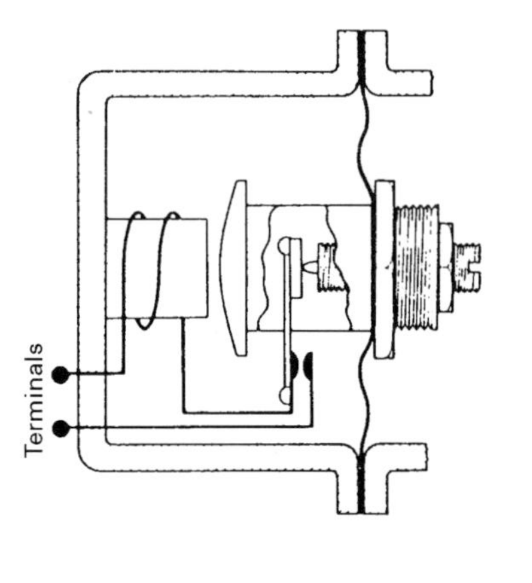

Complete the wiring diagram of the simple twin-horn circuit shown below:

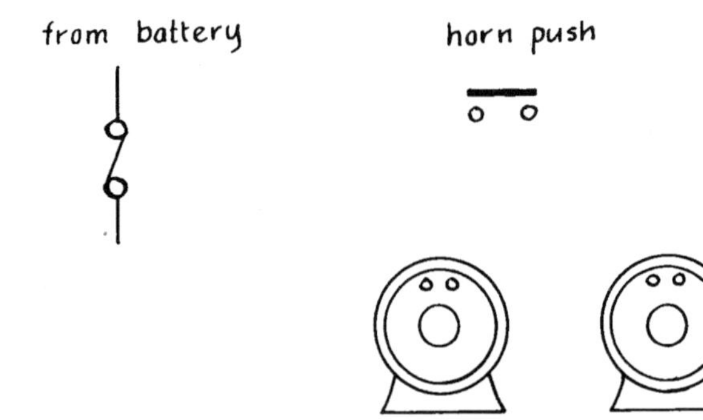

Relay Switches

As in many electrical circuits on a vehicle, a relay is used in the horn circuit. Why is this?

....

....

....

Complete the wiring of the three-terminal relay shown and explain its principle of operation:

Indicate terminal connections.

....

....

....

....

....

....

....

....

....

Sketch a wiring diagram of a twin-horn circuit to include a relay switch:

Air Horns

Air horns are usually purchased from a motor accessory shop and fitted to replace the standard manufacturer's horn on a vehicle. Basically the system consists of a trumpet (or trumpets) and electrically driven air pump.

Examine a vehicle equipped with an air horn system or examine a kit. Sketch the complete system below and describe its operation (show electrical and pneumatic connections):

....

....

....

....

....

....

....

....

....

BI-METAL FUEL AND TEMPERATURE GAUGES

These gauges are fitted on most modern vehicles. The current supply to both fuel and temperature gauge is controlled by a voltage stabiliser. The sender units in both fuel tank and cooling system have resistors that vary depending on either the amount of fuel in the tank or the engine temperature.

Explain the principle of operation of the:

Voltage stabiliser

Bi-metal gauge

Engine cooling system temperature transmitter

Complete this wiring diagram to show the internal wiring symbols of the voltage stabiliser, gauge units and temperature and fuel gauge transmitter units:

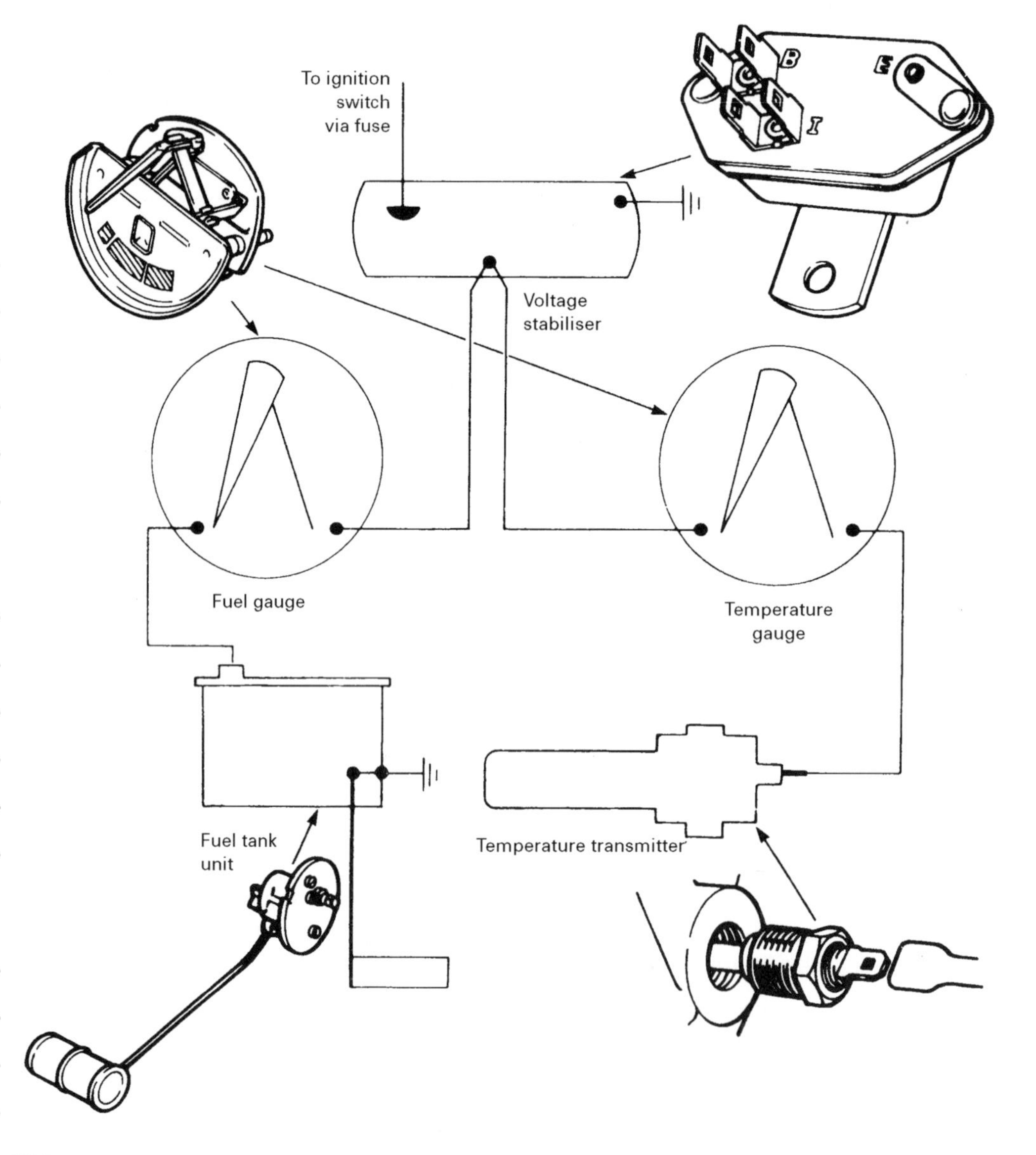

How does the thermal type gauge compare with earlier 'cross-coil' type gauges?

....

....

....

....

An alternative to the thermal (bi-metal) voltage stabiliser is the electronic type shown below.

Describe the operation of this voltage stabiliser:

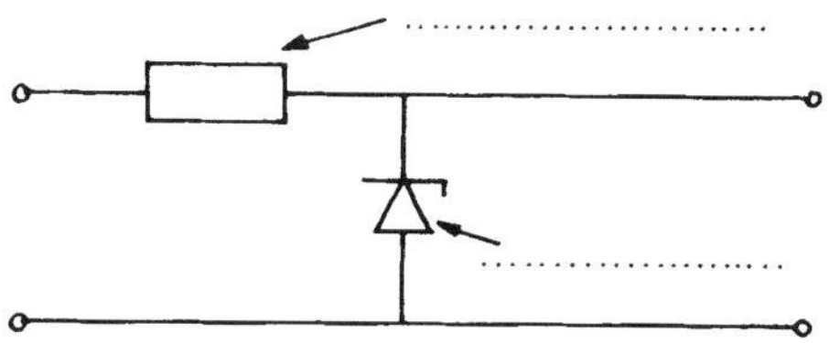

....

....

....

....

....

....

SENSORS

Fluid Level Sensors

Low engine oil level can be indicated by the use of a HOTWIRE sensor located inside the 'dipper' end of the dipstick. Complete the sketch to show this type of sensor and explain how it works:

VEHICLE...

LEVEL INDICATED

....

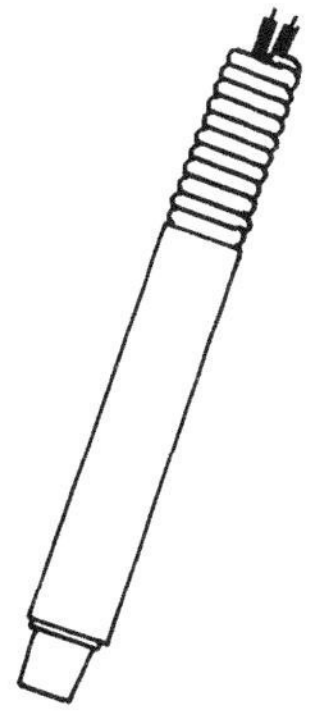

....

....

....

....

....

....

....

....

....

Pressure Sensors

State the purpose of the sensor shown below and describe how it operates:

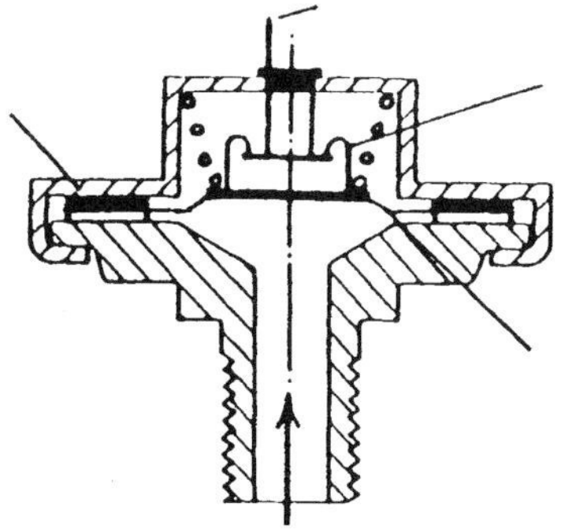

....

....

....

....

....

An indication of oil pressure can also be provided by a thermal type pressure sensor in conjunction with a thermal gauge.

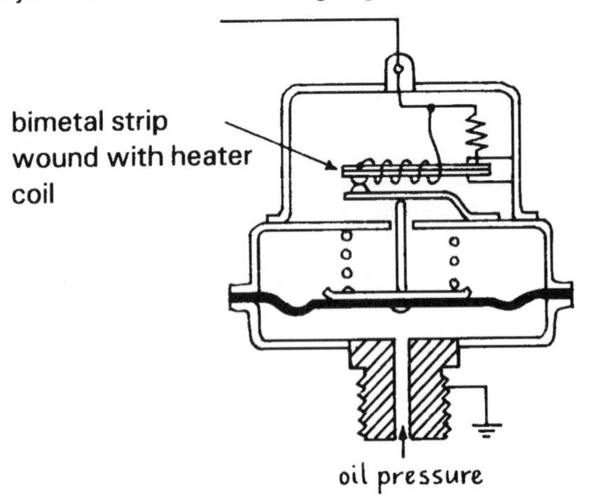

OPERATION (thermal)

....

....

....

....

....

....

Turbo Boost Sensor

A pressure-sensing transducer can be incorporated into the turbo charging system to operate, in conjunction with ECU, wastegate, solenoid valves, etc., to control boost pressure in relation to engine operating requirements.

Sketch and describe one type of pressure sensor which could be used for this purpose:

....

....

....

....

....

....

....

SPEEDOMETER/TACHOGRAPH

The speedometer/tachograph can be driven by an electric motor at the rear of the instrument. A transducer, which is a quenched oscillator pulse generator driven from the gearbox, provides current pulses which are fed to the instrument ECU. The speed of the electric motor is regulated by the ECU in response to the signal from the speed transducer. Describe, with the aid of a simple diagram, the operation of the pulse generator.

Make a simple diagram to show a pulse generator:

ELECTRICALLY OPERATED WINDOWS

Side windows can be operated by electric motors connected to a geared linkage which moves the windows up or down, depending on the motor direction of rotation. A DC permanent magnet motor is used to power each window drive mechanism.

How is the direction of the motor reversed for up or down window operation?

The circuit below shows the main front/rear relays which are energised (fuse 16, wire GY) when the ignition is switched on.

Describe the operation of the system during up and down operation of the driver's door windows. What is the purpose of the 'isolation switch' in the driver's door switch pack?

WINDOW OPERATION

Up

Down

Isolation switch

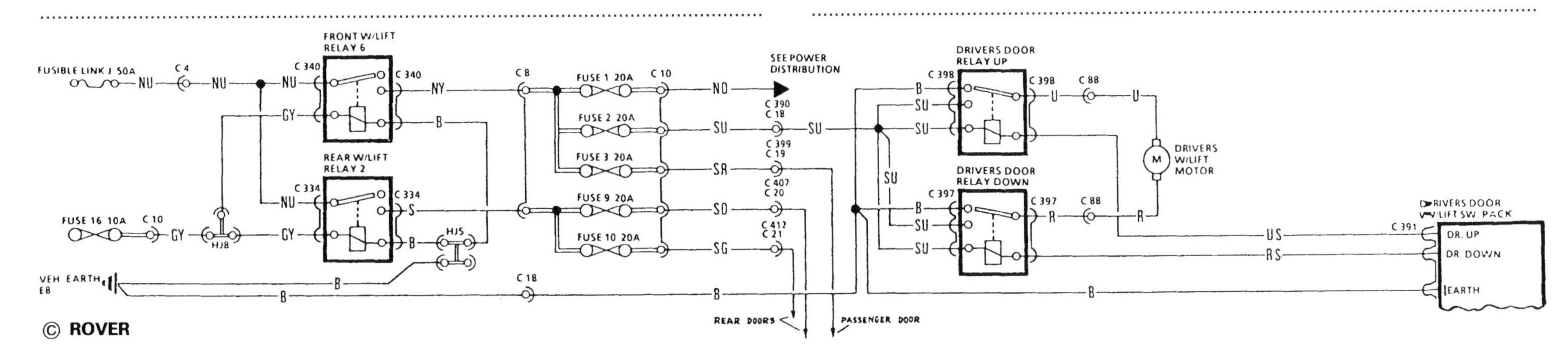

© ROVER

What is the difference between semi-automatic and fully automatic (or 'one-shot') window switching systems?

....

....

....

....

Examine an electric window mechanism and make a simple sketch below to illustrate the motor drive mechanism:

CENTRAL DOOR LOCKING

A central door locking system enables all doors, boot or tailgate to be locked or unlocked simultaneously when the key is turned in the door lock. Electric motors or solenoids in each door and boot or tailgate are activated to operate the door locking mechanism. A front door lock and control is shown below; label the drawing.

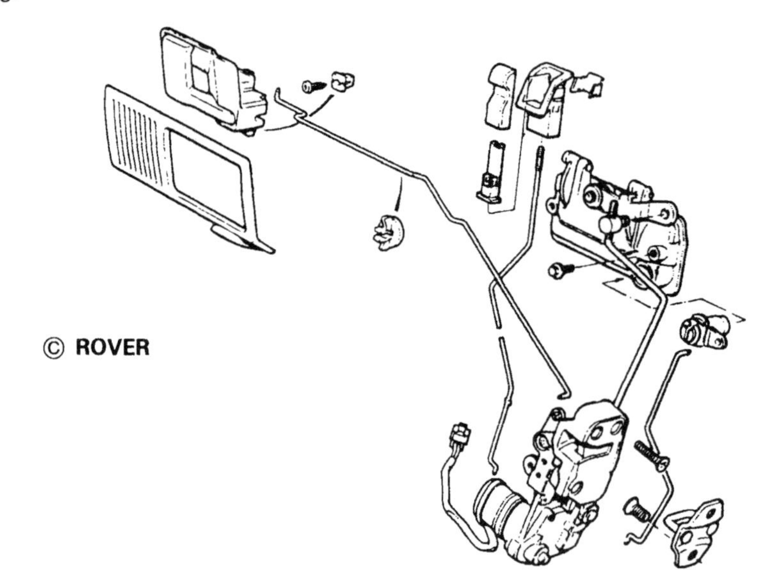

DIAGNOSTICS: ELECTRIC WINDOW/CDL – SYMPTOMS, FAULTS AND CAUSES

State probable causes for each symptom listed below. Each cause will suggest any corrective action required.

SYMPTOMS	FAULTS	PROBABLE CAUSES
Restricted window movement; door lock inoperative	Mechanical failure	....
Window or door lock inoperative	Mechanical failure	....

A complete central door locking electrical circuit is shown below. The ECU provides a current pulse to activate the locks. Study the circuit and describe the operation of the system during door lock and unlock.

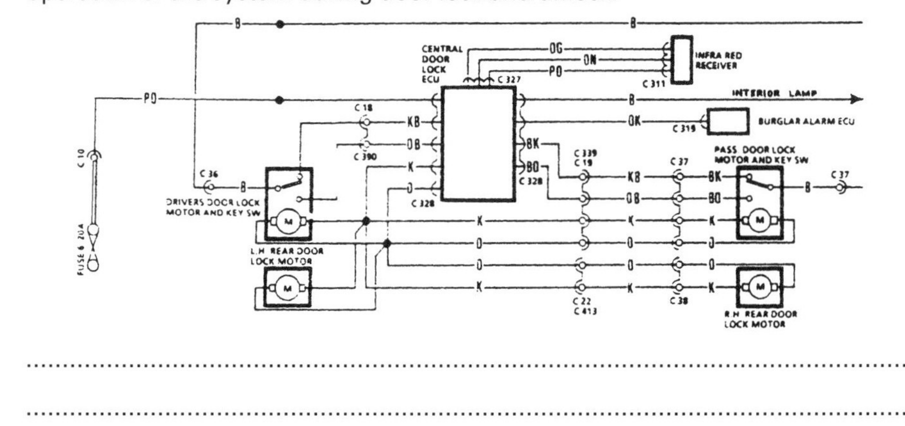

....

....

....

....

....

....

INFRA-RED LOCKING/UNLOCKING

The central door locking circuit can be energised using the door key or remotely using a hand-held infra-red transmitter.

The main components in an infra-red system are: transmitter, receiver/sensor units, and ECU (see CDL circuit).

The transmitter consists of a coded integrated circuit, batteries, switch, emitter diode and function diode.

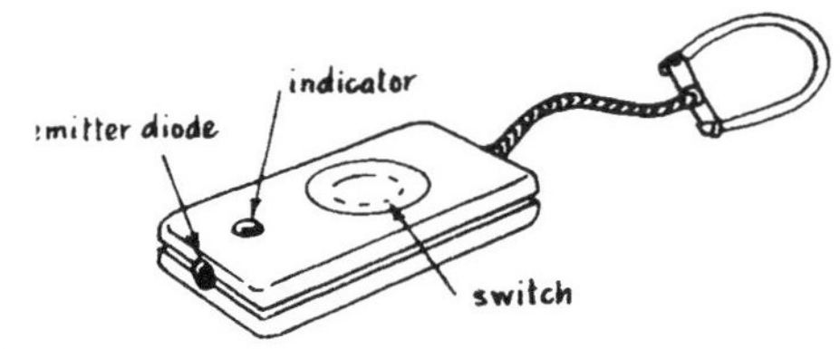

Note: some remote control units use a miniature radio transmitter/receiver rather than infra-red, in order to activate the system.

State where the receiver/sensor units are located and describe the operation of the system:

....

....

....

....

....

....

....

ELECTRONIC COMPONENTS

Complete the table:

COMPONENT	SYMBOL	FUNCTION	M.V. APPLICATION	COMPONENT	SYMBOL	FUNCTION	M.V. APPLICATION
transistor				thermistor			
Zener diode				thyristor			
L.E.D.				diode			

DIAGNOSTICS: AUXILIARY ELECTRIC – SYMPTOMS, FAULTS AND CAUSES

State probable causes for each symptom listed below. Each cause will suggest any corrective action required.

	SYMPTOMS	FAULTS	PROBABLE CAUSES
WINDSCREEN WIPER/WASH	Incorrect parking and swept area	Play in linkage	....
	Abnormal noise; arm and blade judder; blown fuse	Worn or partial seizure of drive gear	....
	Blade lift	Weak wiper arm springs	....
	Screen smearing	Worn wiper blade rubbers; oily deposits on screen	....
	Insufficient or no water spray	Blocked pipe or jet	....
HORN	Low or high note; buzzing	Incorrect tone quality	....
	Horn inoperative; intermittent operation; continuous operation	Circuit fault	....
TEMP. GAUGE	Low or high reading; no or intermittent operation	Incorrect gauge reading	....
ALARM	Incorrect delay; incorrect sensitivity; continuous operation	Short circuit; incorrect adjustment; faulty unit	....

Chapter 14

Enhance Vehicle System Features

Enhance vehicle system features 246
In car entertainment (ICE) 247
Aerials and antennas 249
Interference suppression 250
Fault diagnosis (ICE) 251
Anti-theft systems 252
Mobile communications 253
Towbars 254
Suspension 254
Testing electronic components 255
Electron theory 256
Production of electricity 257
Effects of an electric current 257
OHM's law 258
Electrical circuits 259
Investigation 259
Measuring instruments 259
Parallel circuits 260
Series circuits 261
Permanent and electro-magnets 262
Investigation 262
Lines of force 262
Energy and power 263
Mutual induction 263

ENHANCE VEHICLE SYSTEM FEATURES

To ENHANCE is to improve in value, desirability or attractiveness. The 'standard specification' of a vehicle can be enhanced to provide a vehicle more suited to the customer's requirements; this is customising.

Vehicle manufacturers offer a range of OPTIONAL equipment over and above the standard specification for a particular model. Models at the top end of the range will include as standard many of the optional features available on basic models. In addition to this, vehicle manufacturers and specialist manufacturers and suppliers provide a huge range of ACCESSORIES which can be purchased and fitted as and when required. Very often the governing factor is cost.

List the main items in the technical specification for a vehicle.

ENGINE

TRANSMISSION

STEERING

....

....

....

....

....

....

....

....

....

....

Specific enhancement considerations are:

- in car entertainment
- communication equipment
- anti-theft systems
- safety fitments
- tow bars
- lamps
- wheels and tyres
- performance modifications, e.g. suspension parts
- manufacturer's modifications

A range of typical equipment, standard or optional, is shown right.
(ALFA ROMEO)

Tick those you consider to be 'top options'.

EXTERIOR
Tinted Windows
Electric Front Windows
Electric Rear Windows
Alfa Code Immobiliser and Alarm System
Central Locking with Remote Control
Heated, Electrically Adjusted Door Mirrors
Electric Sunroof
Metallic Paint / Iridescent Paint
15" Alloy Wheels
Selespeed 16" Alloy Wheels with 205/55 WR 16 tyres
INTERIOR
Thermostatically Controlled Automatic Climate Control System
Stereo Radio / Cassette with Six Speaker System
Stereo radio / CD Player with Six Speaker System
Titanium Effect Console / Leather Steering Wheel
Carbon Effect Console / Leather Steering Wheel
Wood Rimmed Steering Wheel and Effect Console
Steering Wheel Reach and Height Adjustment
Driver Seat Height Adjustment and Self-adaptive Lumbar Support
Cloth / Velour Trim
Rev Counter
Analogue Clock
Check Control Panel
Front and Rear Courtesy Lights
Front Armrest
Rear Armrest with Ski Tunnel
Boot and Fuel Filler with Internal Release

INTERIOR
Front Seat Back Pockets
Momo Leather Upholstery (5 Colour Variants)
Instrument Panel Lighting Adjustable to 3 Levels
Metallic Grey Instrument and Control Background
Heated Rear Window
Exterior Temperature Indicator
Boot and Fuel Filler with Internal Release
Lockable Glove Compartment
SAFETY / MECHANICAL
ABS + EBD (Electronic Brake Distribution)
Driver's Airbag
Front Side Airbags
Passenger Airbag
Height-adjustable Front Seat-belts with Pre-tensioners
Power Assisted Steering
Third Brake Light
Fire Prevention System (FPS)
Rear Door Child Locks
Headlight Alignment Adjustment
Front and Rear Foglights
Headlight Washers
Laminated Windscreen
Height-adjustable Front and Rear Head-restraints
3rd Rear Head-restraint plus 3rd Inertia Reel Seatbelt
6-speed Gearbox/Synchronised Reverse
Selespeed Gearbox
Q-system Gearbox
AUDIO
Alpine Radio with Cassette*
Clarion Radio with Single CD Player*
* Includes 2 front tweeters, 2 front woofers, 2 rear speakers, aerial control unit.

IN CAR ENTERTAINMENT (ICE)

The radio cassette player is standard equipment in most of today's cars. However, the range of equipment available for ICE, and the industry involved in it, is considerable. Typical of modern systems would be a radio/cassette head unit with CD control and a CD autochanger.

RADIO/CASSETTE HEAD UNIT

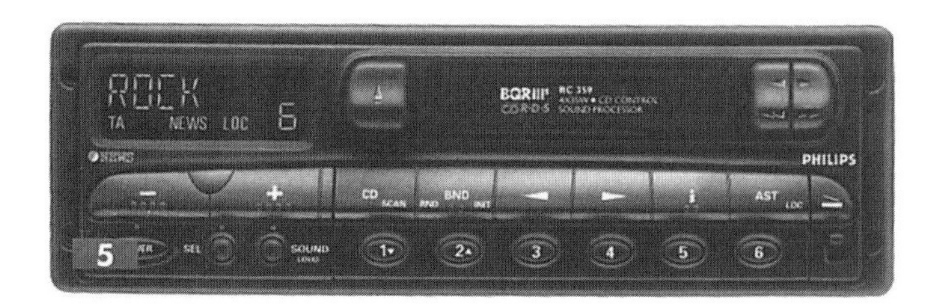

© PHILIPS

CD CHANGER

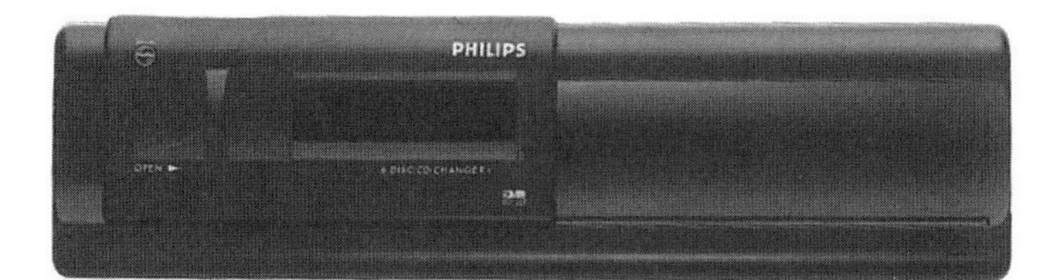

The CD changer is usually installed in the boot horizontally or vertically to suit the location.

Speakers

Modern ICE systems use four or more speakers. It is important to ensure that the IMPEDANCE (ratio of voltage to current in an AC circuit) of the speaker system is matched to the amplifier.

The sound quality of a system is largely dictated by the quality of the speakers. Speakers can be TWEETERS or WOOFERS or a combination of both.

TWEETERS are for ....

WOOFERS are for ....

The reception and reproduction of sound by a system relies not only on the quality of the equipment but also on the quality of the installation. Most reception and sound deficiences are due to problems with the installation. Complete the drawing below to show the installation wiring for a radio and list the important points to be aware of when installing a radio.

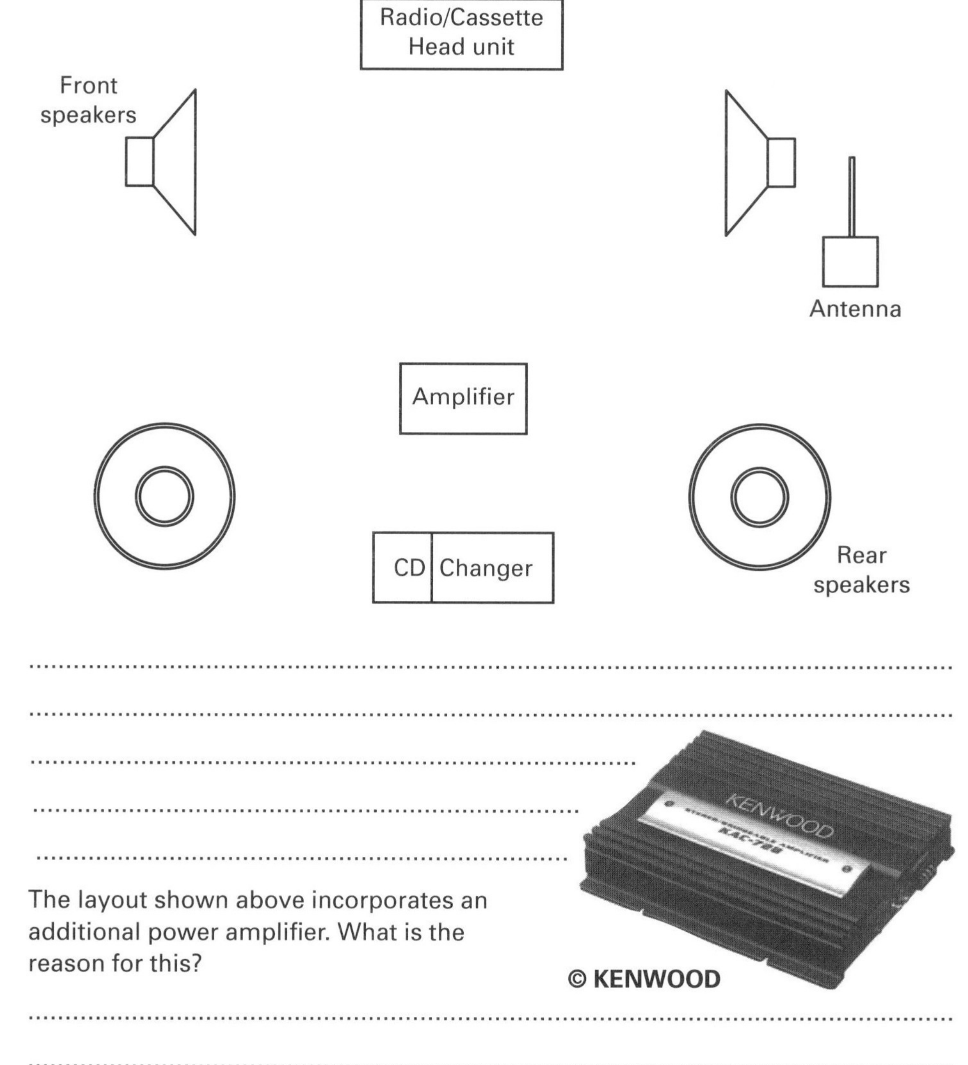

....

....

....

....

....

The layout shown above incorporates an additional power amplifier. What is the reason for this?

© KENWOOD

....

....

IN CAR ENTERTAINMENT

The layout opposite is an example of multi-media equipment available for ICE.

1. System controller/monitor screen/receiver
2. Car navigation unit (DVD)
3. Video player (DVD)
4. CD changer
5. Dolby digital processor
6. Rear monitor
7. Centre speaker
8. Amplifier
9. Speakers.

DVD (Digital Versatile Disc)

DVD has about 7.5 times the capacity of a CD providing: films, video CDs and music CDs to enhance the in car facilities.

© KENWOOD

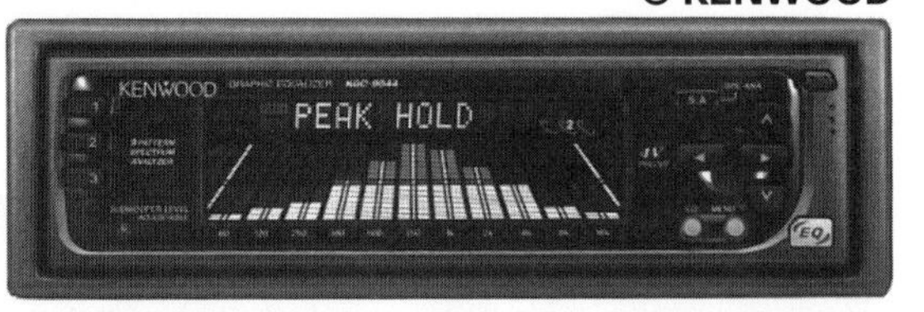

Graphic Equaliser

The GRAPHIC EQUALISER is an ancillary item which enhances ICE equipment. State the function of a graphic equaliser.

....

....

....

....

DAB (Digital Audio Broadcasting)

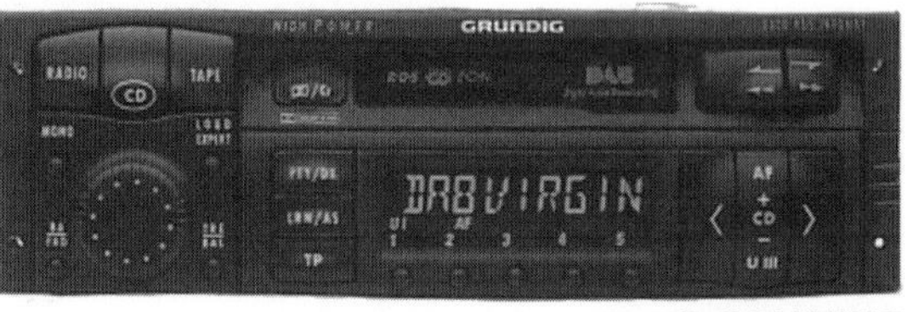

© GRUNDIG

DAB is digital radio. This system offers the advantages of:

Clearer sound quality, more services, no interference, no signal fade (no need to retune during travel).

What is RDS?

....

....

....

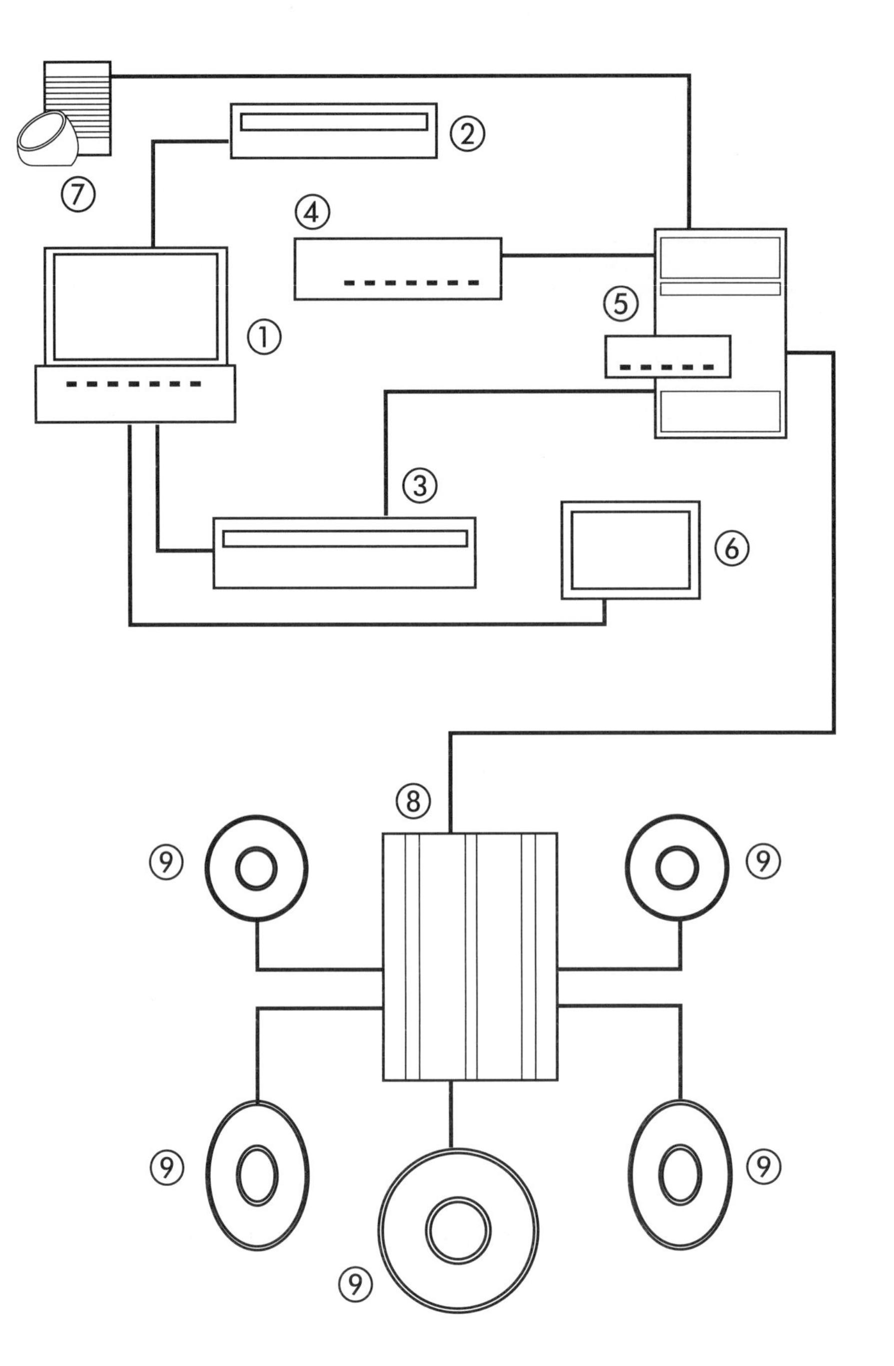

AERIALS AND ANTENNAS

A high percentage of radio service problems can be attributed to inferior or badly fitted aerials. The ideal aerial is a fixed-length antenna but the practicalities of a telescopic aerial make it more suitable (against vandals, car washing, etc.).

Name two other types of aerial:

....

....

Siting
State the factors to consider when choosing a place on the vehicle to position the aerial:

....

....

....

....

Describe the procedure for fitting a mast type aerial on to the car body and complete the drawing to show an installation:

....

....

....

....

....

....

....

....

body panel

What is meant by 'aerial trimming'?

....

....

....

Electrically Operated Aerials

A permanent magnet electric motor can be used to power the drive mechanism for extension or retraction of the aerial mast. A separate two-way switch or a switch incorporated into the radio will operate the motor in opposite directions of rotation. Add the circuit to the aerial assembly shown and describe the drive mechanism for the aerial:

....

....

....

....

....

Twin System

Two antennas, one front wing, one rear screen can be connected via a computer to a receiver which has two tuners. How does this system operate and what are its advantages?

....

....

....

....

....

....

INTERFERENCE SUPPRESSION

Radio interference is caused by high-frequency waves being fed to the receiver, resulting in impaired reception. Interference can reach the receiver by both conduction and radiation.

Describe briefly how frequency interference can reach the receiver via each route.

Conduction interference:

....

....

....

....

....

....

....

Radiation interference:

....

....

....

....

....

....

High-frequency interference waves are produced by electrical components on a vehicle in which sparking or arcing occurs during operation. The main source of interference on a petrol-engined vehicle is the ignition system. List other sources of interference on a motor vehicle:

....

....

....

....

A further source of interference is the build-up of an electro-static charge in body panels, resulting in electrical discharges to adjacent panels. Methods of suppressing interference include the use of: resistors, capacitors, chokes, screens and straps.

How is the high tension ignition circuit suppressed?

....

....

....

....

Show, by sketching, the installation of a capacitor on the components below:

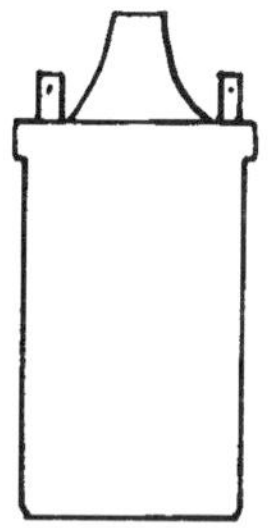

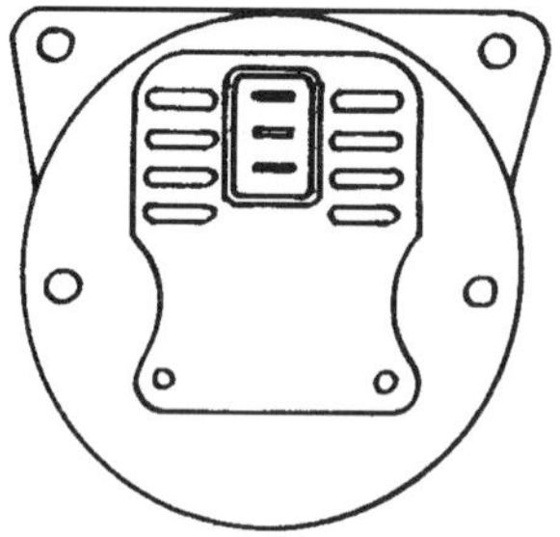

Give examples of the use of CHOKES in electrical circuits for suppression purposes:

....

....

....

....

Screening and Straps

Screening means creating a barrier between the source of interference and the aerial. Interference waves generated inside an effective screen cannot be radiated through it. Metal body parts such as the bonnet have a screening effect.

Metallic screening of the ignition system is an effective but expensive form of suppression.

What does this involve?

....

....

....

....

Earthing straps are an essential part of the interference suppression system. State the reason for these and give examples of places on a vehicle where they would be used:

....

....

....

....

....

....

Certain types of bodywork present particular screening problems. Why is this?

....

....

State the legal requirements relating to suppression on vehicles:

....

....

....

....

FAULT DIAGNOSIS (ICE)

Faults occurring in ICE systems can typically be caused by:

1. poor quality installation
2. physical damage during service, e.g. damage to wiring and connectors or water ingress.
3. main component failure.

Checks on electrical circuitry would include: battery condition, security and routing of cables, security and cleanliness of connections, earth continuity, fuses, supply voltage at components.

State the checks to be made on:

Aerial ....

....

Speakers ....

....

....

Main components (radio/cassette, CD changer)

....

....

....

State the main causes of:

Interference ....

....

....

Poor reception ....

....

....

ANTI-THEFT SYSTEMS

Vehicle anti-theft systems range from a simple immobiliser switch in the ignition circuit, to a highly sophisticated electronic alarm system. Remotely activated tracer systems are now being fitted to some vehicles.

Alarm systems differ basically in the method of sensing interference with a vehicle and in the type of alarm warning activated (such as vehicle horn, siren, flashing lights, personal bleeper, etc.).

Describe briefly the following types of alarm system:

Voltage-drop type

....

....

....

....

....

Inertia type

....

....

....

....

....

Ultrasonic type

....

....

....

....

....

Examine one type of alarm system and make a sketch below to show the circuit. Include: switches, sensors, control and warning device.

Describe any adjustments which may have to be made to a system following installation:

....

....

....

An alarm system can be extended to provide specific protection for items such as spot lamps, radio cassette, caravan. This entails the use of a '24 hour loop' system. Describe this:

....

....

....

....

Make a simple diagram to show an IMMOBILISER installation.

MOBILE COMMUNICATIONS

An alternative to the pocket mobile phone is an in-car phone installation which is integral with the vehicle's audio system. The radio speakers are used as phone speakers with a centrally positioned microphone. This system provides 'hands free' telephone conversation.

Hands Free Phone Kit

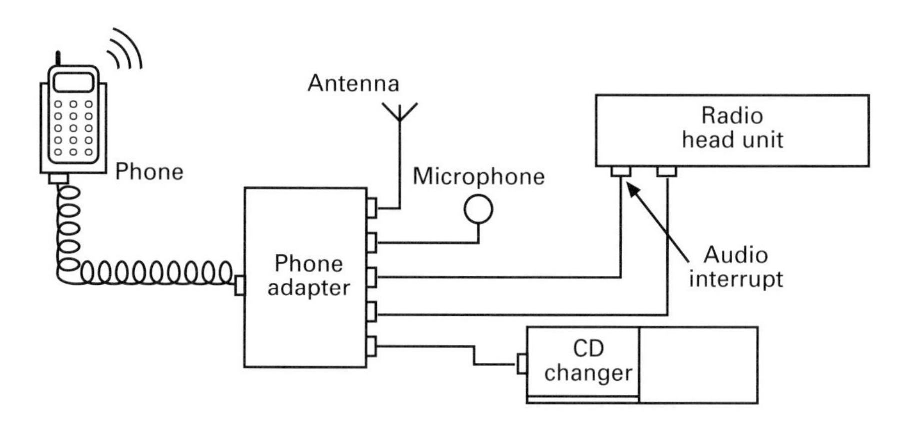

The radio phone is fitted as standard equipment on many top of the range models. Features offered are steering column control for radio/phone switch and scanning various directories on the multi-function monitor screen. With this system a call can be made without taking the hands from the steering wheel.

Using the latest 'plug in' electrical/electronic test equipment it is possible to check out complex circuits and electronic devices in a few seconds.

How might electrical systems be protected against hazards during use or repair?

1. ....

2. ....

3. ....

4. ....

5. ....

6. ....

List general rules for efficiency and any special precautions to be observed when testing, overhauling and repairing the electrical system:

1. ....

2. ....

3. ....

4. ....

5. ....

6. ....

7. ....

8. ....

9. ....

TOWBARS

A towbar is a bracket, usually 'T', 'U' or 'V' formation, which is attached to the rear underframe of a vehicle. The shape and attachment points of a towbar vary according to the vehicle to which it is being fitted. There is, however, a main design consideration for a towbar. What is this?

....

....

....

FIAT ACCESSORY

Make a simple sketch to show a towbar installation. Include the body members to which the towbar is secured.

TOWBAR ATTACHMENT MAKE .... MODEL ....

SUSPENSION

FIAT

This suspension spring is an optional replacement for the standard equipment. When fitted it lowers the suspension giving a lower ride height and sporty appearance to the vehicle.

Uprated suspension springs and dampers are an optional fitment. These make the suspension firmer for sporty driving. (See suspension chapter.)

Tyres and Wheels

The appearance of a car can be enhanced by replacing standard steel wheels with alloy wheels.

However, upgrading tyre and wheel equipment may entail the use of different tyre and wheel specifications, e.g. for a more sporting application.

Give an example of a tyre and wheel upgrade:

Vehicle Make/Model

....

Standard fitment wheel dimensions

....

Wheel upgrade dimensions

....

Standard fitment tyre marking

....

Tyre upgrade marking

....

TESTING ELECTRONIC COMPONENTS

Describe, with the aid of simple sketches, how to test the components on this page.

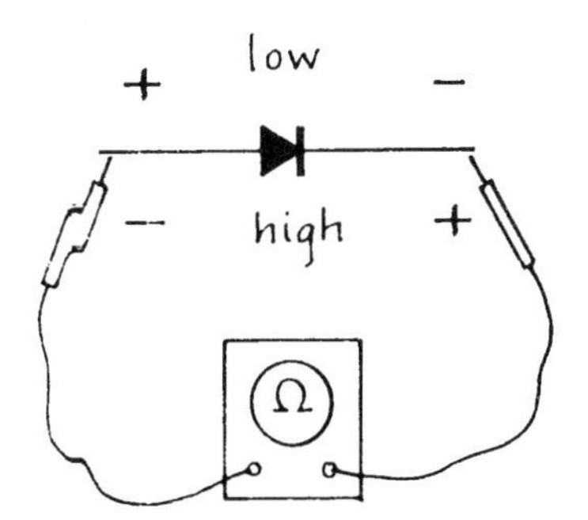

The ohm-meter should show a low reading when connected in one direction and a high reading when connected the opposite way round.

DIODES

THERMISTORS

TRANSISTORS

THYRISTORS

ELECTRON THEORY

The 'Electron Theory' gives a basis by which the flow of electricity can be visualised and understood. Enlarge upon the following points in the Electron Theory:

1. Electricity in a conductor is in the form of electrons.
2. The flow of electric current is a movement of electrons.
3. The electrons may be forced to move by suitably applied magnetism or chemical pressure.

....

....

....

....

....

....

....

Make simple sketches to show the negatively charged electrons, positively charged protons and neutrons of a copper atom, showing why it is a good conductor of electricity:

In terms of electron flow:

a conductor is a material that will ....

....

....

an insulator is a material that will ....

....

....

a resistor is a material that will ....

....

The connections in an electrical circuit are said to be either positive or negative.

The flow of an electric current is said to pass

from .... to ....

and the flow of electrons is said to pass

from .... to ....

State what the following electrical terms mean, and give the units in which they are measured:

Voltage ....

....

....

Current ....

....

Resistance ....

....

....

PRODUCTION OF ELECTRICITY

Electricity can be produced:
chemically, magnetically or thermally.

Chemical means:
Label the simple cell shown.

Copper and zinc plates in dilute sulphuric acid constitute a simple cell.

SIMPLE CELL

The potential difference between the positively charged copper plate and the negatively charged zinc plate is about 1 volt (electromotive force – emf). If a lamp is connected as shown, the emf will cause a current to flow.

Describe briefly, with the aid of simple sketches, production of electrical potential by:

Magnetic means:

Thermal means:

EFFECTS OF AN ELECTRIC CURRENT

The flow of current in an electrical circuit produces three main effects. These are: HEATING, CHEMICAL and MAGNETIC.

Describe the three effects stated above:

Heating

Chemical

Magnetic

Give motor vehicle applications of the three electrical effects:

Heating

Magnetic

Chemical

OHM'S LAW

'Ohm's Law', is the expression that relates voltage, current and resistance to one another.

Ohm's Law states that:

....

....

....

....

Ohm's Law as a formula using symbols can be expressed as:

$$I = \frac{V}{R}$$

where

I = ...

V = ...

R = ...

This simple equation can be used to calculate any one of the three values provided the other two are known.

Resistivity

The unit of resistivity is the OHM and the resistance of a conductor is dependent on:

1. *Material (resistivity)* ....
2. ....
3. ....
4. ....

Resistance (Ω) = $\frac{\quad \times \quad}{\quad}$

State the resistivity of copper: ....

At a constant temperature, the resistance of a wire will vary proportionally to the dimensions of that wire (that is, the length and cross-sectional area).

How is the resistance of most metals affected by temperature change?

....

PROBLEMS

1. What voltage will be required to cause a current flow of 3 A through a bulb having a filament resistance of 4.2 ohms?

2. What will be the total resistance offered by a lighting circuit if a current of 11 A flows under a pressure of 13 V?

3. Two 12 V headlamp bulbs each have a resistance of 2.4 ohms. Calculate the current flowing in each bulb and the total current flowing in the circuit.

4. A cable has a resistance of 0.4 Ω and an area of 15 mm^2. If the area is increased to 90 mm^2 what would be the resistance?

5. If the resistance of a wire is 0.009 Ω when it is 6 m long, what will be its resistance when it is 72 m long?

6. A copper cable has a CSA of 105 mm^2. Calculate the resistance of a 30 m length of the cable. (Resistivity of copper = 1.72×10^{-8}).

ELECTRICAL CIRCUITS

To allow an electrical current to flow an electric circuit must consist of:

(a) A source of supply.

(b) A device that will use the supply to do useful work.

(c) Electrical conducting materials that will transfer the electric current from the supply source to the consuming device, and then return it to the supply source.

On a motor vehicle TWO sources of electrical supply are:

1. ....

2. ....

Name FIVE different types of devices that consume the current to do useful work:

1. ....

2. ....

3. ....

4. ....

5. ....

What is meant by the term electrical conductor?

....

....

What is meant by the term electrical insulator?

....

....

Name SIX electrical conductors and SIX insulators:

Conductors	Insulators
1.	1. ....
2.	2. ....
3.	3. ....
4.	4. ....
5.	5. ....
6.	6. ....

INVESTIGATION

Connect the following components to build an electrical circuit: battery, ammeter, switch, light bulb.

Using conventional symbols draw the circuit diagram. Indicate using arrows the conventionally accepted direction of current flow:

State the amount of current flowing in the circuit shown ....

The bulb wattage was ....

MEASURING INSTRUMENTS

An .... measures the amount of current flow. This instrument when used, must always be connected into the circuit in ....

When checking the voltage (or potential difference) across components in a circuit a .... is used. This must be connected across the terminals of the components being tested, that is in ....

To check the resistance of an electrical component, for example coil, the meter used is called an .... In what way does this meter differ from the other two? ....

....

....

Check the resistance of the following:

coil ...., field winding ...., small light bulbs ....

PARALLEL CIRCUITS

With the aid of diagrams state the basic laws of parallel circuits.

Voltage ....

....

....

....

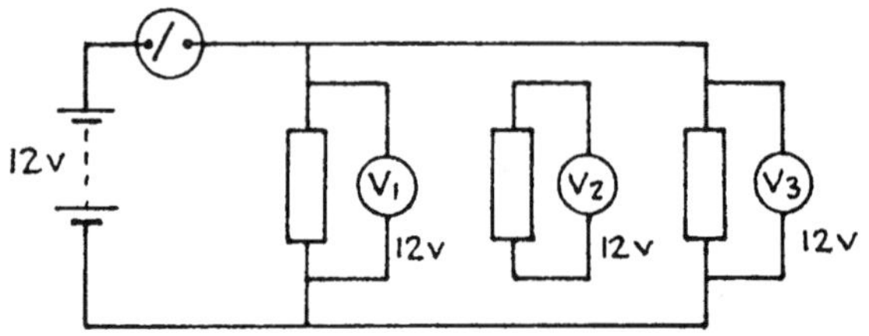

Current ....

....

....

....

....

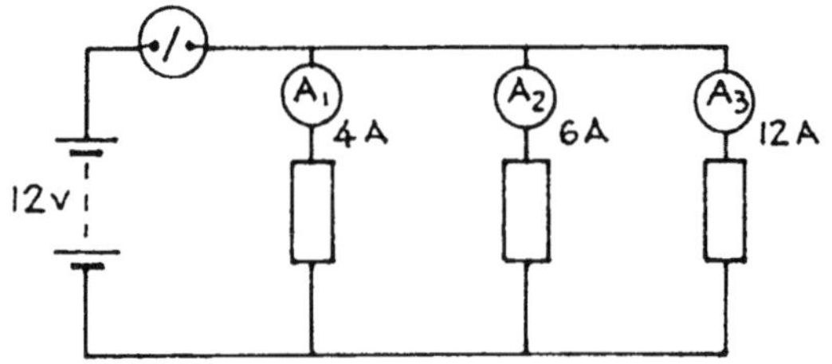

Resistance ....

....

....

....

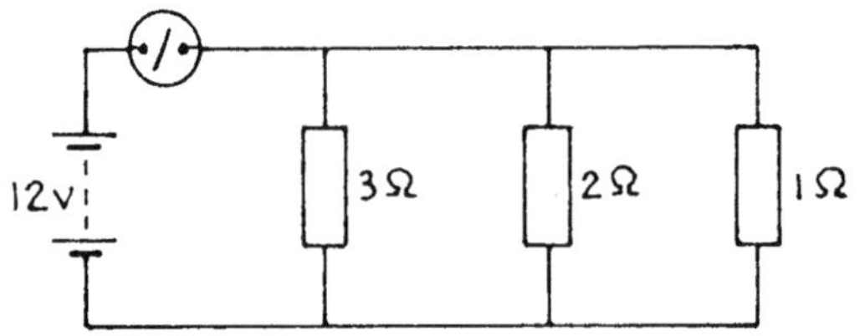

....

....

....

PROBLEMS

1. Three conductors are placed in a parallel circuit, their resistances being 2, 3 and 4 Ω. What current will flow in each when connected to a 12 V system?

2. Two resistors of 20 and 5 Ω are connected in a 12 V parallel circuit. Calculate the total current flow.

3. Two resistors of 8 and 6 Ω are connected in parallel. What voltage would be required to cause a current flow of 7 A?

4. Four resistors of 6, 8, 10 and 12 Ω are connected in parallel to a 12 V circuit. Calculate the current flowing and the total resistance of the circuit.

5. Three resistors of 3, 5 and 6 Ω are connected to a 12 V battery. Calculate the total circuit resistance.

SERIES CIRCUITS

With the aid of diagrams state the basic laws of series circuits.

Voltage ....

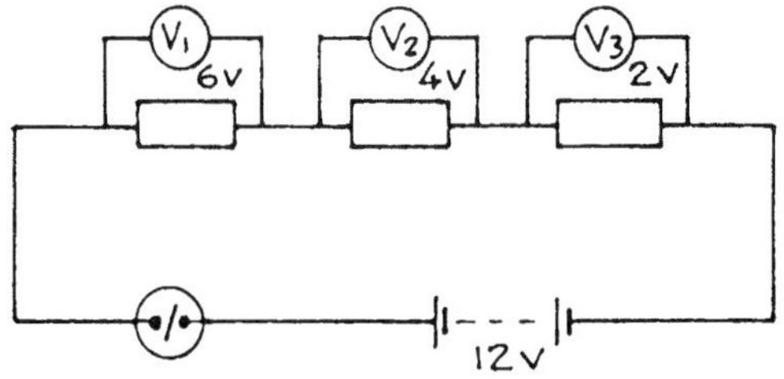

....

....

....

....

V1 6V V2 4V V3 2V 12V

Current ....

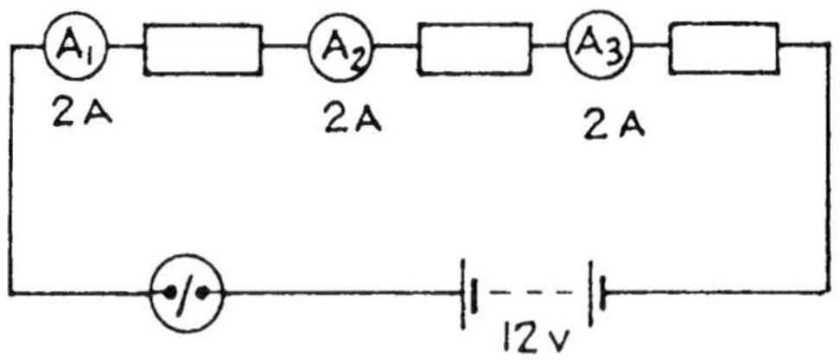

....

....

....

....

A1 2A A2 2A A3 2A 12V

Resistance ....

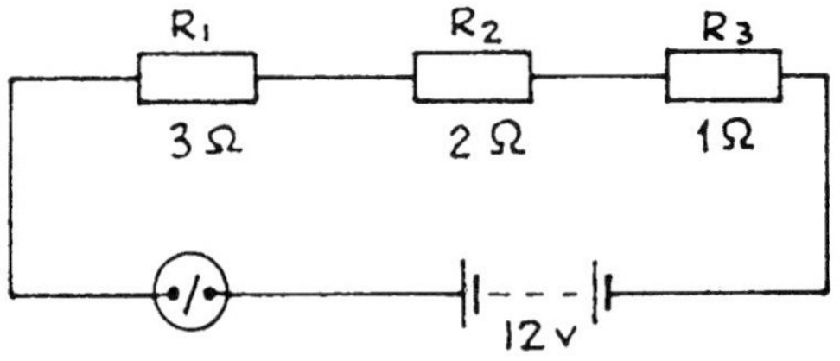

....

....

....

....

....

....

....

....

R1 3 Ω R2 2 Ω R3 1Ω 12V

PROBLEMS

1. Two resistors of 1.75 Ω and 4.25 Ω are connected in series. What voltage would be required to cause a current of 2.5 A to flow in the circuit?

2. Four resistors of 6, 8, 10 and 12 Ω are connected in series to a 12 V circuit.
   Calculate the total resistance of the circuit and the current flowing in each resistor.

3. Four resistors of equal value are placed in series and connected to a 110 V supply. A current of 5 A then flows.
   Calculate the value of each resistor and the voltage across each resistor.

4. Three resistors are wired in series and when connected to a 12 V supply a current of 6 A flows in the circuit.
   If two of the resistors have values of 0.5 and 0.8 Ω calculate the value of the third resistor.

5. Three resistors of 2, 4 and 6 Ω are connected in series to a 12 V battery.
   Calculate the current flowing in the circuit and the voltage across each resistor.

PERMANENT AND ELECTRO-MAGNETS

There are two forms of magnets, permanent and electro-magnets.

What is the difference between a permanent and an electro-magnet?

....

....

....

These two forms of magnetism lead to the important relationship between magnetism and electricity:

....

....

INVESTIGATION

To produce an electro-magnet:

Equipment

Coil of insulated wire and iron bar suitable for passing through centre of coil. Resistance. Battery. Screwdriver or bar for checking magnetism.

Show sketch of apparatus used.

Tests

1. Test bar for magnetism
2. Pass current through coil
3. Test bar for magnetism
4. Attempt to pull bar out of coil. Switch off current.
5. Test bar for magnetism.

Effects of test were:

1. ....
2. ....
3. ....
4. ....
5. ....

What forms the basic construction of an electro-magnet?

....

The core of the electro-magnet may be moving as shown or stationary (for example, the ignition coil). In either case it is made from soft iron. Why is such a material used?

....

....

....

....

List FOUR motor vehicle components where an electro-magnet which produces movement is used:

1. ....
2. ....
3. ....
4. ....

LINES OF FORCE

Magnets act through lines of force. These lines of force stretch between the ends of a magnet and create a magnetic field. The two ends of a magnet are

called .... one end being the ....

and the other the ....

Using small bar magnets, a sheet of paper and iron filings, show the effects of the lines of force when the magnets are held in the positions shown:

(A) [n s]   (B) [n s] [n s]   (C) [s n] [n s]

The effects created by the magnets lead to the statements:

Like magnetic poles ....   Shown by sketch ....

Unlike magnetic poles ....   Shown by sketch ....

ENERGY AND POWER

Electricity is a form of energy:

ENERGY (J) = VOLTS × AMPS × TIME (s)

or Q =

Power is the rate of doing work (energy per unit time):

Power (J/s) (1J/s = 1 Watt)

Electric power is measured in ....

Efficiency (η) is expressed as a percentage.

Efficiency in terms of energy and power is:

$$\frac{\text{Energy output}}{\text{Energy input}} \times 100 \text{ or } \frac{}{} \times 100$$

Calculate the total energy consumed when a current of 3 amps flows in a 12 V ignition coil for 5 seconds.

A starter motor is operated for 30 seconds and the current flow is 200 amps. Calculate the energy consumed and the input power if the terminal voltage is 10 volts.

The power output of a starter motor is 1800 W. If the terminal voltage is 10 V and the current flow during starter operation is 240 amps, calculate the efficiency of the starter motor.

MUTUAL INDUCTION

A voltage can be INDUCED into a circuit by varying the current flow in an adjacent but separate circuit. This property is called MUTUAL INDUCTION.

Label the drawing and describe how the simple transformer shown can STEP-UP input voltage.

....

....

....

....

....

....

....

How does the ratio of turns on primary and secondary windings affect the 'step-up' voltage?

....

....

....

State one motor vehicle application of this principle:

....